VMware vSphere 企业级网络和存储实战

何坤源 著

人民邮电出版社
北京

图书在版编目（ＣＩＰ）数据

VMware vSphere企业级网络和存储实战 / 何坤源著
. -- 北京：人民邮电出版社，2017.12
ISBN 978-7-115-47059-1

Ⅰ．①V… Ⅱ．①何… Ⅲ．①虚拟处理机 Ⅳ．
①TP317

中国版本图书馆CIP数据核字(2017)第249794号

内 容 提 要

本书针对 VMware vSphere 虚拟化架构在生产环境中的实际需求，分 9 章介绍了如何在企业级虚拟化环境中对网络和存储进行安装、配置、管理、维护。

全书以实战操作为主，理论讲解为辅，通过搭建各种物理环境，详细介绍了在企业生产环境中常用的 CiscoNexus 系列设备、DELL MD 存储和 VMware Virtual SAN 如何快速部署，并通过大量实例，迅速提高读者的动手能力。

本书通俗易懂，具有很强的可操作性，不仅适用于 VMware vSphere 虚拟化架构管理人员，而且对其他虚拟化平台管理人员也具有参考作用。

◆ 著　　何坤源
　　责任编辑　王峰松
　　责任印制　焦志炜

◆ 人民邮电出版社出版发行　北京市丰台区成寿寺路 11 号
邮编 100164　电子邮件 315@ptpress.com.cn
网址 http://www.ptpress.com.cn
固安县铭成印刷有限公司印刷

◆ 开本：787×1092　1/16
印张：31.75　　　　　　　2017 年 12 月第 1 版
字数：751 千字　　　　　　2024 年 7 月河北第 7 次印刷

定价：108.00 元

读者服务热线：(010)81055410　印装质量热线：(010)81055316
反盗版热线：(010)81055315
广告经营许可证：京东市监广登字 20170147 号

致 谢

感谢 VMware 公司的叶毓睿老师，虽然我们认识的时间不长，但对于本书，叶老师给出了不少好的建议；感谢 Cisco 公司的禹果老师，是他带领我进入了 Nexus 网络技术的殿堂；感谢 DELL 公司的曾毓老师，始终怀念当年一起备考 CCIE DC 时对于各种技术问题的探讨和研究；感谢张冬老师以及秦柯老师在百忙之中为本书写推荐语；最后感谢我的同事黎叔、晶哥、邱、罗罗、薇姐、倩姐、钒哥、栋哥、小唐等在非技术层面提供的帮助，我们不仅仅是好同事，感谢你们。

何坤源
2017 年 9 月

作者简介

何坤源，业界知名讲师，黑色数据网络实验室创始人，持有 CCIE（RS/DC）、VCP-DCV（4/5/6）、H3CSE、ITIL 等证书，主讲 VMware 虚拟化、oVirt 虚拟化以及数据中心网络、存储等课程，担任多家企业、学校的 IT 咨询顾问。从 2006 年开始，将工作重心转向虚拟化、数据中心以及灾难备份中心建设。2008 年创建 Cisco 路由交换远程实验室，2009 年创建虚拟化远程实验室，2015 年创建云计算平台、数据中心远程实验室。到目前为止，已经参与了多个企业虚拟化建设和改造项目，在虚拟化的设计、设备选型、运营维护等方面积累了丰富的经验。

何坤源编写的《VMware vSphere 5.0 虚拟化架构实战指南》《Linux KVM 虚拟化架构实战指南》《VMware vSphere 6.0 虚拟化架构实战指南》等图书已被多所高校选为教材，累计印刷超过 2 万册。同时，他编写的《VMware vSphere 5.0 虚拟化架构实战指南》和《Linux KVM 虚拟化架构实战指南》两种书已输出版权到台湾。

序一

认识何坤源老师并不久。何老师在申请加我为 QQ 好友时，提及他撰写了多种 vSphere 书籍，令我感到非常惊讶。据何老师介绍，他编写的《VMware vSphere 5.0 虚拟化架构实战指南》以及《Linux KVM 虚拟化架构实战指南》已输出版权到台湾。而且，这几种书同时被国内多所高职院校选为教材。这意味着，何老师为 VMware 虚拟化在中国的普及做出了不小的贡献。

当他提出请我为新书《VMware vSphere 企业级网络和存储实战》撰写序言时，我有些难为情，虽然我领衔撰写的《软件定义存储：原理、实践与生态》也再版了几次，但远远不如何老师，其《VMware vSphere 5.0 虚拟化架构实战指南》一书重印居然高达 12 次，《VMware vSphere 6.0 虚拟化架构实战指南》一书重印也高达 7 次。

何老师在一家企业的信息技术部上班，平时工作也比较繁忙。从我个人曾经的经历知道，写书是一件非常繁重的工作，有时甚至让人抓狂，感觉结束的时间似乎遥遥无期。而让我由衷地敬佩的是，何老师笔耕不辍，先后出版了中文简体和繁体版共 7 种书！

利用每天睡前的一小段时间，连续数日，我简单通读了《VMware vSphere 企业级网络和存储实战》的样稿，感觉这本书特别适合初学者，能够帮助他们快速入门，获得直观的认识。这本书的可操作性强，不过，如果能够再多一些基础的理论阐述，就更丰满了。

这本书分为 9 章，介绍了如何在企业级虚拟化架构中安装、配置、管理、维护网络和存储。读完这本书之后，其实我还有另外一层的思考，我发现何老师的实验环境比较有限，主流的传统外置存储例如 EMC VNX、HDS、NetApp 在他的书中都未出现，于是我就好奇地提出这个问题，他告诉我，实际上在中小企业或小微企业上虚拟化的，很少甚至不可能用到针对大中型企业的主流存储设备。所以，他用了 Open-E、DELL MD 3600F 等存储设备。为了用上 NAS 存储，甚至采用免费或开源的 NAS 软件，如 Nexentastor 和 FreeNAS 等。

世上无难事，只怕有心人！真心地祝愿读者能从这本书的学习中获益！

叶毓睿
VMware 存储架构师
《软件定义存储：原理、实践与生态》作者
微信公众号"乐生活与爱 IT"作者
2017 年 9 月

序二

近年来，云计算、大数据、人工智能等新技术的出现正在改变着社会生活的方方面面，各种技术创新也带给客户前所未有的体验。

本书的作者正是一位埋头苦干、静心学习新技术的工程师。最早接触他，是网络方面的特长。作者多年来长期从事运维工作，积累了大量的实战经验，虽然对于分公司运维人员来说，CCIE 这种层面的技术不是必须掌握的，但是作者依然在 2009 年通过了 CCIE 路由交换认证，也是当时新华保险全系统第 2 位通过 CCIE 认证的工程师，这让笔者感到吃惊。2016 年，作者再次通过了 CCIE 数据中心认证，这让笔者感觉到作者在技术这条道路上的不懈追求。

不久前笔者收到作者的邀请为本书作序，当看完样稿后，内心是兴奋的。VMware vSphere 虚拟化架构是云计算平台的核心，也是整个行业领先的解决方案，几乎所有的世界 500 强企业都部署有 VMware 公司的产品。纵观市面同类书籍，就如作者所说：介绍网络配置的书籍非常多，无论是 Cisco 的也好，华为的也好，大多是基于认证体系的培训教材，并没有针对虚拟化平台进行讲解。存储类书籍相对于网络书籍少了很多，而且大多是基于存储理论的讲解，不涉及具体的配置或很少配置。本书可谓填补了这一领域的空白，具有很强的实战性。

总体来说，本书主要有两大特点。

第一，定位准确、条理清晰。虚拟化数据中心的核心，一是虚拟化系统，二是网络，三是存储。本书以最新版本的 VMware vSphere 6.5 为基础，详细介绍了 ESXi 主机以及 vCenter Server 的基本部署，随后介绍了基于 VMware vSphere 虚拟化系统网络、存储方面的配置，整体思路非常清晰。

第二，具有很强的实战性。本书用大量的篇幅介绍了虚拟化架构网络方面，包括物理网络以及软件定义网络的配置，同时介绍了各种主流存储以及软件定义存储的配置，引入了大量生产环境使用的物理设备，这对于中小企业以及小微企业来说具有很高的参考价值。

可以肯定的是，本书对于从事虚拟化运维的技术人员非常有帮助。祝本书作者能够继续在这个领域不断探索，有更多更好的心得和经验与大家分享。

及戈
新华人寿保险股份有限公司信息技术部总经理
2017 年 9 月

序三

科技与保险行业的结合越来越重要,"平台动力""分享经济"等成为热点,越来越多的人关注并参与进来。前端场景的打造需要基础架构强有力的支撑,而基础架构更多的是基于规范的、规律的、成熟的技术。

很高兴看到坤源发表新作,从工作中找出规律,从实践中总结经验,规范化、系统化、标准化,并以出书的方式将知识进行传承。坤源此前已经出版了多种关于虚拟化实战方面的书籍,这些书重点介绍了虚拟化平台的构建以及各种高级特性如何配置使用,而《VMware vSphere 企业级网络和存储实战》这本书重点介绍的则是企业级网络、存储方面的配置以及优化,与以往的书籍不太一样。笔者尝试寻找了一下市面上是否有类似的书籍,结果发现没有。从这个角度来说,本书是一个创新。

从笔者阅读的样稿来看,本书的内容十分丰富,从基本的 VMware vSphere 6.5 安装部署开始,详细介绍了在虚拟化环境中如何配置物理网络、软件定义网络、传统存储以及软件定义存储等,除了虚拟化上层应用系统的安装部署未介绍外,对底层的框架结构体系的介绍非常完善。令人高兴的是,本书使用了大量真实的设备进行演示,比如思科的 NEXUS 交换、DELL 存储等,而非模拟器,这对于读者来说具有很强的参考性和复制性。

工作贵在持之以恒地反省、总结、提炼,不断在工作中提升自己,并以师父带徒弟的模式培养新员工。何坤源能够在非常繁忙的工作之余,利用休息时间,投入大量精力在写作上,是值得每个 IT 人学习的。希望他在以后的工作中,继续保持写作的习惯,把好的工作方法、工作经验分享给大家。

最后,祝读者有个愉快的学习过程。

姚仁毅
天安人寿保险股份有限公司信息技术部总经理
2017 年 9 月

序四

记得我是 2002 年开始听说 VMware 的,那时 VMware 还没有企业级产品,只有 VMware Workstation 单机版。当时在我的台式机安装后,给人的感觉真的很神奇啊,原先要装多个系统,就必须安装多引导的软件,而使用 VMware 后一切都解决啦,想用哪个系统,就启动哪个系统,实在是太方便了。有一段时间,我在一家外资公司负责产品和售前,我们的演示版本基本都采用虚拟机安装,真的很方便。

时光飞逝,2012 年我们的数据中心已经在生产服务器上大量使用 VMware 虚拟化技术,这给我们的工作带来了极大的方便性,现在公司里有 60 多套生产系统,开发和测试环境加在一起,大概使用了近 300 台虚拟机,虚机比达到了 1:17,节约了大量硬件购置的成本,运维人员进行维护也非常方便。

虚拟化技术发展到如今,除了系统虚拟化以外,网络虚拟化、存储虚拟化也是发展趋势。本书以最新版本的 VMware vSphere 6.5 为基础,详细介绍了 ESXi 主机以及 vCenter Server 的基本部署,同时也系统介绍了 VMware vSphere 虚拟化系统网络和存储方面的配置,内容深入浅出,条理清晰,是一本市场上少有的实战性技术资料。

在编排上,本书花费了大量的篇幅介绍虚拟化架构网络,同时介绍了各种主流存储以及软件定义存储的配置,并引入了大量生产环境使用的物理设备,对于中小企业以及小微企业来说具有很高的参考价值,对于有志于学习虚拟化技术的从业者来说也是一本难得的学习教材。

认识坤源还是我在生命人寿负责"生命动力项目"的时候。2011 年 10 月,我带领西区推广小组到成都分公司,让我吃惊地看到:一个房间里面全是电脑和网络设备,何坤源在搭建各种网络,进行测试。再次见到已是 2014 年,听说他撰写了一系列有关网络、虚拟化技术的图书,就一直想着"先读为快"。今天何坤源先生的新书《VMware vSphere 企业级网络和存储实战》又要出版了,我怀着无比高兴的心情为此书一序,既为何坤源先生孜孜不倦的努力成果而感到高兴,同时又为本书给广大读者即将带来的价值而感到高兴。

徐斌
利安人寿保险股份有限公司信息技术部总经理
2017 年 9 月于南京

自序

从 2013 年开始，笔者写作了多本关于虚拟化方面的技术书，通过这些书认识了许多正在学习或使用虚拟化架构的朋友。

在日常的交流中笔者发现，虚拟化涉及的知识点非常多，除了各种虚拟化架构本身的技术之外，还包括网络以及存储的规划与配置。一般情况下，网络的配置与管理由网络管理员来完成，存储的配置与管理由存储管理员来完成。从目前国内的实际情况看，作为 IT 运维人员，特别是中小企业的 IT 运维人员，其岗位多数不能细分，所以无论是虚拟化架构还是网络和存储，这些工作可能都由相同的 IT 运维人员来完成。

再来看看市面上出版的 IT 类书籍，介绍网络配置的非常多，无论是关于 Cisco 的技术图书，还是关于华为的技术图书，大多是基于认证体系的培训教材，并没有针对虚拟化平台进行讲解。存储类书籍相对于网络技术类书籍少了很多，大多是基于存储理论的讲解，而不涉及具体的配置或很少涉及配置。

对于正在学习虚拟化架构或者对网络、存储不太了解的朋友来说，他们需要一本综合配置的书籍。而笔者的生产环境中部署有 VMware vSphere 虚拟化、Cisco CATALYST 系列交换机、Cisco Nexus 系列交换机、Open-E 系列存储、DELL MD 3600F 系列存储等设备，因此产生了写作一本基于虚拟化平台网络以及存储配置的专业书的想法。

从 2016 年 8 月动笔至 2017 年 7 月完成，花费了整整 1 年的时间进行本书的写作。最初的想法是基于 VMware vSphere 6.0 架构来完成本书，然而 VMware 公司在 2016 年 11 月发布了 VMware vSphere 6.5 版本，新版本提供了不少新的功能，特别是 vSAN 部分，因此笔者又重新调整架构，重写了部分章节的内容。希望这本书对读者朋友构建企业级网络和存储有所帮助。

黑色数据　何坤源
2017 年 7 月于成都

前 言

软件定义数据中心是最近几年非常热门的话题。在软件定义服务器市场成熟后，各大厂商纷纷把目光转向了软件定义存储、软件定义网络、超融合等领域。

无论是传统数据中心还是新型的虚拟化数据中心，从目前情况来看，处于基础架构位置的网络和存储还不可能完全被软件定义所取代。作者一直认为，基础架构的网络和存储作为数据中心的核心，依旧扮演着非常重要的角色。

本书的重点是介绍如何在虚拟化数据中心中部署网络和存储。希望通过这本书，能够让 IT 技术人员在部署过程中得到一定程度的参考和指引。

本书一共分为 9 章，采用循序渐进的方式带领大家掌握基于 VMware vSphere 虚拟化架构的企业级网络和存储在生产环境中的部署。

由于作者水平有限，本书涉及的知识又很多，书中难免有不妥和错误之处，欢迎大家与作者进行交流。有关本书的任何问题、意见或建议，可以发邮件到 heky@vip.sina.com 与作者联系，也可与本书编辑（wangfengsong@ptpress.com.cn）联系。

以下是作者的技术交流平台：

技术交流网站：www.bdnetlab.com（黑色数据网络实验室）；

技术交流 QQ：44222798；

技术交流 QQ 群：240222381。

目　录

第1章　实战环境搭建 ... 1
1.1　物理设备及拓扑介绍 ... 1
1.1.1　实战环境物理设备配置 ... 1
1.1.2　实战环境拓扑 ... 2
1.2　虚拟化平台及其他系统介绍 ... 2
1.2.1　虚拟化平台介绍 ... 2
1.2.2　其他系统介绍 ... 3
1.3　本章小结 ... 3

第2章　部署VMware vSphere 6.5 ... 4
2.1　VMware vSphere 6.5 虚拟化介绍 ... 4
2.1.1　什么是VMware vSphere ... 4
2.1.2　VMware vSphere的用途 ... 4
2.1.3　VMware vSphere的优势 ... 5
2.1.4　VMware vSphere 6.5 的新特性 ... 5
2.2　部署VMware ESXi 6.5 ... 8
2.2.1　部署ESXi 6.5 硬件要求 ... 8
2.2.2　部署ESXi 6.5 ... 9
2.3　部署VMware vCenter Server 6.5 ... 18
2.3.1　部署vCenter Server 6.5 要求 ... 18
2.3.2　部署vCenter Server 6.5 ... 19
2.4　创建使用虚拟机 ... 37
2.4.1　创建使用Windows虚拟机 ... 37
2.4.2　创建使用Linux虚拟机 ... 47
2.5　本章小结 ... 50

第3章　部署VMware vSphere 基本网络 ... 51
3.1　VMware vSphere 网络介绍 ... 51
3.1.1　ESXi主机通信原理介绍 ... 51
3.1.2　ESXi主机网络组件介绍 ... 52
3.1.3　ESXi主机网络VLAN实现方式 ... 55
3.1.4　ESXi主机网络NIC Teaming ... 56
3.1.5　ESXi主机TCP/IP协议堆栈 ... 60
3.2　配置使用标准交换机 ... 61
3.2.1　创建运行虚拟机流量标准交换机 ... 61
3.2.2　创建基于VMkernel流量标准交换机 ... 65
3.2.3　标准交换机多VLAN配置 ... 71
3.2.4　标准交换机NIC Teaming配置 ... 77
3.2.5　标准交换机其他策略配置 ... 83
3.2.6　TCP/IP协议堆栈配置 ... 85
3.3　配置使用分布式交换机 ... 101
3.3.1　创建分布式交换机 ... 101
3.3.2　将ESXi主机添加到分布式交换机 ... 104
3.3.3　分布式交换机多VLAN配置 ... 110
3.3.4　迁移虚拟机到分布式交换机 ... 113
3.3.5　分布式交换机LACP配置 ... 120
3.3.6　分布式交换机策略配置 ... 131
3.4　本章小结 ... 136

第4章　部署Nexus 1000V分布式交换机 ... 137
4.1　Nexus 1000V介绍 ... 137

4.1.1 虚拟化架构面临的网络
      问题 ……………………… 137
4.1.2 Nexus 1000V 基本介绍 …… 138
4.1.3 Nexus 1000V 架构介绍 …… 139
4.2 部署 Nexus 1000V VSM ………… 140
  4.2.1 部署 VSM 前的准备工作 … 140
  4.2.2 部署 VSM ……………………… 141
  4.2.3 VSM 常用命令 ……………… 152
4.3 部署 Nexus Port-Profile ………… 156
  4.3.1 部署 Port-Profile 前的准备
        工作 ……………………… 156
  4.3.2 部署 Port-Profile ………… 157
  4.3.3 Port-Profile 常用命令 …… 159
4.4 部署 Nexus 1000V VEM ………… 160
  4.4.1 部署 VEM 前的准备工作 … 161
  4.4.2 部署 VEM ……………………… 162
  4.4.3 VEM 常用命令 ……………… 177
  4.4.4 VEM 常见故障排除 ……… 183
4.5 虚拟机使用 Nexus 1000V ……… 186
  4.5.1 迁移虚拟机到 Nexus 1000V
        交换机 …………………… 186
  4.5.2 Nexus 1000V 安全策略
        配置 ……………………… 194
4.6 部署使用 VXLAN ………………… 204
  4.6.1 VXLAN 基础知识介绍 …… 204
  4.6.2 配置 VXLAN ………………… 206
4.7 本章小结 ………………………… 222

第 5 章 部署 Nexus N5K&N2K
        交换机 ……………………… 223
5.1 Nexus N5K&N2K 交换机介绍 … 223
  5.1.1 Nexus 系列交换机介绍 …… 223
  5.1.2 Nexus N5K 介绍 …………… 225
  5.1.3 Nexus N2K 介绍 …………… 226
  5.1.4 Nexus NXOS 基本命令行
        介绍 ……………………… 227
5.2 配置使用 Nexus FEX …………… 232
  5.2.1 Nexus FEX 技术介绍 …… 232
  5.2.2 配置 Nexus FEX …………… 235
5.3 配置使用 Nexus vPC …………… 241
  5.3.1 Nexus vPC 技术介绍 …… 241
  5.3.2 配置 Nexus vPC …………… 243
5.4 虚拟化架构使用 N5K&N2K …… 248
  5.4.1 N5K&N2K 连接设计 ……… 248
  5.4.2 ESXi 主机应用配置 ……… 249
5.5 本章小结 ………………………… 261

第 6 章 部署存储服务器 ……………… 262
6.1 VMware vSphere 支持的存储
    介绍 ……………………………… 262
  6.1.1 常见存储类型 ……………… 262
  6.1.2 FC SAN 存储介绍 ………… 263
  6.1.3 FCoE 介绍 ………………… 265
  6.1.4 iSCSI 存储介绍 …………… 266
  6.1.5 NFS 介绍 …………………… 266
6.2 部署使用 Open-E 存储
    服务器 …………………………… 266
  6.2.1 Open-E 存储服务器介绍 … 266
  6.2.2 Open-E 存储服务器安装 … 267
  6.2.3 生产环境部署 Open-E 存储服
        务器建议 ………………… 279
6.3 部署使用 DELL MD 3620 存储
    服务器 …………………………… 280
  6.3.1 DELL MD 3620F 存储
        服务器介绍 ……………… 280
  6.3.2 DELL MD 3620F 存储
        服务器基本操作 ………… 282
  6.3.3 生产环境部署 DELL MD 存储
        服务器建议 ……………… 289
6.4 本章小结 ………………………… 289

第 7 章 部署使用 FC SAN 存储 ……… 290
7.1 FC SAN 存储介绍 ……………… 290
  7.1.1 FC SAN 基本概念 ………… 290
  7.1.2 FC SAN 的组成 …………… 290
  7.1.3 FC 协议介绍 ……………… 291
  7.1.4 FC 拓扑介绍 ……………… 291
  7.1.5 FC 端口介绍 ……………… 292

7.1.6　WWN/FCID 介绍 …………… 294
7.1.7　FC 数据通信介绍 …………… 295
7.1.8　VSAN 介绍 ………………… 295
7.1.9　ZONE 介绍 ………………… 296
7.1.10　NPV/NPIV 介绍 …………… 296
7.2　FCoE 存储介绍 ………………… 297
7.2.1　FCoE 存储组成 ……………… 298
7.2.2　FCoE 协议介绍 ……………… 298
7.3　配置 DELL MD 系列企业级
　　存储 ……………………………… 299
7.3.1　DELL MD 存储磁盘配置 … 299
7.3.2　DELL MD 存储映射配置 … 307
7.3.3　DELL MD 存储其他配置 … 315
7.4　配置 Cisco MDS 系列企业级
　　存储交换机 ……………………… 326
7.4.1　基本命令行介绍 …………… 326
7.4.2　配置 VSAN ………………… 329
7.4.3　配置 ZONE ………………… 332
7.4.4　配置多台 FC 交换机级联 … 333
7.4.5　配置 NPV/NPIV …………… 344
7.5　配置 ESXi 主机使用 FC 存储 … 349
7.5.1　配置 ESXi 主机 SANBOOT
　　　启动 ………………………… 349
7.5.2　配置 ESXi 主机使用 FC 共享
　　　存储 ………………………… 357
7.5.3　ESXi 主机在 FC 存储下
　　　高级特性使用 ……………… 362
7.6　配置 ESXi 主机使用 FCoE
　　存储 ……………………………… 367
7.6.1　配置 FCoE 存储准备
　　　工作 ………………………… 367
7.6.2　配置 FCoE 交换机 ………… 373
7.6.3　使用 FCoE 存储 …………… 378
7.7　实验 FC 设备配置信息 ………… 383
7.7.1　Cisco MDS 交换机配置
　　　信息 ………………………… 383
7.7.2　Cisco Nexus 交换机配置
　　　信息 ………………………… 387
7.8　本章小结 ………………………… 390

第 8 章　部署使用 iSCSI 存储 ……… 391
8.1　iSCSI 协议介绍 ………………… 391
8.1.1　SCSI 协议介绍 ……………… 391
8.1.2　iSCSI 协议基本概念 ……… 392
8.1.3　iSCSI 协议名字规范 ……… 392
8.2　配置 Open-E 存储服务器 ……… 392
8.2.1　配置 Open-E 存储磁盘 …… 392
8.2.2　配置 Open-E 存储 iSCSI
　　　选项 ………………………… 396
8.2.3　配置 Open-E 存储负载
　　　均衡 ………………………… 399
8.3　配置 ESXi 主机使用 iSCSI
　　存储 ……………………………… 403
8.3.1　配置 ESXi 主机启用 iSCSI
　　　存储 ………………………… 403
8.3.2　配置 ESXi 主机绑定 iSCSI
　　　流量 ………………………… 413
8.3.3　配置使用 ESXi 主机高级
　　　特性 ………………………… 421
8.3.4　配置 ESXi 主机启用 iSCSI 安全
　　　特性 ………………………… 423
8.4　生产环境使用 iSCSI 存储
　　讨论 ……………………………… 434
8.4.1　生产环境选择 iSCSI 存储还是
　　　FC 存储 …………………… 434
8.4.2　生产环境 iSCSI 存储网络
　　　设计 ………………………… 435
8.5　本章小结 ………………………… 435

第 9 章　部署使用 Virtual SAN ……… 436
9.1　Virtual SAN 存储介绍 ………… 436
9.1.1　软件定义存储介绍 ………… 436
9.1.2　什么是 Virtual SAN ……… 437
9.1.3　Virtual SAN 功能介绍 …… 438
9.1.4　Virtual SAN 常用术语 …… 440
9.1.5　Virtual SAN 存储策略
　　　介绍 ………………………… 440
9.2　部署 Virtual SAN 6.5 …………… 442

9.2.1 使用 Virtual SAN 要求 …… 442
9.2.2 配置 Virtual SAN 所需网络 …… 443
9.2.3 启用 Virtual SAN …… 458
9.2.4 配置 Virtual SAN 存储策略 …… 465
9.2.5 配置 Virtual SAN 去重和压缩 …… 467
9.2.6 配置 Virtual SAN 纠删码 … 469
9.2.7 配置 Virtual SAN 故障域 … 474
9.2.8 配置 Virtual SAN 延伸集群 …… 478
9.2.9 配置 Virtual SAN 为 iSCSI 目标服务器 …… 482
9.2.10 配置 Virtual SAN 性能服务 …… 488
9.3 生产环境使用 Virtual SAN 讨论 …… 491
9.3.1 Virtual SAN 是否能代替传统存储 …… 491
9.3.2 生产环境使用 Virtual SAN 主机数量 …… 491
9.3.3 生产环境使用 Virtual SAN 网络要求 …… 491
9.3.4 生产环境使用 Virtual SAN 硬件兼容性要求 …… 491
9.4 本章小结 …… 492

第 1 章 实战环境搭建

本书重点介绍基于 VMware vSphere 虚拟化架构构建企业级网络以及存储，因此在实战操作讲解中会针对大量的物理设备进行，希望对初学者以及运维人员有所参考。本章介绍实战环境使用的物理设备以及虚拟化平台。

本章要点
- 物理设备及拓扑介绍
- 虚拟化平台及其他介绍

1.1 物理设备及拓扑介绍

为保证实战操作更具参考价值和可复制性，同时最大程度地还原企业生产环境真实应用，作者使用了全物理设备构建本书的实战环境。

1.1.1 实战环境物理设备配置

实战环境使用多台物理服务器安装 VMware ESXi 6.5，使用 DELL MD 3600F 构建 FC SAN 存储系统，使用 Open-E 系统构建 IP SAN 存储（iSCSI 存储），使用 Cisco Nexus 系列交换机。所使用设备的详细配置如表 1-1-1 所列。

表 1-1-1　　实战环境硬件配置

设备名称	CPU 型号	内存	硬盘	备注
ESXi07-ESXi11 服务器	XEON L5520×2	64GB	64GB SSD + 128GB SSD	用于 vSAN 全闪存
ESXi12-ESXi15 服务器	XEON L5620×2	64GB	无	SANBOOT 引导 QLOGIC QLE2460 4GB HBA 卡
FC SAN 存储系统	DELL MD3620F	4GB（缓存）	双 RAID 控制器 15K 146GB×4 + 10K 300GB×8	
iSCSI 存储服务器	XEON 5420×2	8GB	1TB SATA×4	Open-E 系统

续表

设备名称	CPU 型号	内存	硬盘	备注
Cisco Nexus 交换机	Cisco MDS9124-01，16 端口激活 Cisco MDS9124-02，16 端口激活 Cisco Nexus 5010P-01，提供 20 个 10GE 以太网口以及 8 个 FC 口 Cisco Nexus 5010P-02，提供 20 个 10GE 以太网口 Cisco Nexus 5548UP-01 提供 32 个 10GE 以太网口（可切换为 FC 接口） Cisco Nexus 2248TP-01，提供 48 个 1GE 以太网口 Cisco Nexus 2248TP-02，提供 48 个 1GE 以太网口			
Cisco Catalyst 交换机	Cisco Catalyst 4506，提供 48 个 1GE 以太网口			

1.1.2 实战环境拓扑

本书的实战环境由于使用了大量的物理设备，因此整体的架构比较复杂（整体拓扑如图 1-1-1 所示）。

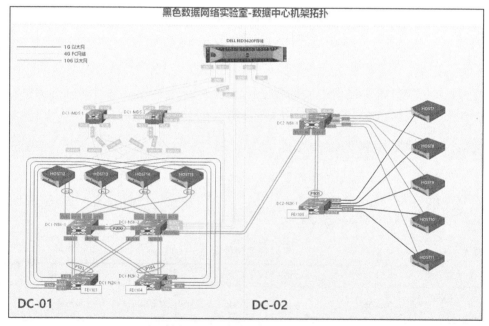

图 1-1-1 实战环境设备拓扑

1.2 虚拟化平台及其他系统介绍

1.2.1 虚拟化平台介绍

本书的实战操作主要使用 VMware 最新发布的 VMware vSphere 6.5 版本。由于第三方软件更新的原因，某些章节的实战操作会使用 VMware vSphere 的其他版本。

1.2.2　其他系统介绍

企业生产环境除使用了 VMware vSphere 虚拟化平台外，还会使用多种系统，比如常见的有 DNS。在实战环境中，使用 Windows 2008 R2 构建了 AD、DNS 服务器，用于提供活动目录以及 DNS 服务。

存储部分除主要使用的 DELL MD3620F 存储外，还使用 Open-E 构建企业级 DIY 存储，用于提供 NFS、ISCSI、FC 等服务。

1.3　本章小结

本章介绍了实战环境物理设备配置以及拓扑，使用大量的物理设备来还原企业真实生产环境。需要说明的是，对于有生产环境的读者，可以亲自动手实验；对于没有生产环境的读者，日常模拟器几乎不能使用，建议仔细阅读本书的操作部分，以能够在一定程度上提高动手能力。

第 2 章　部署 VMware vSphere 6.5

2016 年 11 月，VMware 公司开放 VMware vSphere 6.5 版本下载，这意味着 vSphere 新一代版本的正式发布。作为 VMware 软件定义数据中心核心组件，VMware vSphere 6.5 提供了不少新的功能，其中 WEB 管理端提供 HTML5 是一个重大的改进，同时其整合进 vSphere 的 vSAN 也升级到 6.5 版本。本章介绍如何在生产环境的物理服务器上部署 VMware vSphere 6.5。

本章要点
- VMware vSphere 6.5 虚拟化介绍
- 部署 VMware ESXi 6.5
- 部署 VMware vCenter Server 6.5
- 创建使用虚拟机

2.1　VMware vSphere 6.5 虚拟化介绍

VMware vSphere 是 VMware 公司开发的虚拟化平台，是 VMware 软件定义数据中心的基础。VMware vSphere 6.5 是为新一代数据中心应用而打造的，可用作软件定义的数据中心的核心基础架构。

2.1.1　什么是 VMware vSphere

VMware vSphere 是业界领先的虚拟化平台，能够通过虚拟化纵向扩展和横向扩展应用、重新定义可用性以及简化虚拟数据中心，最终实现高可用、恢复能力强的按需基础架构，这是任何云计算环境的理想基础。同时，VMware vSphere 可以降低数据中心成本，增加系统和应用正常运行时间，显著简化 IT 运行数据中心的方式。

VMware vSphere 的两个核心组件是 ESXi 和 vCenter Server。ESXi 是用于创建和运行虚拟机及虚拟设备的虚拟化平台。vCenter Server 是管理平台，充当连接到网络的 ESXi 主机的中心管理员，可用于将多个 ESXi 主机加入池中并管理这些资源。

2.1.2　VMware vSphere 的用途

VMware vSphere 的用途主要分为以下几个方面。

1. 虚拟化应用

提供增强的可扩展性、性能和可用性，使用户能够虚拟化应用。

2. 简化虚拟数据中心的管理

通过功能强大且简单直观的工具，管理虚拟机的创建、共享、部署和迁移。

3. 数据中心迁移和维护

执行工作负载实时迁移和数据中心维护，而无需中断应用。

4. 为虚拟机实现存储转型

使外部存储阵列更多地以虚拟机为中心来运行，从而提高虚拟机运维的性能和效率。

5. 灵活选择云计算环境的构建和运维方式

使用 VMware vSphere 和 VMware 产品体系或开源框架（例如 OpenStack 和 VMware Integrated OpenStack 附加模块），可以构建和运维适合生产环境需求的云计算环境。

2.1.3 VMware vSphere 的优势

VMware vSphere 的优势主要分为以下几个方面。

1. 通过提高利用率和实现自动化获得高效率

可实现 15∶1 或更高的整合率，将硬件利用率从 5%～15%提高到 80%甚至更高，而且无需牺牲性能。

2. 在整个云计算基础架构范围内最大限度地增加正常运行时间

减少计划外停机时间并消除用于服务器和存储维护的计划内停机时间。

3. 大幅降低 IT 成本

使资金开销降幅高达 70%，运营开销降幅高达 30%，从而为 VMware vSphere 上运行的每个应用降低 20%～30% 的 IT 基础架构成本。

4. 兼具灵活性和可控性

快速响应不断变化的业务需求而又不牺牲安全性或控制力，并且为 VMware vSphere 上运行的所有关键业务应用提供零接触式基础架构以及内置的可用性、可扩展性和性能保证。

5. 可自由选择

使用基于标准的通用平台，既可利用已有 IT 资产又可利用新一代 IT 服务，并且通过开放 API 与来自全球领先技术提供商体系的解决方案集成，增强 VMware vSphere。

2.1.4 VMware vSphere 6.5 的新特性

相对于 VMware vSphere 6.0 来说，VMware vSphere 6.5 提供了更多的特性，可以提供高度可用、弹性和按需定制的云基础架构，从而运行、保护并管理多种应用程序。

如图 2-1-1、图 2-1-2 所示为 VMware vSphere 6.5 的部分新功能。

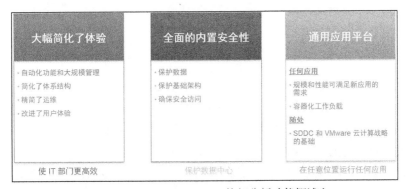

图 2-1-1　VMware vSphere 6.5 的部分新功能概述之一

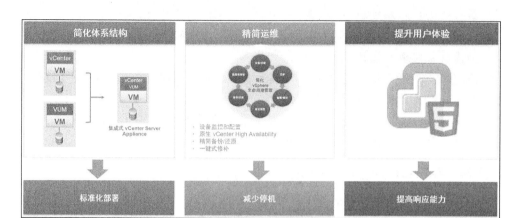

图 2-1-2　VMware vSphere 6.5 的部分新功能概述之二

1. 标准化 vCenter Server 部署并简化管理

从 VMware vSphere 6.5 版本开始，VMware vCenter Appliance 作为"首选"vCenter Server，支持企业扩展、高可用性和备份，提供全面的设备管理和监控。

VMware vCenter Appliance 简化了 vSphere 生命周期管理，通过加快 vCenter Server 部署速度提高了业务敏捷性，同时因没有 Windows 或 MS SQL 许可要求而降低了总体拥有成本。如图 2-1-3 所示为 VMware vCenter Appliance 6.5 支持的主机数量以及虚拟机数量等。

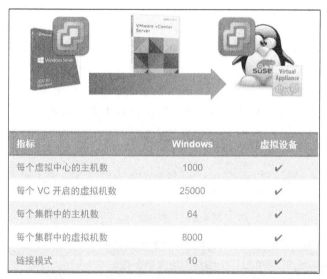

图 2-1-3　VMware vCenter Appliance 6.5 具有的新特性

2. vSphere Client（HTML5）支持

VMware vSphere 6.5 版本 Web Client 管理端开始提供 HTML5 的支持。由于是刚发布的第一个支持版本，因此目前只具备基础的管理功能（如图 2-1-4 所示）。

3. 提供全面的内置安全性

VMware vSphere 6.5 加强了内置的安全性管理，提供了虚拟机加密、vMotion 传输加密等新的安全特性（如图 2-1-5 所示）。

2.1 VMware vSphere 6.5 虚拟化介绍

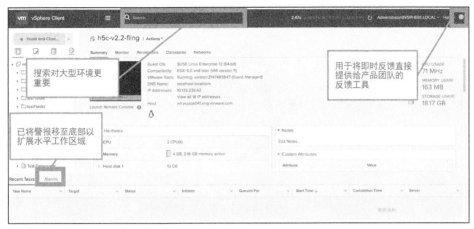

图 2-1-4　vSphere Client（HTML5）支持

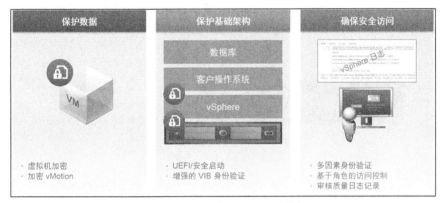

图 2-1-5　提供全面的内置安全性

4．通用应用平台支持

VMware vSphere 6.5 版本能够更好地支持 vSphere 家族产品，常见的 vRealize、NSX、Virtual SAN 以及 Horizon 等产品能够更好地为软件定义数据中心服务（如图 2-1-6 所示）。

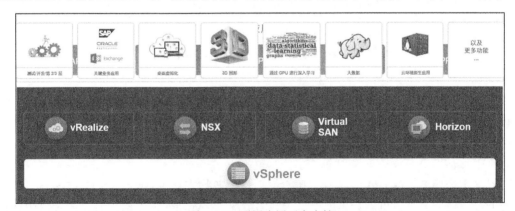

图 2-1-6　通用应用平台支持

5．vSphere Integrated Containers

VMware vSphere 6.5 版本提供了对主流容器的支持，兼容 Docker 接口等，可以方便开发人员便捷地开发出对应的程序（如图 2-1-7 所示）。

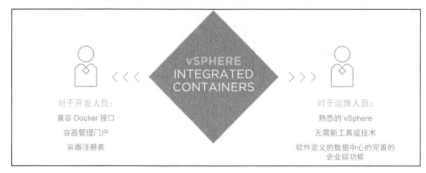

图 2-1-7　vSphere Integrated Containers

6．自由选择如何搭建和使用云平台

VMware vSphere 6.5 版本通过软件定义存储以及软件定义网络，能够更好地打造私有云、公有云以及混合云（如图 2-1-8 所示）。

图 2-1-8　自由选择如何搭建和使用云平台

整体来说，与 VMware vSphere 6.0 版本相比，VMware vSphere 6.5 版本有多种创新，但尚不是跨越式的升级。关于 VMware vSphere 6.5 版本的其他新功能可以查阅 VMware 官方网站。

2.2　部署 VMware ESXi 6.5

VMware ESXi 6.5 的安装过程与 VMware ESXi 6.0 基本一样，同时 VMware ESXi 6.5 提供了更多硬件支持，主流的服务器均可完成安装。

2.2.1　部署 ESXi 6.5 硬件要求

目前主流服务器的 CPU、内存、硬盘、适配器等均支持 VMware ESXi 6.5 安装。需要注意的是，使用兼容机可能会出现无法安装的情况。VMware 官方推荐的部署 ESXi 6.5 的硬件标准如下。

1．处理器

ESXi 6.5 支持 2006 年 9 月后发布的 64 位 x86 处理器，这其中包括了多种多核处理器。

需要说明的是，CPU 必须能够支持硬件虚拟化（Intel VT-x 或 AMD RVI）技术。

2. 内存

ESXi 6.5 要求物理服务器至少具有 4GB 或以上内存，生产环境至少推荐 8GB 以上，这样才能满足虚拟机的基本运行。

3. 适配器

ESXi 6.5 要求物理服务器至少具有 2 个 1GE 以上的适配器，对于使用 vSAN 软件定义存储的环境推荐 10GE 以上的适配器。

4. 存储适配器

SCSI 适配器、光纤通道适配器、聚合的网络适配器、iSCSI 适配器或内部 RAID 控制器。

5. 硬盘

ESXi 6.5 支持主流的 SATA、SAS、SSD 硬盘安装，同时也支持 SD 卡、U 盘等非硬盘介质安装。需要说明的是，使用 USB 和 SD 设备容易对 I/O 产生影响，安装程序不会在这些设备上创建暂存分区。

对于硬件方面的详细要求，可以参考 VMware 官方网站的《VMware 兼容性指南》。VMware 的官方网址为 http://www.vmware.com/resources/compatibility。

2.2.2 部署 ESXi 6.5

准备好安装介质后，就可以开始部署 VMware ESXi 6.5。本节实战操作将通过 DELL 远程管理卡安装 VMware ESXi 6.5 系统。

第 1 步，选择"ESXi-6.5.0-4564106-standard Installer"（如图 2-2-1 所示），按【Enter】键开始安装 VMware ESXi 6.5。其中，4564106 代表 VMware ESXi 6.5 版本号。

图 2-2-1　部署 VMware ESXi 6.5 之一

第 2 步，系统开始加载安装文件（如图 2-2-2 所示）。需要注意的是，如果物理服务器硬件不支持或 BIOS 相关参数未打开虚拟化支持，可能会出现错误提示，无法继续安装 VMware ESXi 6.5。

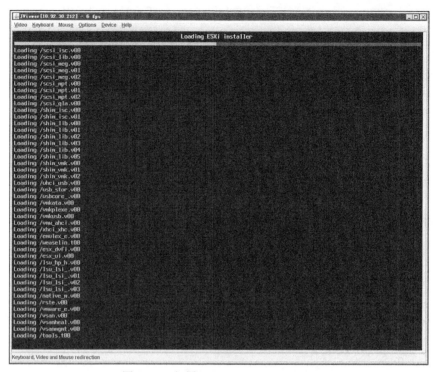

图 2-2-2　部署 VMware ESXi 6.5 之二

第 3 步，进入 VMware ESXi 6.5 基本文件加载界面（如图 2-2-3 所示）。

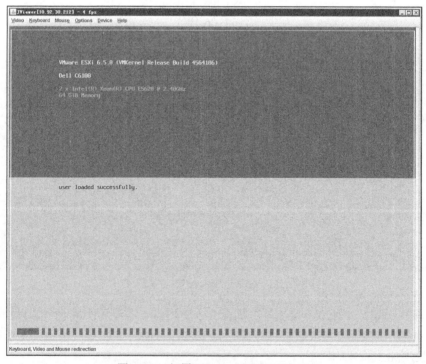

图 2-2-3　部署 VMware ESXi 6.5 之三

第 4 步，加载文件完成后会出现如图 2-2-4 所示界面，按【Enter】键开始安装 VMware ESXi 6.5。

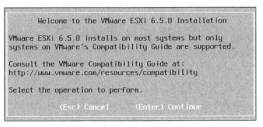

图 2-2-4　部署 VMware ESXi 6.5 之四

第 5 步，系统出现 "End User License Agreement（EULA）"界面，也就是最终用户许可协议（如图 2-2-5 所示），按【F11】键接受 "Accept and Continue"，接受许可协议。

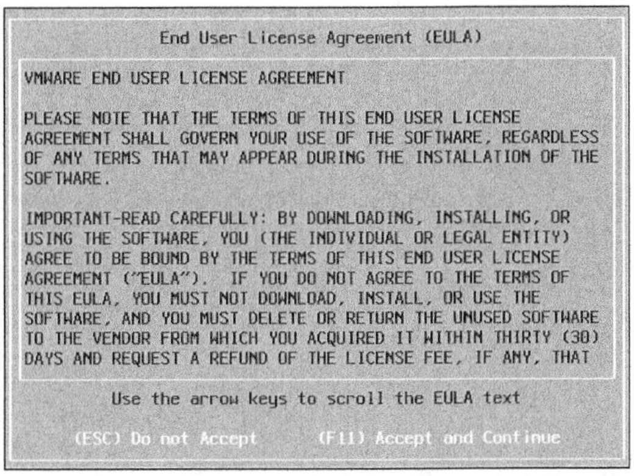

图 2-2-5　部署 VMware ESXi 6.5 之五

第 6 步，系统对服务器硬盘进行扫描（如图 2-2-6 所示）。

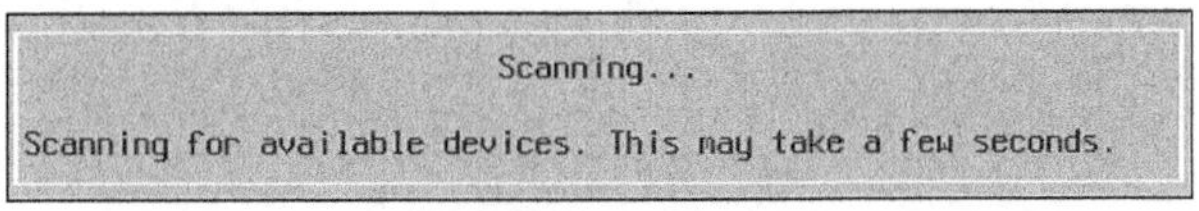

图 2-2-6　部署 VMware ESXi 6.5 之六

第 7 步，系统提示选择安装 VMware ESXi 6.5 的硬盘，ESXi 支持 U 盘以及 SD 卡安装，选择服务器使用的 SanDisk U 盘进行安装（如图 2-2-7 所示），按【Enter】键继续安装。

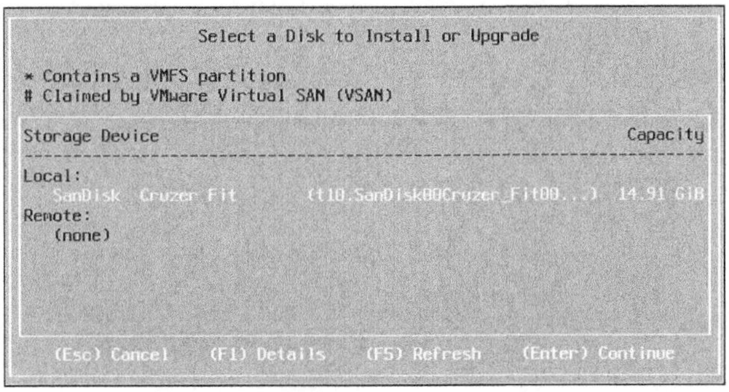

图 2-2-7　部署 VMware ESXi 6.5 之七

第 8 步，提示选择键盘类型，选择"US Default"，默认美国标准（如图 2-2-8 所示），按【Enter】键继续。

图 2-2-8　部署 VMware ESXi 6.5 之八

第 9 步，系统提示配置 root 用户的密码（如图 2-2-9 所示），根据实际情况输入，按【Enter】键继续。

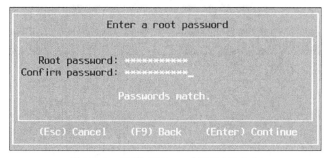

图 2-2-9　部署 VMware ESXi 6.5 之九

第 10 步，系统提示 VMware ESXi 6.5 将安装在刚才选择的 SanDisk U 盘（如图 2-2-10 所示），按【F11】键开始安装。

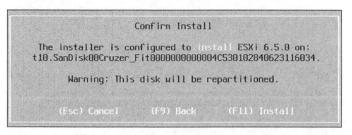

图 2-2-10　部署 VMware ESXi 6.5 之十

第 11 步，开始安装 VMware ESXi 6.5（如图 2-2-11 所示）。

图 2-2-11　部署 VMware ESXi 6.5 之十一

第 12 步，安装的时间取决于服务器的性能，等待一段时间后即可完成 VMware ESXi 6.5 的安装（图 2-2-12 所示），按【Enter】键重启服务器。

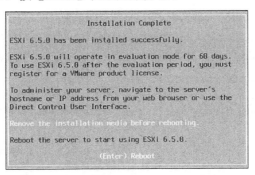

图 2-2-12　部署 VMware ESXi 6.5 之十二

第 13 步，服务器重启完成后，进入 VMware ESXi 6.5 正式界面（如图 2-2-13 所示）。

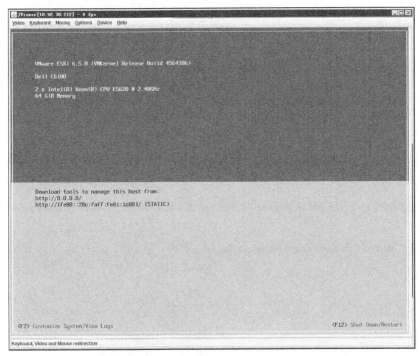

图 2-2-13　部署 VMware ESXi 6.5 之十三

第 14 步，按【F2】键输入 root 用户密码进入主机配置模式（如图 2-2-14 所示）。

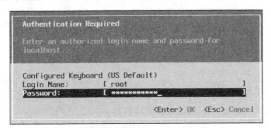

图 2-2-14　部署 VMware ESXi 6.5 之十四

第 15 步，选择"Configure Management Network"配置管理网络（如图 2-2-15 所示），按【Enter】键继续。

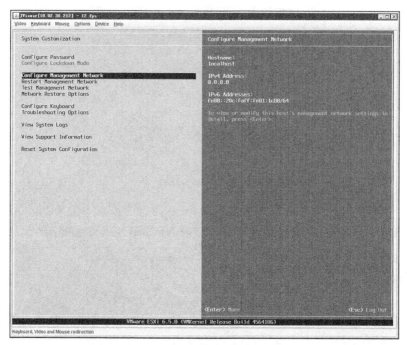

图 2-2-15　部署 VMware ESXi 6.5 之十五

第 16 步，选择"Network Adapters"对适配器进行配置（如图 2-2-16 所示），按【Enter】键继续。

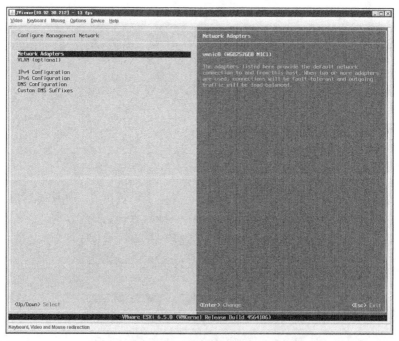

图 2-2-16　部署 VMware ESXi 6.5 之十六

第 17 步，默认情况一般使用 vmnci0（如图 2-2-17 所示）。如果需要调整管理适配器，可以通过空格键进行选择，按【Enter】键继续。

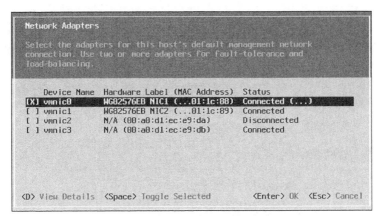

图 2-2-17　部署 VMware ESXi 6.5 之十七

第 18 步，因为实战环境交换机端口默认模式为 TRUNK，所以需要配置 VLAN ID。输入相应的 VLAN ID（如图 2-2-18 所示），按【Enter】键继续。

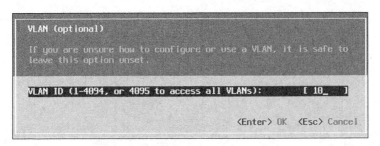

图 2-2-18　部署 VMware ESXi 6.5 之十八

第 19 步，选择 "IPv4 Configuration" 对 IP 进行配置，按【Enter】键进入配置界面。选择 "Set static IPv4 address and network configuration"，配置静态地址、子网掩码、默认网关（如图 2-2-19 所示），按【Enter】键完成配置。

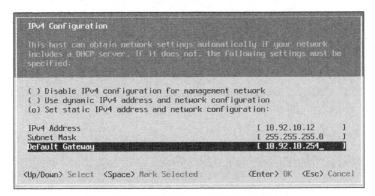

图 2-2-19　部署 VMware ESXi 6.5 之十九

第 20 步，系统询问是否确定修改管理网络配置（如图 2-2-20 所示），确定按【Y】键继续。

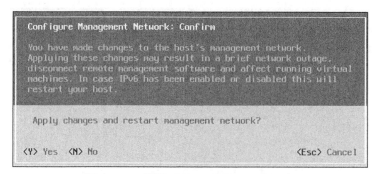

图 2-2-20　部署 VMware ESXi 6.5 之二十

第 21 步，ESXi 主机 IP 修改配置完成（如图 2-2-21 所示）。

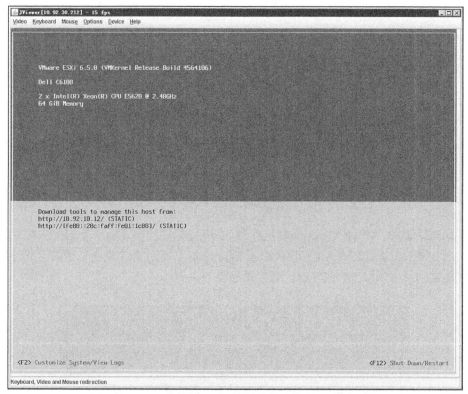

图 2-2-21　部署 VMware ESXi 6.5 之二十一

第 22 步，运行安装好的 VMware vSphere Client，输入 IP 地址，用户名 root，密码为安装 ESXi 主机时所设置的密码（如图 2-2-22 所示），单击"登录"按钮。

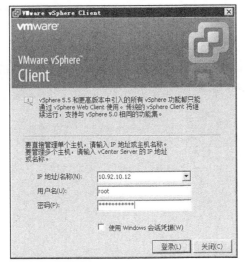

图 2-2-22　部署 VMware ESXi 6.5 之二十二

第 23 步，系统出现 "安全警告"（如图 2-2-23 所示），勾选 "安装此证书并且不显示针对'10.92.10.12'的任何安全警告（N）"，单击 "忽略" 按钮。

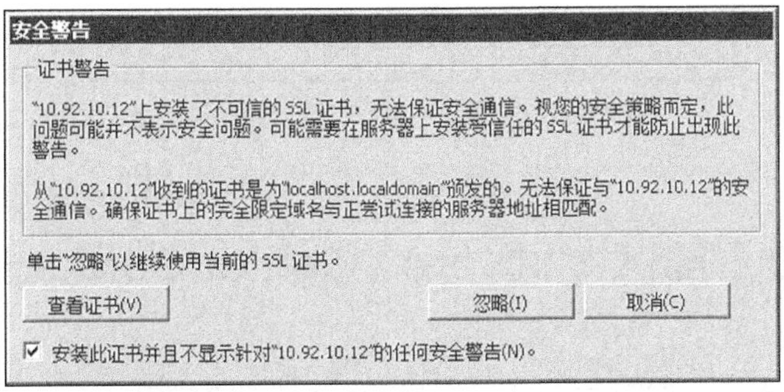

图 2-2-23　部署 VMware ESXi 6.5 之二十三

第 24 步，由于 VMware vSphere Client 客户端工具之前连接过相同 IP 的 ESXi 主机，所以会出现一个 "安装警告" 提示使用新证书替换原证书（如图 2-2-24 所示），单击 "是" 按钮。

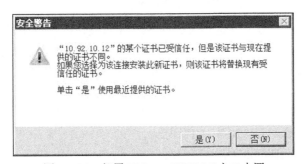

图 2-2-24　部署 VMware ESXi 6.5 之二十四

第 25 步，成功使用 vSphere Client 工具登录 ESXi 6.5 主机（如图 2-2-25 所示）。

图 2-2-25　部署 VMware ESXi 6.5 之二十五

2.3　部署 VMware vCenter Server 6.5

从 VMware vSphere 6.0 版本开始，VMware 一直推荐使用基于 Linux 版本的 vCenter Server。vCenter Server 6.5 版本不仅优化了对 Linux 的支持，而且发布了从 Windows 迁移到 Linux 版本的工具。

2.3.1　部署 vCenter Server 6.5 要求

从 vCenter Server 5.5 版本开始，vCenter Server 对硬件以及操作系统提出了新的要求。比如 Windows 版 vCenter Server 6.0，内存如果小于 8GB 会终止安装；Linux 版 vCenter Server 6.5，其配置内存超过 10GB。vCenter Server 6.5 对系统以及硬件的要求如下。

1. 操作系统要求

在 Windows 下安装 vCenter Server 6.5，需要使用以下操作系统：

（1）Windows Server 2008 Service Pack 2

（2）Windows Server 2008 R2

（3）Windows Server 2012

（4）Windows Server 2012 R2（推荐）

需要说明的是，Windows Server 2008 Service Pack 1 以及 Windows Server 2003 系统不再支持 vCenter Server 6.5 的安装。

2. CPU 要求

在 Winddows 下安装 vCenter Server 6.5，推荐使用 4 个或以上的 CPU。

3. 内存要求

在 Winddows 下安装 vCenter Server 6.5，需要配置 8GB 或以上内存，低于这个要求，安装会被终止。

2.3.2　部署 vCenter Server 6.5

由于 VMware 极力推荐使用 Linux 版本 vCenter Server 6.5，因此实战环境将部署 Linux 版本 vCenter Server 6.5。

第 1 步，使用光驱挂载或解压 VCSA 6.5 ISO 文件。此处使用挂载的方式，双击"installer"图标（如图 2-3-1 所示）。

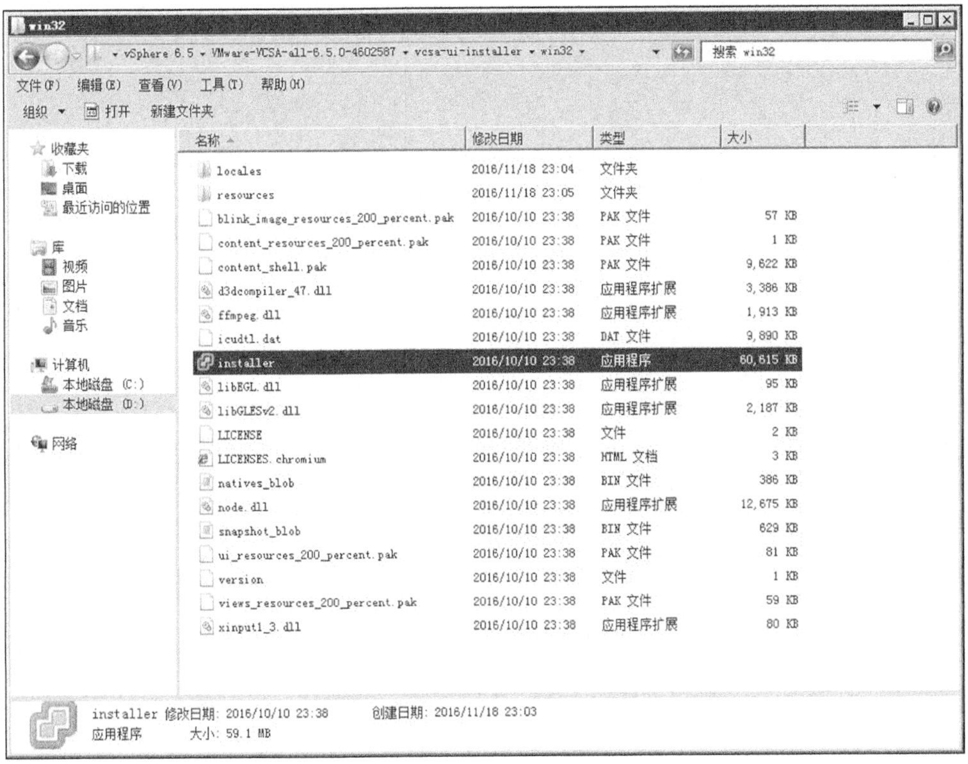

图 2-3-1　部署 vCenter Server 6.5 之一

第 2 步，进入 vCenter Server Appliance 6.5 Installer 向导，有"Install""Upgrade""Migrate""Restore"四个选项，分别为安装、升级、迁移以及重置（如图 2-3-2 所示）。

第 3 步，选择"Install"进入部署向导（如图 2-3-3 所示），单击"Next"按钮。

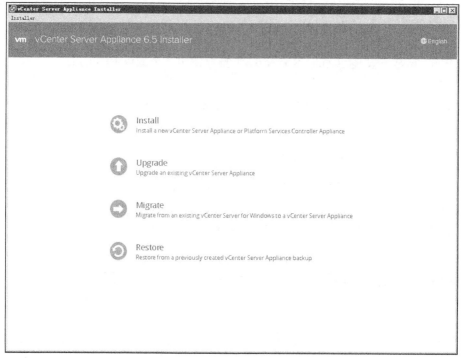

图 2-3-2　部署 vCenter Server 6.5 之二

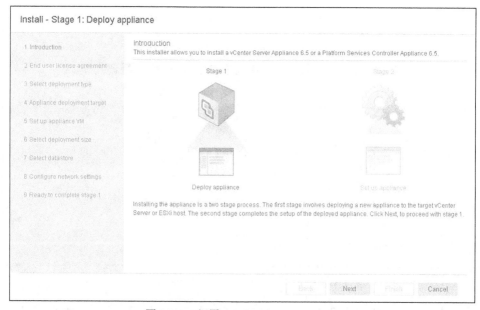

图 2-3-3　部署 vCenter Server 6.5 之三

第 4 步，接受"VMware 最终用户许可协议"，勾选"I accept the terms of the license agreement"（如图 2-3-4 所示），单击"Next"按钮。

第 5 步，选择使用"vCenter Server with an Embedded Platform Services Controller"模式（如图 2-3-5 所示），单击"Next"按钮。

2.3 部署 VMware vCenter Server 6.5

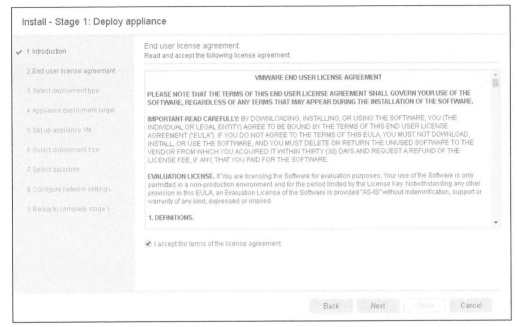

图 2-3-4　部署 vCenter Server 6.5 之四

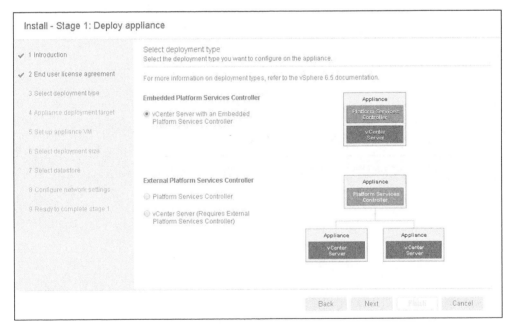

图 2-3-5　部署 vCenter Server 6.5 之五

第 6 步，设置 VCSA 6.5 需要安装的 ESXi 主机相关信息，输入设备的名称、root 用户密码（如图 2-3-6 所示），单击"Next"按钮。

第 7 步，系统出现证书警告提示（如图 2-3-7 所示），接受并继续应用单击"Yes"按钮。

第 8 步，设置 vCenter Server 虚拟机名字以及密码（如图 2-3-8 所示），单击"Next"按钮。

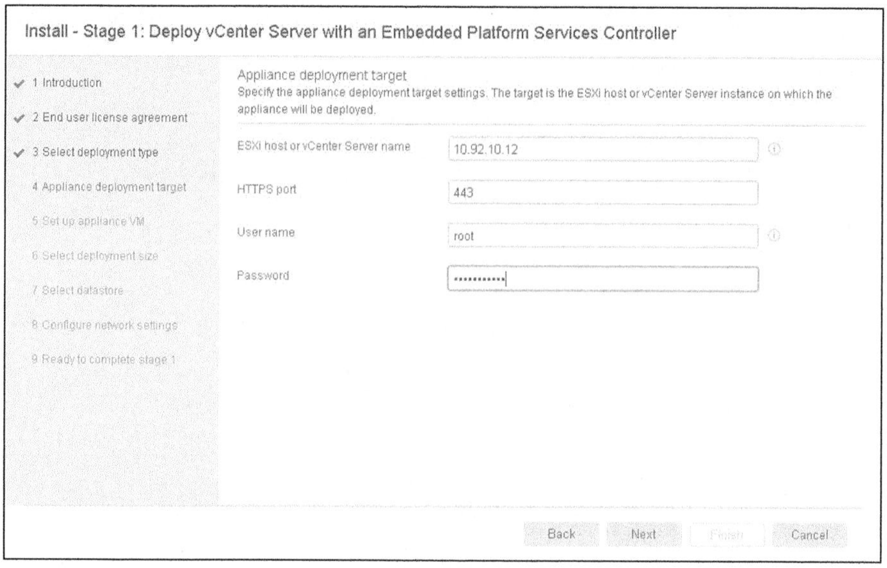

第 2 章 部署 VMware vSphere 6.5

图 2-3-6　部署 vCenter Server 6.5 之六

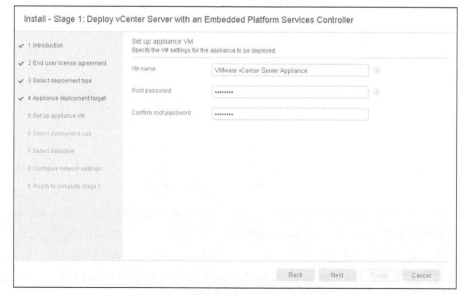

图 2-3-7　部署 vCenter Server 6.5 之七

图 2-3-8　部署 vCenter Server 6.5 之八

第 9 步，选择 vCenter Server 管理 ESXi 主机以及虚拟机规模，不同规模需要配置不同的 vCPU 以及内存（如图 2-3-9 所示），单击 "Next" 按钮。

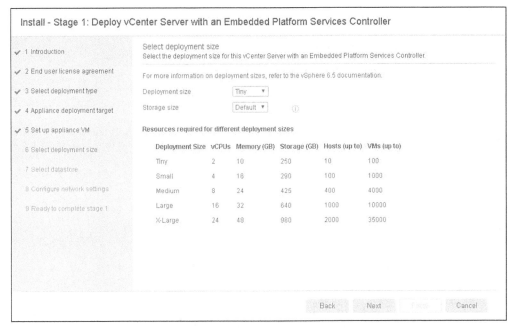

图 2-3-9　部署 vCenter Server 6.5 之九

第 10 步，选择 vCenter Server 虚拟机存放的存储（如图 2-3-10 所示），单击 "Next" 按钮。

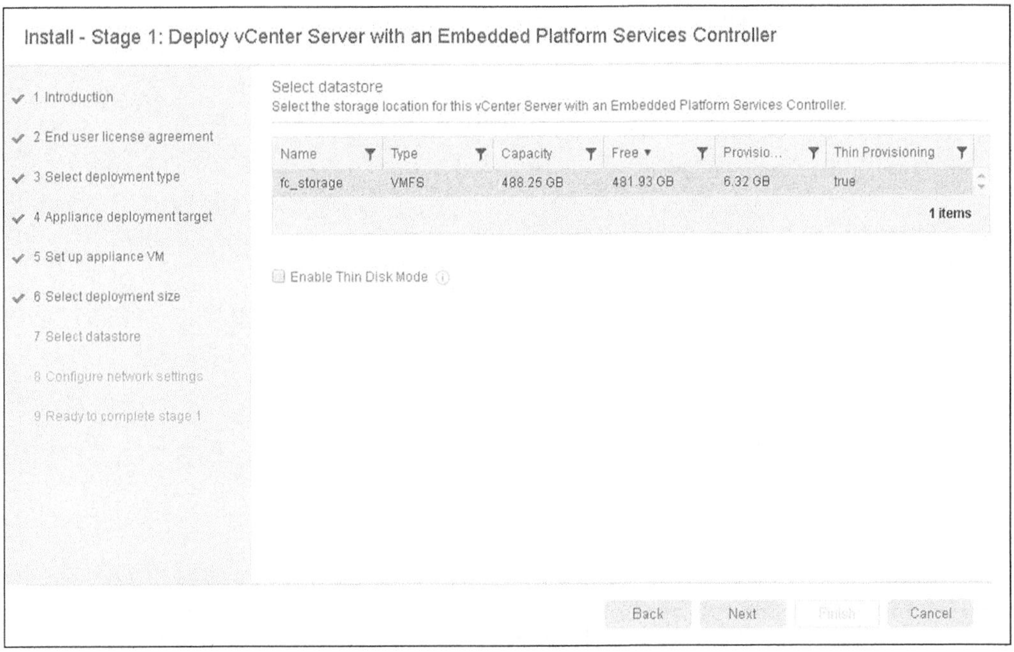

图 2-3-10　部署 vCenter Server 6.5 之十

第 11 步，设置 vCenter Server 虚拟机网络相关信息（如图 2-3-11 所示），单击"Next"按钮。

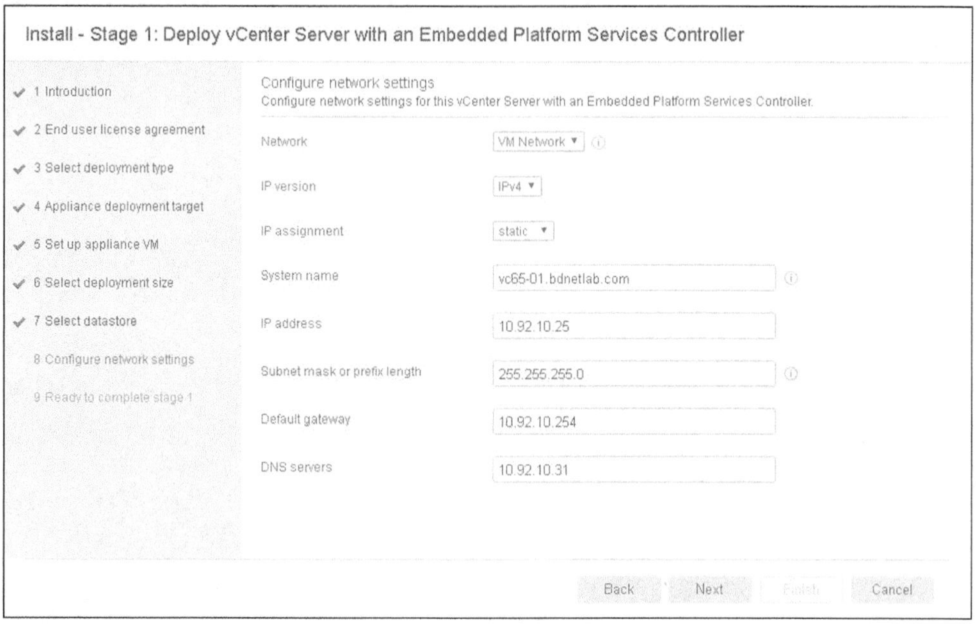

图 2-3-11　部署 vCenter Server 6.5 之十一

第 12 步，完成基本的参数设置，确认所有参数设置正确（如图 2-3-12 所示），单击"Finish"按钮开始部署。

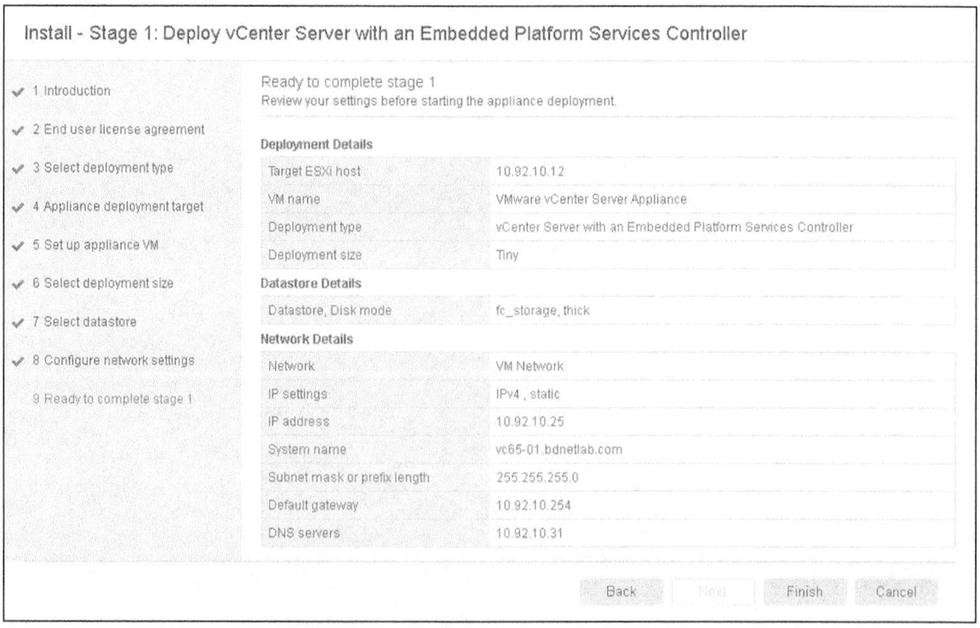

图 2-3-12　部署 vCenter Server 6.5 之十二

第 13 步，系统开始自动部署 vCenter Server 6.5（如图 2-3-13 所示）。

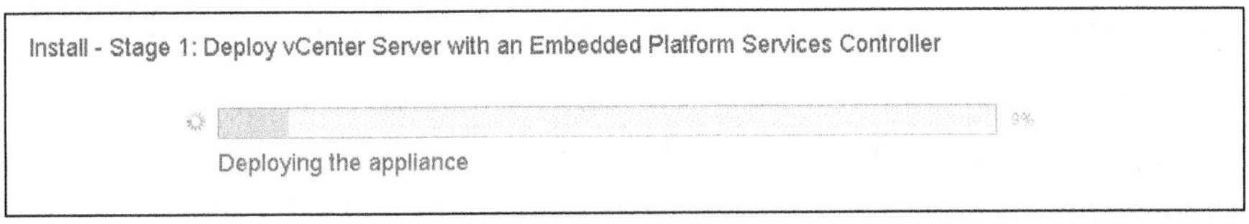

2.3 部署 VMware vCenter Server 6.5

图 2-3-13　部署 vCenter Server 6.5 之十三

第 14 步，完成 vCenter Server 6.5 部署。如果在部署过程存在问题，会给出相应的提示（如图 2-3-14 所示）。

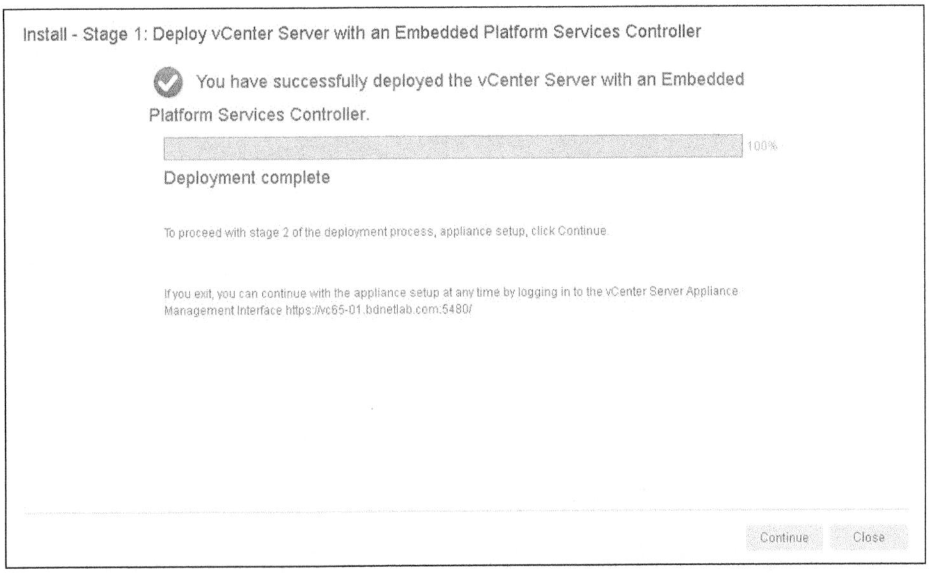

图 2-3-14　部署 vCenter Server 6.5 之十四

第 15 步，进入 vCenter Server 6.5 后续设置（如图 2-3-15 所示），单击"Next"按钮。

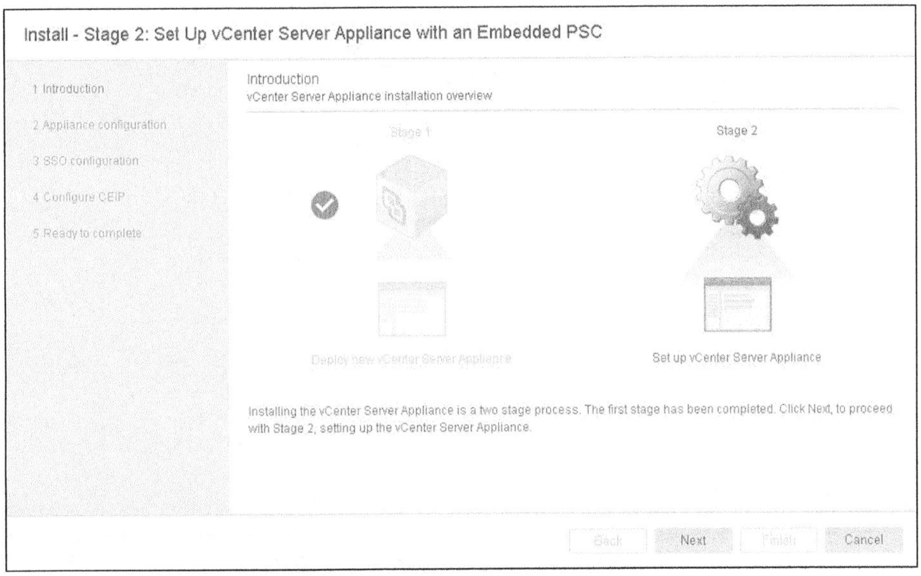

图 2-3-15　部署 vCenter Server 6.5 之十五

第 16 步，设置 vCenter Server 虚拟机时间同步。如果环境中有 NTP 时间服务器，可以指定 NTP 服务器；如果没有，应选择"Synchronize time with the ESXi host"与 ESXi 主机同步（如图 2-3-16 所示），单击"Next"按钮。

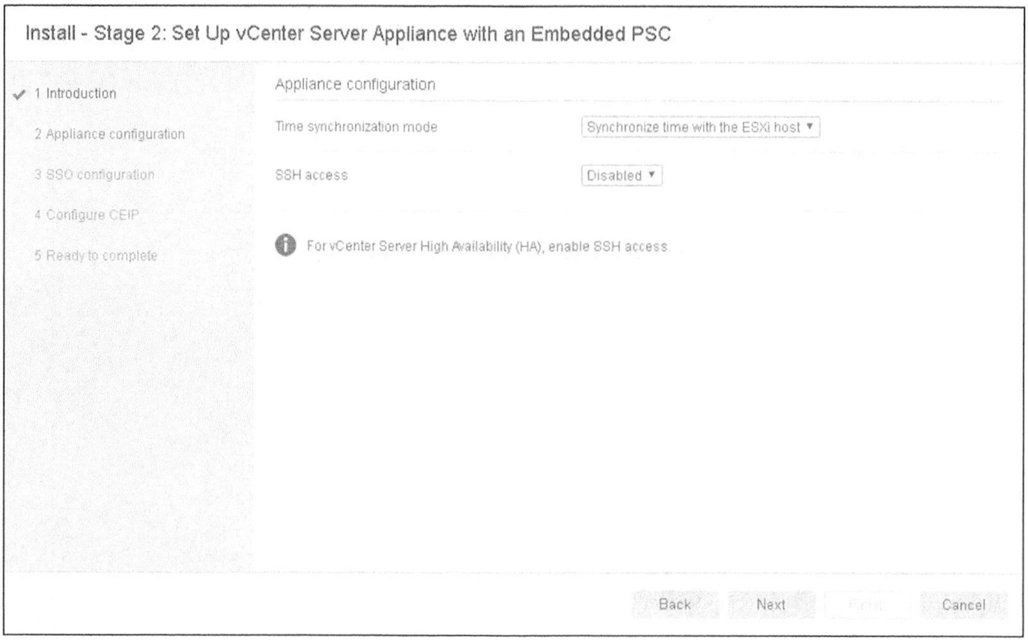

图 2-3-16　部署 vCenter Server 6.5 之十六

第 17 步，配置 SSO 相关信息（如图 2-3-17 所示），单击"Next"按钮。

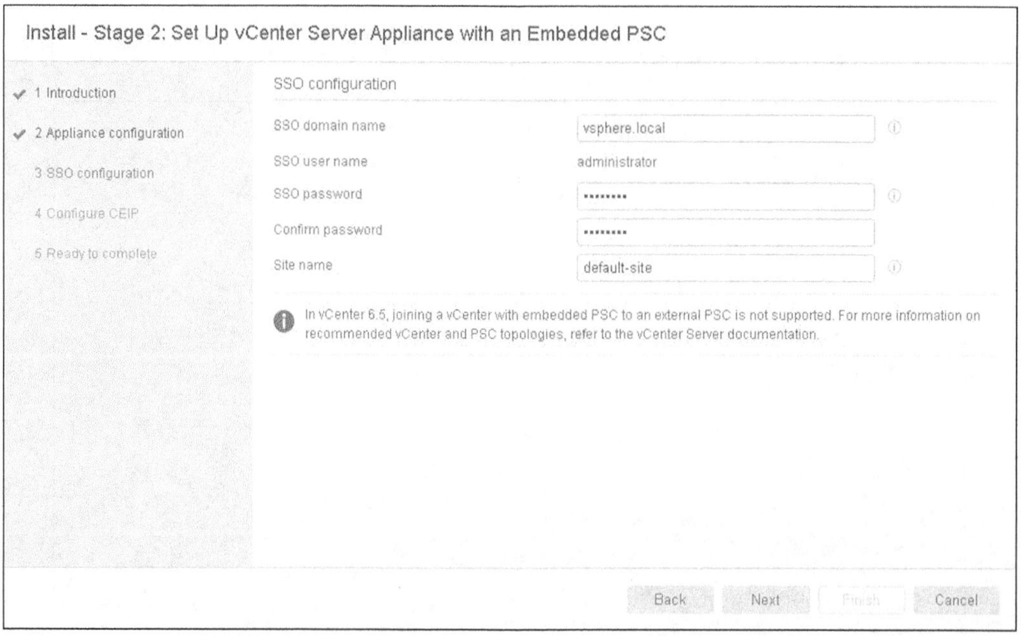

图 2-3-17　部署 vCenter Server 6.5 之十七

第 18 步，提示是否加入 VMware 客户体验改善计划，一般情况下取消勾选（如图 2-3-18 所示），单击"Next"按钮。

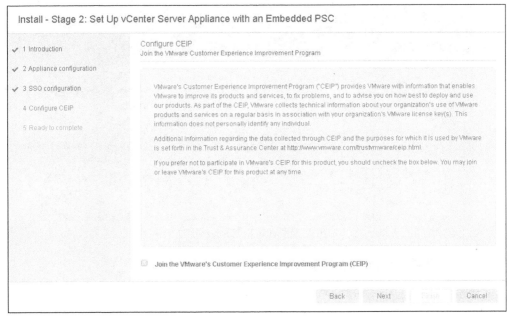

图 2-3-18　部署 vCenter Server 6.5 之十八

第 19 步，完成基本的参数设置，确认所有参数设置正确（如图 2-3-19 所示），单击"Finish"按钮开始部署。

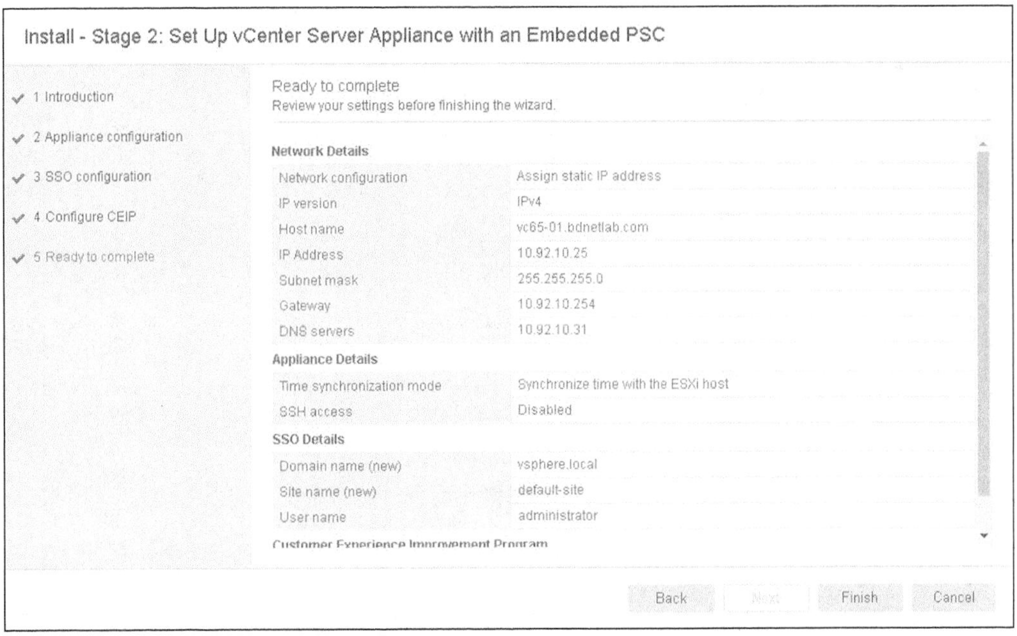

图 2-3-19　部署 vCenter Server 6.5 之十九

第 20 步，警告在部署过程中不能暂停或中止（如图 2-3-20 所示），单击"OK"按钮。

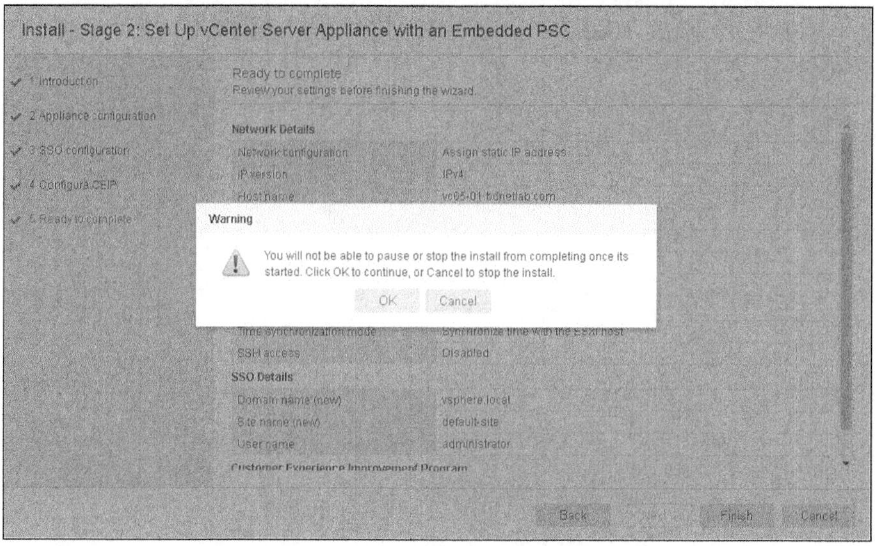

28　第 2 章　部署 VMware vSphere 6.5

图 2-3-20　部署 vCenter Server 6.5 之二十

第 21 步，系统自动部署并启动相关应用服务（如图 2-3-21 所示）。

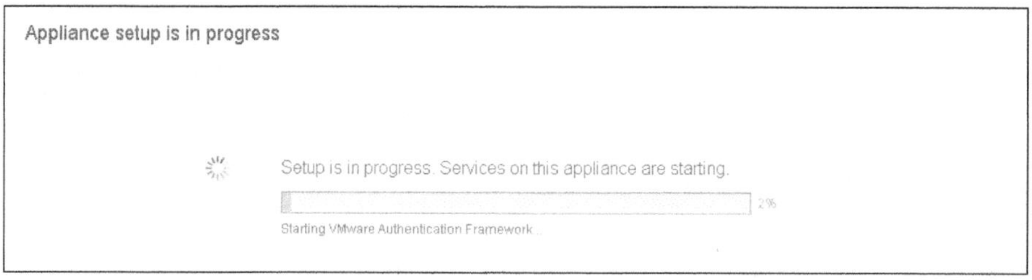

图 2-3-21　部署 vCenter Server 6.5 之二十一

第 22 步，完成部署并启动相关服务（如图 2-3-22 所示）。

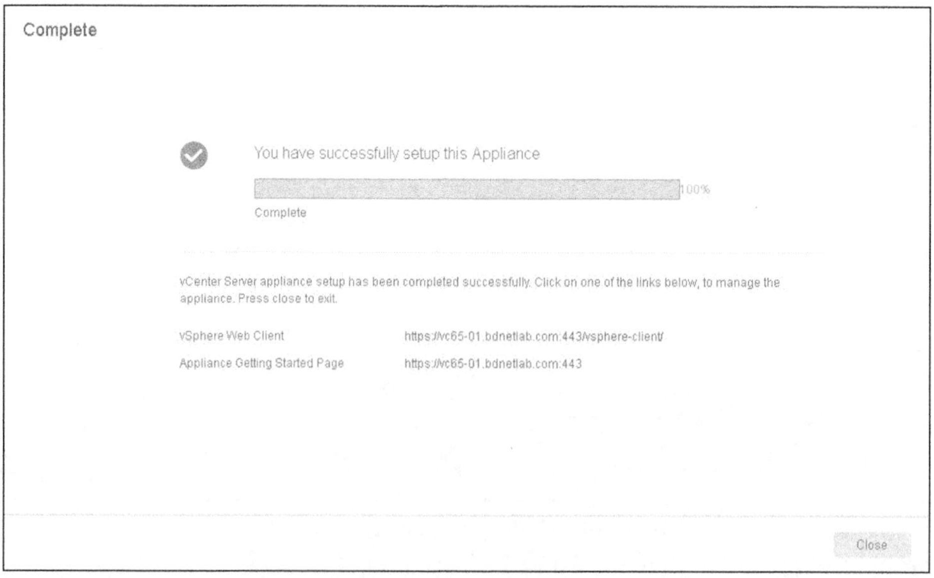

图 2-3-22　部署 vCenter Server 6.5 之二十二

第 23 步，使用浏览器访问 vCenter Server。通过图 2-3-23 可以看到，vCentere Server 6.5 提供了传统的 Web Client 以及 HTML5 两种访问方式。

图 2-3-23　部署 vCenter Server 6.5 之二十三

第 24 步，选择传统的 Web Client 登录 vCenter Server，输入用户名以及密码（如图 2-3-24 所示），单击"登录"按钮。

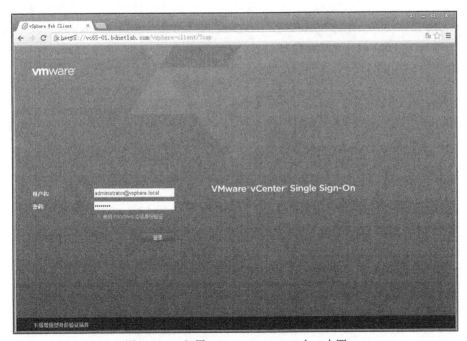

图 2-3-24　部署 vCenter Server 6.5 之二十四

第 25 步，成功登录 vCenter Server 6.5（如图 2-3-25 所示）。

图 2-3-25　部署 vCenter Server 6.5 之二十五

第 26 步，使用 HTML5 方式登录后的界面如图 2-3-26 所示。

图 2-3-26　部署 vCenter Server 6.5 之二十六

第27步，在 vCenter Server 上单击右键，选择"新建数据中心"（如图 2-3-27 所示）。

图 2-3-27　部署 vCenter Server 6.5 之二十七

第28步，输入新建数据中心的名称（如图 2-3-28 所示），单击"确定"按钮。

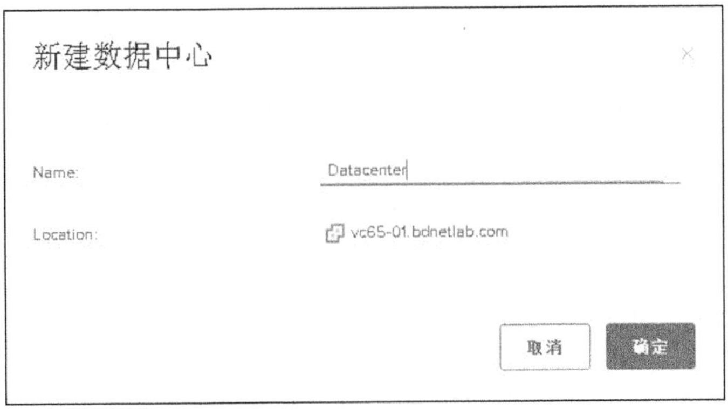

图 2-3-28　部署 vCenter Server 6.5 之二十八

第29步，数据中心创建后，选择"添加主机"，将安装好的 ESXi 6.5 主机加入 vCenter Server 管理（如图 2-3-29 所示）。

第30步，输入需要添加的 ESXi 主机名或 IP 地址（如图 2-3-30 所示），单击"NEXT"按钮。

图 2-3-29　部署 vCenter Server 6.5 之二十九

图 2-3-30　部署 vCenter Server 6.5 之三十

第 31 步，输入 ESXi 主机用户名以及密码（如图 2-3-31 所示），单击"NEXT"按钮。

第 32 步，系统出现安全警示，提示 vCenter Server 的证书无法验证该证书（如图 2-3-32 所示），单击"是"按钮使用新证书替换原证书。

第 33 步，确认需要添加的主机参数设置正确（如图 2-3-33 所示），单击"NEXT"按钮。

2.3 部署 VMware vCenter Server 6.5

图 2-3-31 部署 vCenter Server 6.5 之三十一

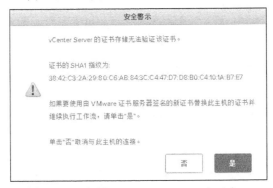

图 2-3-32 部署 vCenter Server 6.5 之三十二

图 2-3-33 部署 vCenter Server 6.5 之三十三

第 34 步，系统提示向 ESXi 主机分配许可证。如果没有正式许可，可以使用 60 天的评估版本（如图 2-3-34 所示）。单击"NEXT"按钮。

图 2-3-34　部署 vCenter Server 6.5 之三十四

第 35 步，基于安全考虑，系统提示是否使用锁定模式（如图 2-3-35 所示），单击"NEXT"按钮。

图 2-3-35　部署 vCenter Server 6.5 之三十五

第 36 步，设置 ESXi 主机存放的位置（如图 2-3-36 所示），单击"NEXT"按钮。

图 2-3-36　部署 vCenter Server 6.5 之三十六

第 37 步，确认添加主机的相关参数设置正确（如图 2-3-37 所示），单击"FINISH"按钮。

图 2-3-37　部署 vCenter Server 6.5 之三十七

第 38 步，成功将 ESXi 主机添加到 vCenter Server（如图 2-3-38 所示）。

图 2-3-38　部署 vCenter Server 6.5 之三十八

第 39 步，使用相同的方式将其余 ESXi 主机添加进 vCenter Server（如图 2-3-39 所示）。

图 2-3-39　部署 vCenter Server 6.5 之三十九

第 40 步，查看 vCenter Server 数据中心相关信息，通过图 2-3-40 可以看到已添加的两台 ESXi 主机的状态。

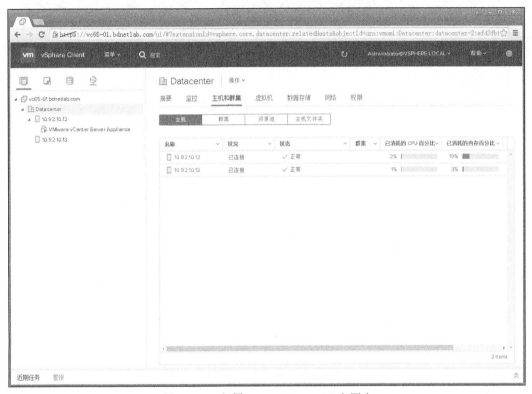

图 2-3-40　部署 vCenter Server 6.5 之四十

2.4　创建使用虚拟机

完成 ESXi 以及 vCenter Server 安装后，就可以创建使用虚拟机。虚拟机正常运行也是整个虚拟化架构正常运行的关键之一。作为虚拟化架构实施人员或者管理人员，必须考虑如何在企业生产环境构建高可用虚拟化环境，以保证虚拟机的正常运行。

VMware vSphere 6.5 对于 Windows 操作系统的支持是非常完善的，从早期的 MS-DOS 到最新的 Windows Server 2016，几乎覆盖了整个 Windows 操作系统。当然 VMware vSphere 6.5 对 Linux 系统的支持也是非常完善的，基本 Redhat、Centos、SUSE 等主流厂商各个版本的 Linux 都能够运行。

2.4.1　创建使用 Windows 虚拟机

第 1 步，使用 Web Client 客户端登录 vCenter Server，在集群或主机上单击右键，选择"新建虚拟机"（如图 2-4-1 所示）。

第 2 章 部署 VMware vSphere 6.5

图 2-4-1　创建 Windows 虚拟机之一

第 2 步，进入新建虚拟机向导，选择"创建新虚拟机"（如图 2-4-2 所示），单击"NEXT"按钮。

图 2-4-2　创建 Windows 虚拟机之二

第 3 步，输入需要新创建虚拟机的名称，同时选择所属的数据中心（如图 2-4-3 所示），单击"NEXT"按钮。

图 2-4-3 创建 Windows 虚拟机之三

第 4 步，由于还未启用集群的高级特性，所以需要指定虚拟机运行的 ESXi 主机（如图 2-4-4 所示），单击"NEXT"按钮。

图 2-4-4 创建 Windows 虚拟机之四

第 5 步，选择虚拟机文件放置的位置（如图 2-4-5 所示），单击"NEXT"按钮。

图 2-4-5 创建 Windows 虚拟机之五

第 6 步，选择 ESXi 主机硬件兼容版本，一般推荐使用最新硬件版本以获取更好的特性，但如果环境中存在其他硬件版本的虚拟机，推荐使用下向兼容的版本（如图 2-4-6 所示），单击"NEXT"按钮。

图 2-4-6 创建 Windows 虚拟机之六

第 7 步，选择虚拟机操作系统类型。系统内置了 Windows、Linux 以及其他多种操作系统，根据实际情况选择（如图 2-4-7 所示）。

图 2-4-7 创建 Windows 虚拟机之七

第 8 步，选择操作系统后，系统会给出一个基本的硬件配置（如图 2-4-8 所示），可以根据实际情况调整硬件配置，单击"NEXT"按钮。

图 2-4-8 创建 Windows 虚拟机之八

第 9 步，确认新建虚拟机的基本配置（如图 2-4-9 所示），单击"FINISH"按钮。

图 2-4-9　创建 Windows 虚拟机之九

第 10 步，新的虚拟机 WIN_2012R2 创建完成（如图 2-4-10 所示）。

图 2-4-10　创建 Windows 虚拟机之十

第 11 步，在新创建的虚拟机单击右键，选择"编辑设置"（如图 2-4-11 所示）。

图 2-4-11　创建 Windows 虚拟机之十一

第 12 步，通过 CD/DVD 驱动器挂载 Windows Server 2012 R2 安装 ISO，特别需要注意勾选"打开电源时连接"选项（如图 2-4-12 所示），单击"确定"按钮。

图 2-4-12　创建 Windows 虚拟机之十二

第 13 步，启动 VMware Remote Console（VMRC）控制台开始安装 Windows 2012 R2 操作系统（如图 2-4-13 所示）。

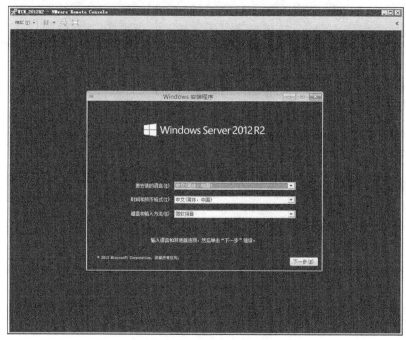

图 2-4-13　创建 Windows 虚拟机之十三

第 14 步，Windows Server 2012 R2 的安装过程与在物理服务器上安装一样，安装的时间取决于 ESXi 主机性能。如图 2-4-14 所示为安装完成后的界面。

图 2-4-14　创建 Windows 虚拟机之十四

第 15 步，通过图 2-4-15 可以看到，Windows 2012 R2 操作系统已安装成功，但 VMware Tools 处于未运行、未安装状态，单击"安装 VMware Tools"。

图 2-4-15　创建 Windows 虚拟机之十五

第 16 步，进入 VMware Tools 安装向导（如图 2-4-16 所示），单击"下一步"按钮。

图 2-4-16　创建 Windows 虚拟机之十六

第 17 步，一般情况下，选择典型安装 VMware Tools（如图 2-4-17 所示），单击"下一步"按钮。

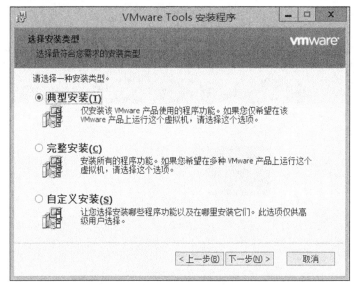

图 2-4-17 创建 Windows 虚拟机之十七

第 18 步，已准备好安装 Windows 系统 VMware Tools（如图 2-4-18 所示），单击 "安装" 按钮。

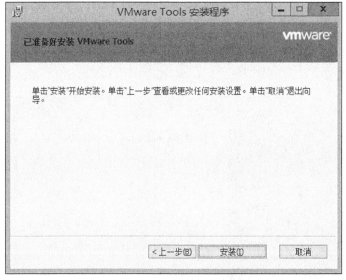

图 2-4-18 创建 Windows 虚拟机之十八

第 19 步，VMware Tools 安装完成（如图 2-4-19 所示），单击 "完成" 按钮。

第 20 步，VMware Tools 安装完成后需要重新启动虚拟机才能生效（如图 2-4-20 所示），单击 "是" 按钮。

第 21 步，重新启动虚拟机后可以看到 VMware Tools 处于正在运行状态（如图 2-4-21 所示）。

图 2-4-19　创建 Windows 虚拟机之十九

图 2-4-20　创建 Windows 虚拟机之二十

图 2-4-21　创建 Windows 虚拟机之二十一

2.4.2　创建使用 Linux 虚拟机

Linux 虚拟机的创建过程与 Windows 基本一致。需要说明的是，CentOS 7.0 以后的版本可以不安装 VMware Tools。

第 1 步，创建名为 CENTOS_7 的虚拟机（如图 2-4-22 所示）。

图 2-4-22　创建 Linux 虚拟机之一

第 2 步，挂载 CENTOS 7 安装 ISO，选择"Install CentOS 7"安装操作系统（如图 2-4-23 所示）。

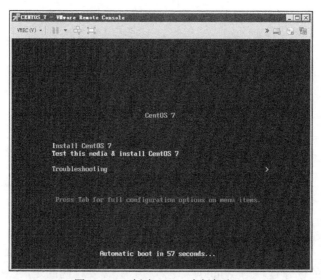

图 2-4-23　创建 Linux 虚拟机之二

第 3 步，安装过程与在物理服务器上安装一样，安装的时间取决于 ESXi 主机性能。如图 2-4-24 所示为安装完成后的界面。

图 2-4-24　创建 Linux 虚拟机之三

第 4 步，通过图 2-4-25 可以看到，CENTOS 7 操作系统已安装成功，VMware Tools 正处于运行状态。

图 2-4-25　创建 Linux 虚拟机之四

2.5 本章小结

本章介绍 VMware vSphere 6.5 虚拟化架构新增功能以及 ESXi 6.5、vCenter Server 6.5 的基本部署。由于本书的重点 vSphere 虚拟化架构企业级网络以及存储配置，仅介绍了基础的部署，如果读者对于数据库以及升级等有需求，可以参考作者已出版的《VMware vSphere 6.0 虚拟化架构实战指南》一书。

需要说明的是，本书的企业级网络及存储配置基本围绕 VMware vSphere 6.5 版本进行介绍，但是由于第三方软件发布更新问题，部分章节会使用到 VMware vSphere 的其他版本。

第 3 章 部署 VMware vSphere 基本网络

网络在 VMware vSphere 环境中相当重要，无论是管理 ESXi 主机还是 ESXi 主机上运行的虚拟机对外提供服务都必须依赖于网络。VMware vSphere 提供了强大的网络功能，其基本的网络配置就是标准交换机以及分布式交换机。在生产环境中，日常的配置除了在 ESXi 主机、vCenter Server 配置外，还涉及多种物理交换机的配置，我们先掌握基本的网络配置后再进行物理交换机配置延伸学习。本章介绍如何在 ESXi 主机上配置使用标准交换机以及如何在 vCenter Server 上配置使用分布式交换机等。

本章要点
- VMware vSphere 网络介绍
- 配置使用标准交换机
- 配置使用分布式交换机

3.1 VMware vSphere 网络介绍

VMware vSphere 网络是 ESXi 主机管理以及虚拟机外部通信的关键。如果配置不当，可能影响网络的性能，情况严重时将导致服务全部停止。

3.1.1 ESXi 主机通信原理介绍

ESXi 主机通过模拟出一个 Virtual Switch（虚拟交换机）的主机内虚拟机对外进行通信，其功能相当于一台传统的二层交换机，图 3-1-1 显示了 ESXi 主机的通信原理。

安装完 ESXi 主机后，会默认创建一个虚拟交换机，物理适配器作为虚拟标准交换机的上行链路接口与物理交换机连接对外提供服务。在图 3-1-1 中，左边有 4 台虚拟机，每台虚拟机配置 1 个虚拟适配器，这些虚拟适配器连接到虚拟交换机的端口，然后通过上行链路接口连接到物理交换机，虚拟机就可以对外提供服务。如果上行链路接口没有对应的物理适配器，那么这些虚拟机就成为一个网络孤岛，无法对外提供服务。

第 3 章 部署 VMware vSphere 基本网络

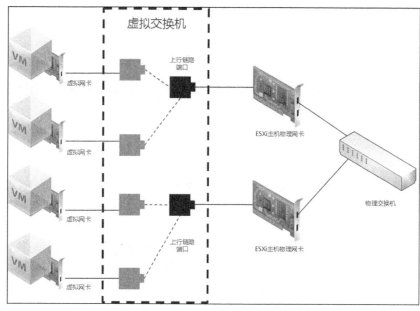

图 3-1-1　ESXi 主机通信原理

3.1.2　ESXi 主机网络组件介绍

介绍了 ESXi 主机通信原理后，下面对 ESXi 主机所涉及的网络组件进行简要的介绍。

1. Standard Switch（标准交换机）

Standard Switch，中文名为标准交换机，简称 vSS，是由 ESXi 主机虚拟出来的交换机。在安装完成 ESXi 后，系统会自动创建一个标准交换机 vSwitch0，这个虚拟交换机主要是提供管理、虚拟机与外界通信等功能。在生产环境中，一般会根据应用需要，创建多个标准交换机对各种流量进行分离、提供冗余以及负载均衡。图 3-1-2 显示了一台 ESXi 主机上的虚拟标准交换机，除了默认的 vSwitch0 以外，还创建了 vSwitch1 用于 iSCSI 以及 vSwitch2 用于 vMotion。在生产环境中应该根据实际情况创建多个标准交换机。

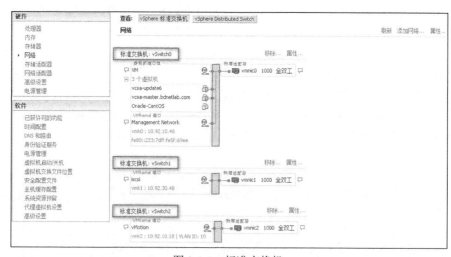

图 3-1-2　标准交换机

2. Distributed Switch（分布式交换机）

Distributed Switch，中文名为分布式交换机，简称 vDS，是 VMware 从 4.0 版本后推出的新一代网络交换机。vDS 是个横跨多台 ESXi 主机的虚拟交换机，减化了管理人员的配置。如果使用 vSS，需要在每台 ESXi 主机进行网络的配置，在 ESXi 主机数量较少的情况下，是比较适用的。如果 ESXi 主机数量较多，vSS 就不适用了，因为会极大增加管理人员的工作量。此时，使用 vDS 是更好的选择。图 3-1-3 显示了一台 ESXi 主机上的分布式交换机，上行链路接口有 3 个，分布式交换机上创建了多个端口组并划分了 VLAN，可以根据生产环境的实际情况在分布式交换机创建多个端口组用于各种用途。

图 3-1-3　分布式交换机

3. vSwitch Port（虚拟交换机端口）

vSwitch Port，中文名为虚拟交换机端口。ESXi 主机上创建的 vSwitch 相当于一个传统的二层交换机。默认情况下，一个 vSwitch 的端口数为 120，在 ESXi 6.5 版本中，最多可以设置 4088 个端口。图 3-1-4 显示了 vSwitch0 默认的端口数为 120。

图 3-1-4　虚拟交换机端口

4. Port Group（端口组）

Port Group，中文名为端口组。在一个 vSwitch 中，可以创建一个或多个 Port Group，并且针对不同的 Port Group 进行 VLAN 以及流量控制等方面的配置，然后将虚拟机划入不同的 Port Group，这样可以提供不同优先级的网络使用率。图 3-1-5 显示了一台 ESXi 主机上分布式交换机上创建的多个端口组。在生产环境中，可以创建多个端口组满足不同的应用。

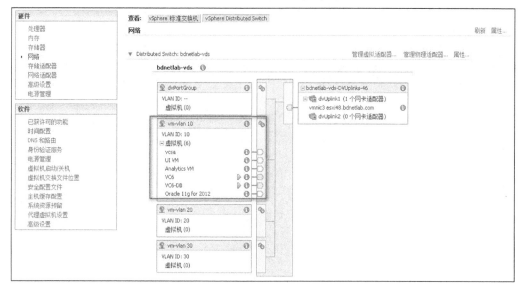

图 3-1-5　端口组

5. Virtual Machine Port Group（虚拟机端口组）

Virtual Machine Port Group，中文名为虚拟机端口组。在 ESXi 系统安装完成后系统自动创建的 vSwitch0 上默认创建一个虚拟机端口组，用于虚拟机外部通信使用。图 3-1-6 显示了一台刚安装完成的 ESXi 主机上的虚拟机端口组，同时这个交换机上还存在管理网络。在生产环境中，一般会将管理网络与虚拟机端口组进行分离配置。

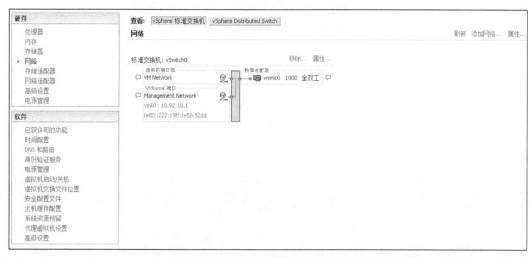

图 3-1-6　虚拟机端口组

3.1 VMware vSphere 网络介绍

6. VMkernel Port（VMware 端口）

VMkernel Port 在 ESXi 主机网络中是一个特殊的端口，VMware 对其的定义为运行特殊的流量端口，比如管理流量、iSCSI 流量、NFS 流量、vMotion 流量等。与虚拟机端口组不同的是，VMkernel Port 必须配置 IP 地址。图 3-1-7 显示了一台 ESXi 主机上的 VMkernel Port 配置，其中 vSwitch0 的 VMkernel Port 运行管理流量，vSwitch1 的 VMkernel Port 运行 iSCSI 流量，vSwitch2 的 VMkernel Port 运行 vMotion 流量。

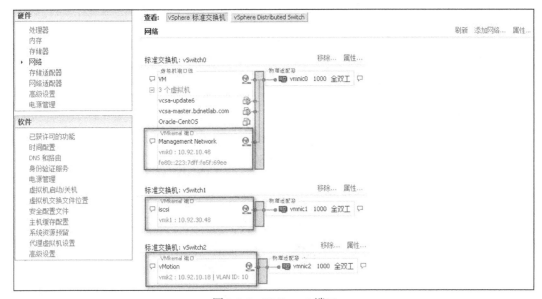

图 3-1-7　Vmkernel 端口

3.1.3 ESXi 主机网络 VLAN 实现方式

在生产环境中，VLAN 的使用相当普遍，ESXi 主机的标准交换机以及分布式交换机都支持 802.1Q 标准。当然，与传统的支持方式也有一定差异。ESXi 主机网络比较常用的实现方式有以下两种。

1. External Switch Tagging（EST 模式）

External Switch Tagging，简称 EST 模式。这种模式的好处是将 ESXi 主机物理适配器对应的物理交换机端口划入 VLAN 即可，ESXi 主机不需额外配置。图 3-1-8 显示了 EST 模式下 VLAN 实现方式，这种模式下只需将 ESXi 主机物理适配器对应的物理交换机端口划入 VLAN，该端口就会传递相应的 VLAN 信息。

2. Virtual Switch Tagging（VST 模式）

Virtual Switch Tagging，简称 VST 模式。这种模式要求 ESXi 主机物理适配器对应的物理交换机端口模式必须为 TRUNK 模式，同时 ESXi 主机需要启用 TRUNK 模式，以便端口组接受相应的 VLAN Tag 信息。图 3-1-9 显示了 VST 模式下 VLAN 实现方式，这种模式下首先需要配置物理交换机端口模式为 TRUNK，然后在 ESXi 主机网络对应的端口组下配置对应的 VLAN 信息。

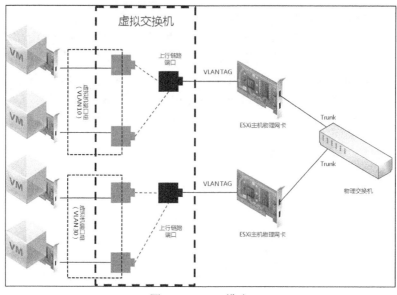

图 3-1-8　EST 模式

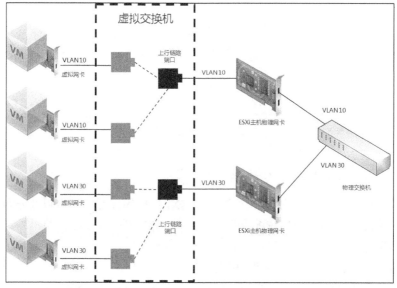

图 3-1-9　VST 模式

3.1.4　ESXi 主机网络 NIC Teaming

如果 ESXi 主机的虚拟交换机只使用一个物理适配器，那么就存在单点故障，当这个物理适配器出现故障时整个网络将中断，ESXi 主机服务就全部停止。所以，对于虚拟交换机来说，负载均衡是必须考虑的事情，当一个虚拟交换机有多个物理适配器的时候，就可以形成负载均衡。多物理适配器情况下负载均衡的实现主要有以下几种方式。

1. Originating Virtual Port ID（源虚拟端口 ID）

Originating Virtual Port ID，基于源虚拟端口负载均衡，ESXi 主机网络默认的负载均衡方式。这种方式下，系统会将虚拟机适配器与虚拟交换机所属的物理适配器进行对应和绑定，绑定后虚拟机流量始终走虚拟交换机分配的物理适配器，不管这个物理适配器流量是否过载，除非当分配的这个物理适配器故障后才会尝试走另外活动的物理适配器。也就是说，基于源虚拟端口负载均衡不属于动态的负载均衡方式，但可以实现冗余功能。

图 3-1-10 显示了基于源虚拟端口负载均衡。在这种模式下，虚拟机通过算法与 ESXi 主机物理适配器进行绑定，虚拟机 01 和虚拟机 02 与 ESXi 主机物理适配器 vmnic0 进行绑定，虚拟机 03 和虚拟机 04 与 ESXi 主机物理适配器 vmnic1 进行绑定，无论网络流量是否过载，虚拟机只会通过绑定的适配器对外进行通信。当虚拟机 03 和虚拟机 04 绑定 ESXi 主机物理适配器 vmnic1 出现故障时，虚拟机才会使用 ESXi 主机物理适配器 vmnic0 对外进行通信，如图 3-1-11 所示。

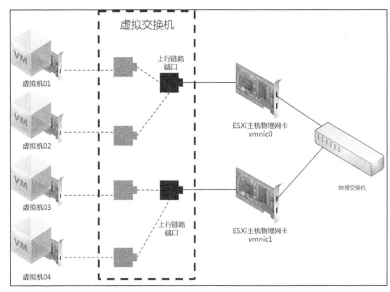

图 3-1-10　基于源虚拟端口负载均衡之一

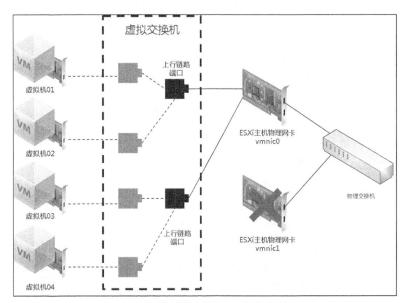

图 3-1-11　基于源虚拟端口负载均衡之二

2. Source MAC Hash（源 MAC 地址哈希算法）

Source MAC Hash，基于源 MAC 地址哈希算法负载均衡，这种方式与基于源虚拟端口负载均衡方式相似，如果虚拟机只使用一个物理适配器，那么它的源 MAC 地址不会发生任何变化，系统分配物理适配器以及绑定后，虚拟机流量始终走虚拟交换机分配的物理适配器，不管这个物理适配器流量是否过载，只有当分配的这个物理适配器出现故障后才会尝试走另外活动的物理适配器。基于源 MAC 地址哈希算法负载均衡的另外一种实现方式是虚拟机使用多个虚拟适配器，以便生成多个 MAC 地址，这样虚拟机就可能绑定多个物理适配器以实现负载均衡。

图 3-1-12 显示了基于源 MAC 地址负载均衡，虚拟机如果只有一个 MAC 地址的情况下，与基于源虚拟端口负载均衡相同，虚拟机 01 和虚拟机 02 与 ESXi 主机物理适配器 vmnic0 进行绑定，虚拟机 03 和虚拟机 04 与 ESXi 主机物理适配器 vmnic1 进行绑定，那么无论网络流量是否过载，虚拟机只会通过绑定的适配器对外进行通信。当虚拟机 03 和虚拟机 04 绑定 ESXi 主机物理适配器 vmnic1 出现故障时，虚拟机才会使用 ESXi 主机物理适配器 vmnic0 对外进行通信，如图 3-1-13 所示。

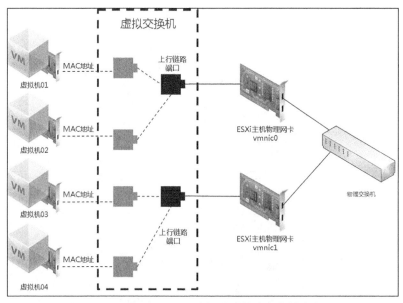

图 3-1-12　基于源 MAC 地址哈希算法负载均衡之一

基于源 MAC 地址负载均衡存在另外一种方式，即虚拟机多 MAC 地址模式，也就是说虚拟机有多个虚拟适配器，图 3-1-14 中的虚拟机 02 和虚拟机 03 有两个适配器，意味着虚拟机有两个 MAC 地址。在这样的模式下，通过基于源 MAC 地址负载均衡算法，虚拟机可能使用不同的 ESXi 主机物理适配器对外通信。

3. IP Base Hash（基于 IP 哈希算法）

IP Base Hash，基于 IP 哈希算法的负载均衡，这种方式与前两种负载均衡方式完全不一样，IP 哈希是基于源 IP 地址和目标 IP 地址计算出一个散列值，源 IP 地址和不同目标 IP 地址计算的散列值不一样，当虚拟机与不同目标 IP 地址通信时使用不同的散列值，这个散列值就会走不同的物理适配器，这样就可以实现动态的负载均衡。在 ESXi 主机网络上使用基于 IP 哈希算法的负载均衡，还必须满足一个前提，就是物理交换机必须支持链路聚合协议（Link Aggregation Control Protocol）以及 Cisco 私有的端口聚合协议（Port Aggregation Protocol），同时要求端口必须处于同一物理交换机（如果使用 Cisco Nexus 交换机 Virtual Port Channel 功能，不需要端口处于同一物理交换机）。

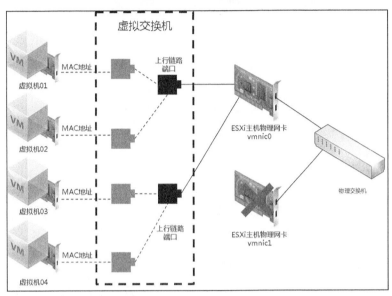

图 3-1-13　基于源 MAC 地址哈希算法负载均衡之二

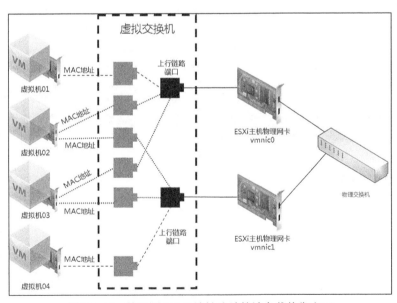

图 3-1-14　基于源 MAC 地址哈希算法负载均衡之三

图 3-1-15 显示了基于 IP 哈希算法的负载均衡，由于虚拟机源 IP 地址和不同目标 IP 地址计算的散列值不一样，所以虚拟机就不存在绑定某个 ESXi 主机物理适配器的情况，虚拟机 01 至 04 可能根据不同的散列值，选择不同 ESXi 主机物理适配器对外进行通信。需要特别注意的是，如果交换机不配置使用链路聚合协议，那么基于 IP 哈希算法的负载均衡模式无效。

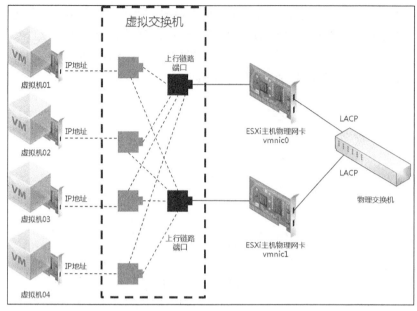

图 3-1-15　源 IP 地址哈希算法负载均衡

3.1.5　ESXi 主机 TCP/IP 协议堆栈

从 VMware vSphere 5.5 版本开始，新增加了一个名为 TCP/IP 协议堆栈的配置参数。TCP/IP 协议堆栈解决了共享路由表和 DNS 的问题，解决了一些环境使用单一网关的问题。

TCP/IP 协议堆栈的出现，在生产环境中可以创建多个 TCP/IP 协议堆栈，每个 TCP/IP 协议堆栈都有各自的路由表和 DNS 配置，这样可以更加灵活地配置网络。图 3-1-16 显示了系统默认的 TCP/IP 堆栈。

图 3-1-16　TCP/IP 协议堆栈

3.2 配置使用标准交换机

虚拟机的通信必须依靠网络，配置使用标准交换机在 VMware vSphere 虚拟化环境中相当重要。本节介绍如何配置使用标准交换机。

3.2.1 创建运行虚拟机流量标准交换机

通过 3.1 节的介绍可知，ESXi 主机存在不同的数据流量。本小节介绍如何创建独立的标准交换机运行虚拟机流量。

第 1 步，使用 Web Client 登录 vCenter Server，选择需要配置网络的 ESXi 主机，在"管理"标签中选择"网络"，可以看到默认创建的虚拟交换机 vSwitch0（如图 3-2-1 所示），单击"添加主机网络"。

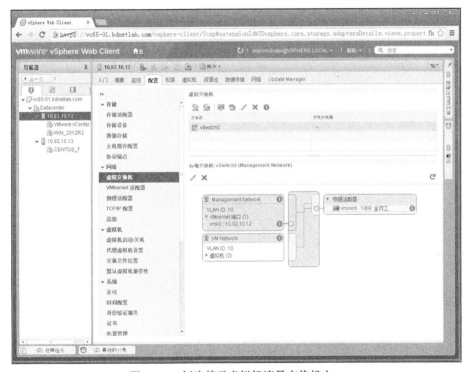

图 3-2-1　创建基于虚拟机流量交换机之一

第 2 步，选择"标准交换机的虚拟机端口组"（如图 3-2-2 所示），单击"下一步"按钮。

第 3 步，选择目标设备，可以使用已有的标准交换机，也可以新建，在生产环境中一般推荐新建交换机来满足不同的需求。选择"新建标准交换机"（如图 3-2-3 所示），单击"下一步"按钮。

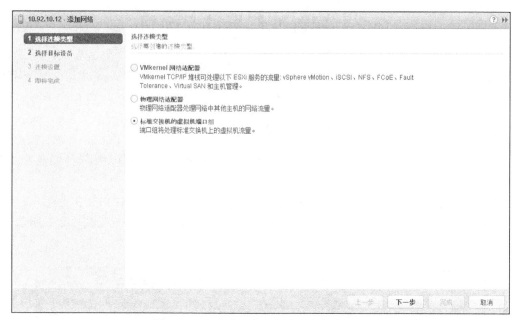

图 3-2-2　创建基于虚拟机流量交换机之二

图 3-2-3　创建基于虚拟机流量交换机之三

第 4 步，单击"添加适配器"按钮，为新创建的标准交换机分配物理适配器（如图 3-2-4 所示）。

图 3-2-4　创建基于虚拟机流量交换机之四

第 5 步，ESXi 主机可以选择的物理适配器为 vmnic1、vmnic2、vmnic3。选择 vmnic1 作为新创建标准交换机的上行链路适配器（如图 3-2-5 所示），单击"确定"按钮。

图 3-2-5　创建基于虚拟机流量交换机之五

第 6 步，确认将 vmnic1 适配器添加到新创建的标准交换机（如图 3-2-6 所示），单击"下一步"按钮。

图 3-2-6　创建基于虚拟机流量交换机之六

第 7 步，输入网络标签的参数，可以理解为虚拟机端口组的名称，一个标准交换机支持多个网络标签，根据实际情况输入（如图 3-2-7 所示），单击"下一步"按钮。

图 3-2-7　创建基于虚拟机流量交换机之七

第 8 步，确认新创建的标准交换机参数设置正确（如图 3-2-8 所示），单击"完成"按钮。

图 3-2-8 创建基于虚拟机流量交换机之八

第 9 步，名为 vSwitch1 的标准交换机创建成功（如图 3-2-9 所示）。

图 3-2-9 创建基于虚拟机流量交换机之九

3.2.2 创建基于 VMkernel 流量标准交换机

VMkernel 作为特殊的端口，可以承载 iSCSI、vMotion、vSAN 等流量，VMkernel 端口

可以在标准交换机和分布式交换机上进行创建。本小节介绍如何创建独立的标准交换机运行 VMkernel 流量。

第 1 步,选择需要配置网络的 ESXi 主机,在"管理"标签中选择"网络",单击"添加主机网络",选择"VMkernel 网络适配器"(如图 3-2-10 所示),单击"下一步"按钮。

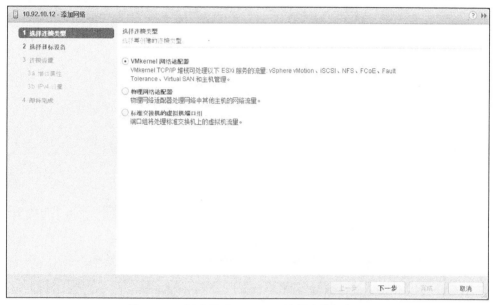

图 3-2-10　创建基于 VMkernel 流量交换机之一

第 2 步,选择目标设备,可以使用已有的标准交换机,也可以新建,在生产环境中一般推荐新建交换机来满足不同的需求。选择"新建标准交换机"(如图 3-2-11 所示),单击"下一步"按钮。

图 3-2-11　创建基于 VMkernel 流量交换机之二

第 3 步，单击"添加适配器"按钮，为新创建的标准交换机分配物理适配器（如图 3-2-12 所示）。

图 3-2-12　创建基于 VMkernel 流量交换机之三

第 4 步，ESXi 主机可以选择的物理适配器为 vmnic2、vmnic3。选择 vmnic2 作为新创建标准交换机的上行链路适配器（如图 3-2-13 所示），单击"确定"按钮。

图 3-2-13　创建基于 VMkernel 流量交换机之四

第 5 步，确认将 vmnic2 适配器添加到新创建的标准交换机（如图 3-2-14 所示），单击"下一步"按钮。

图 3-2-14　创建基于 VMkernel 流量交换机之五

第 6 步，与运行虚拟机流量的标准交换机不同的是，VMkernel 需要配置使用 IP 地址，建议在网络标签处注明该 VMkernel 的具体应用（如图 3-2-15 所示），单击"下一步"按钮。

图 3-2-15　创建基于 VMkernel 流量交换机之六

第 7 步，配置 VMkernel 相关的 IP 地址（如图 3-2-16 所示），单击"下一步"按钮。

图 3-2-16　创建基于 VMkernel 流量交换机之七

第 8 步，确认新创建的标准交换机的相关配置正确（如图 3-2-17 所示），单击"完成"按钮。

图 3-2-17　创建基于 VMkernel 流量交换机之八

第 9 步，名为 vSwitch2 的标准交换机创建成功（如图 3-2-18 所示）。

图 3-2-18 创建基于 VMkernel 流量交换机之九

第 10 步,切换到"VMkernel 适配器",可以看到两个 VMkernel 适配器,vmk0 的适配器网络标签为"Mangement Network",vmk1 的适配器网络标签是刚创建的"VMkernel-iSCSI"(如图 3-2-19 所示)。

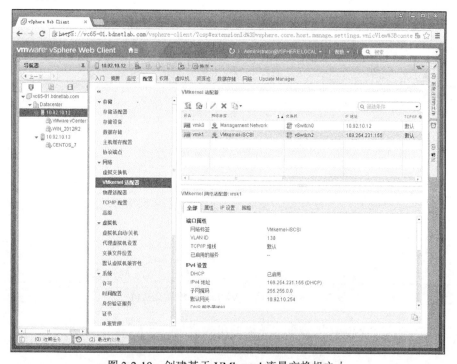

图 3-2-19 创建基于 VMkernel 流量交换机之十

3.2.3 标准交换机多 VLAN 配置

VLAN 作为一种有效减少或隔离广播域的技术在生产环境中的使用非常广泛。VMware vSphere 虚拟化架构中，无论是标准交换机还是分布式交换机，都提供对 VLAN 的支持，比较常用的方式是 VST 模式。也就是说，ESXi 主机物理适配器对应的物理交换机端口必须配置为 TRUNK 模式。

第 1 步，使用 Web Client 登录 vCenter Server，选择需要配置网络的 ESXi 主机，可以看到创建的虚拟交换机 vSwitch1，vSwitch1 上有名为"虚拟机网络"的端口组，VLAN ID 为 20（如图 3-2-20 所示）。

图 3-2-20　标准交换机多 VLAN 配置之一

第 2 步，在生产环境中可能在同一个标准交换机上运行多个 VLAN 的信息，也就需要创建多个 VLAN 的端口组。选择需要配置网络的 ESXi 主机，在"管理"标签中选择"网络"，单击"添加主机网络"，选择"标准交换的虚拟机端口组"（如图 3-2-21 所示），单击"下一步"按钮。

图 3-2-21　标准交换机多 VLAN 配置之二

第 3 步，选择目标设备。由于新建标准交换机需要对应一个物理适配器，而 ESXi 主机物理适配器数量有限，所以可以附加到现在交换机上，选择"现在标准交换机 vSwitch1"（如图 3-2-22 所示），单击"确定"按钮。

图 3-2-22　标准交换机多 VLAN 配置之三

第 4 步，输入网络标签的参数。注意，此处网络标签为 VLAN_10，VLAN ID 为 10（如图 3-2-23 所示），单击"下一步"按钮。

图 3-2-23　标准交换机多 VLAN 配置之四

第 5 步，确认标准交换机参数设置正确（如图 3-2-24 所示），单击"完成"按钮。

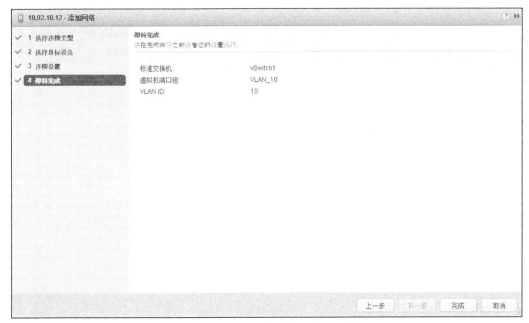

图 3-2-24　标准交换机多 VLAN 配置之五

第 6 步，通过图 3-2-25 可以看到，名为"VLAN_10"且 VLAN ID 为 10 的虚拟机端口组创建成功，名为"虚拟机网络"且 VLAN ID 为 20 的虚拟机端口组也位于同一个标准交换机 vSwitch1，这意味标准交换机 vSwitch1 可以同时接受 VLAN ID 为 10 和 20 的 VLAN Tag 信息。

图 3-2-25　标准交换机多 VLAN 配置之六

第 7 步，按相同的方式创建名为 VLAN_20 且 VLAN ID 为 20 的虚拟机端口组（如图 3-2-26 所示）。

图 3-2-26　标准交换机多 VLAN 配置之七

第 8 步，编辑名为 WIN_2012R2 的虚拟机网络适配器，调整网络适配器 1 到 "VLAN_10"（如图 3-2-27 所示）。

图 3-2-27　标准交换机多 VLAN 配置之八

第 9 步，编辑名为 WIN_2012R2_AD 的虚拟机网络适配器，调整网络适配器 1 到 "VLAN_20"（如图 3-2-28 所示）。

图 3-2-28 标准交换机多 VLAN 配置之九

第 10 步，在名为 WIN_2012R2 的虚拟机上使用命令 "ipconfig" 查看获取的 IP 地址（如图 3-2-29 所示）。

图 3-2-29 标准交换机多 VLAN 配置之十

第 11 步，在名为 WIN_2012R2_AD 的虚拟机上使用命令"ipconfig"查看获取的 IP 地址（如图 3-2-30 所示）。

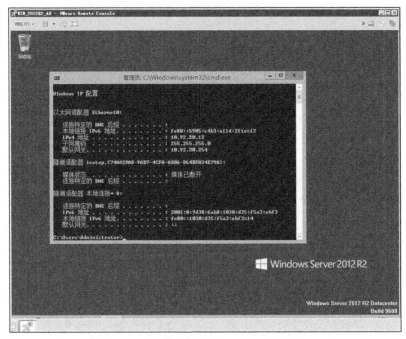

图 3-2-30　标准交换机多 VLAN 配置之十一

第 12 步，在名为 WIN_2012R2 的虚拟机上使用命令"ping 10.92.10.254"检测虚拟机的网络连通性（10.92.10.254 为 VLAN 10 的网关，如图 3-2-31 所示）。

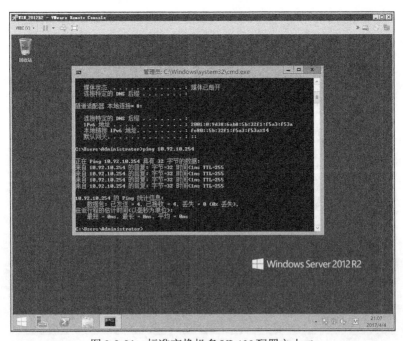

图 3-2-31　标准交换机多 VLAN 配置之十二

第13步，在名为 WIN_2012R2_AD 的虚拟机上使用命令"ping 10.92.20.254"检测虚拟机的网络连通性（10.92.20.254 为 VLAN 20 的网关，如图 3-2-32 所示）。

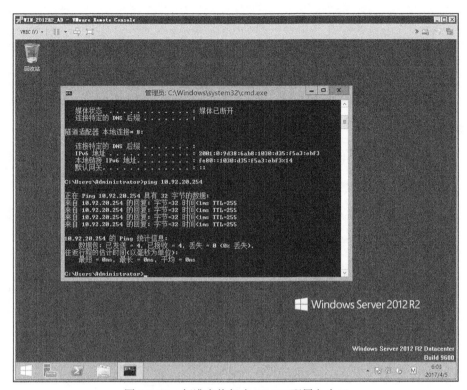

图 3-2-32　标准交换机多 VLAN 配置之十三

3.2.4　标准交换机 NIC Teaming 配置

前面小节所创建的标准交换机都使用了一个物理适配器。对于生产环境来说，标准交换机使用一个物理适配器容易造成单点故障，根据不同的应用，一个标准交换机会使用一个或多个物理适配器，当使用 2 个以上物理适配器时就需要进行 NIC Teaming 配置，以实现负载均衡。

NIC Teaming 的三种模式在 3.1.4 小节已进行了介绍。生产环境可以根据实际情况选择负载均衡的模式。

第1步，使用 Web Client 登录 vCenter Server，选择需要配置网络的 ESXi 主机，在"管理"标签中选择"网络"，可以看到虚拟交换机 vSwitch0 只有一个物理适配器（如图 3-2-33 所示）。要实现 NIC Teaming 必须添加一个物理适配器，单击"管理已连接到选定交换机的物理网络适配器"。

第2步，单击"添加适配器"按钮，为标准交换机增加物理适配器（如图 3-2-34 所示）。

图 3-2-33　标准交换机 NIC Teaming 配置之一

图 3-2-34　标准交换机 NIC Teaming 配置之二

第 3 步，目前可以选择的物理适配器有 vmnic1、vmnic2、vmnic3。选择 vmnic1（如图 3-2-35 所示），单击"确定"按钮。

第 4 步，确认将 vmnic1 适配器添加到 vSwitch0（如图 3-2-36 所示），单击"确定"按钮。

图 3-2-35　标准交换机 NIC Teaming 配置之三

图 3-2-36　标准交换机 NIC Teaming 配置之四

第 5 步，成功为 vSwitch0 配置 2 个物理适配器（如图 3-2-37 所示），单击"编辑设置"配置 NIC Teaming。

第 6 步，默认情况下，负载均衡的方式为"基于源虚拟端口的路由"（如图 3-2-38 所示）。其他参数解释如下：

（1）网络故障检测

网络故障检测分为"仅链路状态"以及"信标探测"。仅链路状态是通过物理交换机的事件来判断，常见的是物理线路断开或物理交换机故障，其缺点是无法判断配置错误；信标探测也会使用链路状态，但它增加了一些其他检测机制，比如由于 STP 阻塞端口、端口 VLAN 配置错误等。

图 3-2-37　标准交换机 NIC Teaming 配置之五

图 3-2-38　标准交换机 NIC Teaming 配置之六

（2）通知交换机

虚拟机启动、虚拟机进行 vMotion 操作、虚拟机 MAC 地址发生变化等情况发生时，物理交换机会收到用反向地址解析协议 RARP 表示的变化通知，物理交换机是否知道取决于通知交换机的设置，设置为是则立即知道，设置为否则不知道，RARP 会更新物理交换机的查询表，并且在故障恢复事件时提供最短延迟时间。

3.2 配置使用标准交换机　　81

（3）故障恢复

这里的故障恢复是指网络故障恢复后的数据流量的处理方式，以图 3-2-28 为例，当 vmnic0 出现故障时，数据流量全部迁移到 vmnic1；当 vmnic0 故障恢复后，可以设置数据流量是否切换回 vmnic0。需要特别注意的是，运行 IP 存储的 vSwitch 推荐将故障恢复配置为"否"，以免 IP 存储流量来回切换。

第 7 步，为了测试负载均衡的效果，需要断开 vmnic0 的网络，在断开网络前查看 vmnic0 适配器的相关信息（如图 3-2-39 所示）。

图 3-2-39　标准交换机 NIC Teaming 配置之七

第 8 步，断开 vmnic0 网络，vSwitch0 只有一个活动的物理适配器（如图 3-2-40 所示）。

图 3-2-40　标准交换机 NIC Teaming 配置之八

第 9 步，在名为 CENTOS_7 的虚拟机上使用命令 "ifconfig" 查看获取 IP 地址（如图 3-2-41 所示）。

图 3-2-41　标准交换机 NIC Teaming 配置之九

第 10 步，在名为 CENTOS_7 的虚拟机上使用命令 "ping 10.92.10.254" 检测虚拟机的网络连通性（10.92.10.254 为 VLAN 10 的网关，如图 3-2-42 所示）。虽然断开其中一个物理适配器，但虚拟机网络并没有中断，也就是负载均衡生效。

图 3-2-42　标准交换机 NIC Teaming 配置之十

第 11 步，也可根据生产的需求将负载均衡的方式调整为"基于 IP 哈希的路由"（如图 3-2-43 所示）。需要注意的是，基于 IP 哈希的路由物理交换机需要相应的配置（有关物理交换机配置的相关内容将在后续章节中讲述）。

3.2 配置使用标准交换机   83

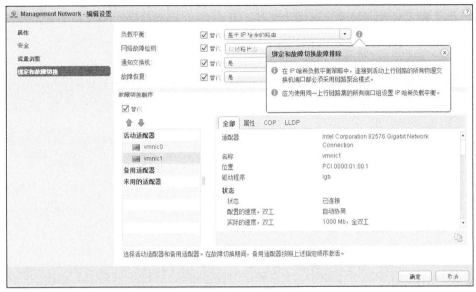

图 3-2-43　标准交换机 NIC Teaming 配置之十一

3.2.5　标准交换机其他策略配置

对于标准交换机来说，策略参数的配置相对简单。策略配置分为基于 vSwitch 全局配置以及基于端口组配置，可以根据生产环境的实际情况进行配置，通常情况下基于端口组进行配置。

1. 基于标准交换机的 MTU 配置

VMware 标准交换机支持修改端口 MTU 值，默认值为 1500（如图 3-2-44 所示）。可以修改为其他参数，但需要物理交换机的支持，建议两端配置的 MTU 值一致。如果与物理交换机不匹配，可能导致网络传输问题。

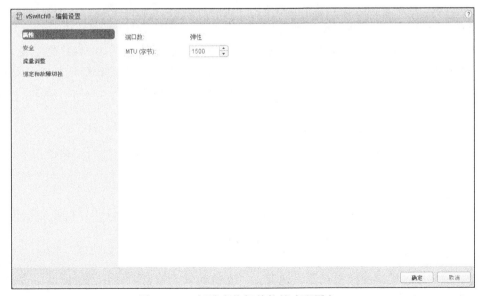

图 3-2-44　标准交换机其他策略配置之一

2. 基于标准交换机的安全配置

VMware 标准交换机提供基本的安全配置，主要包括混杂模式、MAC 地址更改和伪传输（如图 3-2-45 所示）。

图 3-2-45 标准交换机其他策略配置之二

（1）混杂模式

默认为拒绝模式，其功能类似于传统物理交换机，虚拟机进行传输数据通过标准交换机的 ARP 表，仅在源端口和目的端口进行接收和转发，标准交换机的其他接口不会接收和转发。

如果需要对标准交换机上的虚拟机流量进行抓包分析或端口镜像，可以将混杂模式修改为"接受"。修改后，其功能类似于集线器，标准交换机所有端口都可以收到数据。

（2）MAC 地址更改和伪传输

默认为接受，虚拟机在刚创建时会生成一个 MAC 地址，可以理解为初始 MAC 地址。当安装操作系统后可以使用初始 MAC 地址进行数据转发，这时候初始 MAC 地址变为有效 MAC 地址且两者相同。如果通过操作系统修改 MAC 地址，则初始 MAC 地址和有效 MAC 地址就不相同，数据的转发取决于 MAC 地址更改和伪传输状态，状态为"接受"时进行转发，状态为"拒绝"时则丢弃。

3. 基于标准交换机的流量调整

VMware 标准交换机提供了基本的流量调整功能。标准交换机流量调整仅用于出站方向，默认状态为已禁用（如图 3-2-46 所示）。

（1）平均带宽

平均带宽表示每秒通过标准交换机的数据传输量。如果 vSwitch0 上行链路为 1GE 适配器，则每个连接到这个 vSwitch0 的虚拟机都可以使用 1Gbit/s 带宽。

（2）峰值带宽

峰值带宽表示标准交换机在不丢包前提下支持的最大带宽。如果 vSwitch0 上行链路为 1GE 适配器，则 vSwitch0 的峰值带宽即为 1Gbit/s。

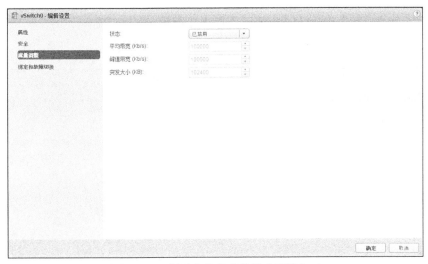

图 3-2-46　标准交换机其他策略配置之三

（3）突发大小

突发大小规定了突发流量中包含的最大数据量，计算方式是"带宽×时间"。在高使用率期间，如果有一个突发流量超出配置值，那么这些数据包就会被丢弃，其他数据包可以传输；如果处理的网络流量队列未满，那么这些数据包后来会被继续传输。

3.2.6　TCP/IP 协议堆栈配置

配置 TCP/IP 协议堆栈整体来说不复杂，但需要具备一些路由方面的基础知识。本节通过几个实验来介绍 TCP/IP 协议堆栈在生产环境中的使用。

1. TCP/IP 协议堆栈的基本配置

第 1 步，使用 Web Client 登录 vCenter Server，在 ESXi 主机的配置中选择"网络"→"TCP/IP 配置"（如图 3-2-47 所示）。默认情况下，主机有 3 个系统堆栈。

图 3-2-47　TCP/IP 协议堆栈的基本配置之一

第 2 步，选择需要编辑的 TCP/IP。除名称不能配置外，其余参数均可配置（如图 3-2-48 所示）。

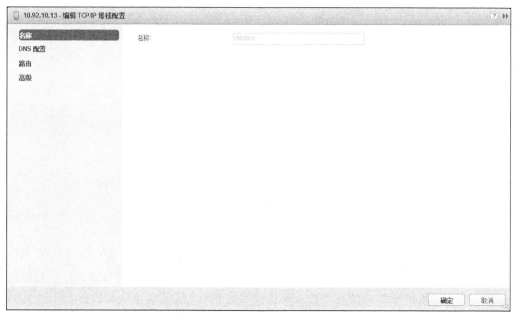

图 3-2-48　TCP/IP 协议堆栈的基本配置之二

第 3 步，可以根据生产环境的需要确定是否修改 DNS 配置（如图 3-2-49 所示）。

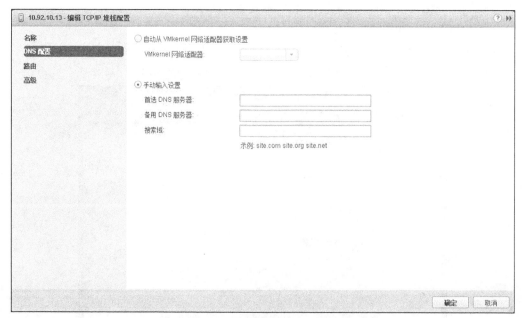

图 3-2-49　TCP/IP 协议堆栈的基本配置之三

第 4 步，最重要的配置，编辑 TCP/IP 堆栈的路由（如图 3-2-50 所示）。需要注意的是，

VMkernel 网关呈灰色状态，无法配置。这个状态可能是由于没有配置 vMotion 的 VMkernel 导致，因此需要配置后再修改网关。具体如何创建，请参考第 5 步的配置。

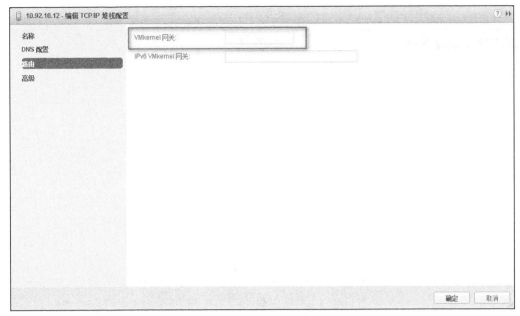

图 3-2-50　TCP/IP 协议堆栈的基本配置之四

第 5 步，当配置好 vMotion 的 VMkernel 后再编辑即处于可编辑状态，同时也可以配置 IPv6 的地址（如图 3-2-51 所示）。

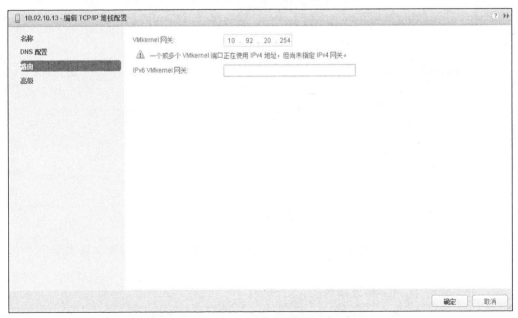

图 3-2-51　TCP/IP 协议堆栈的基本配置之五

第 6 步，TCP/IP 高级选项一般保持默认配置即可（如图 3-2-52 所示）。

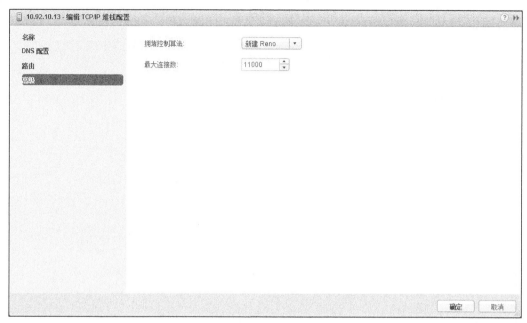

图 3-2-52　TCP/IP 协议堆栈的基本配置之六

第 7 步，IP 地址为 10.92.10.12 的主机配置完 vMotion 的 TCP/IP 堆栈，如图 3-2-53 所示。

图 3-2-53　TCP/IP 协议堆栈的基本配置之七

第 8 步，按照上述方式配置 IP 地址为 10.92.10.13 的主机，配置完成后如图 3-2-54 所示。

2．创建使用 TCP/IP 协议堆栈的 VMkernel

第 1 步，默认情况下 vmk0 的 vMotion 处于已禁用状态（如图 3-2-55 所示），创建一个单独的 VMkernel 用于 vMotion。

3.2 配置使用标准交换机   89

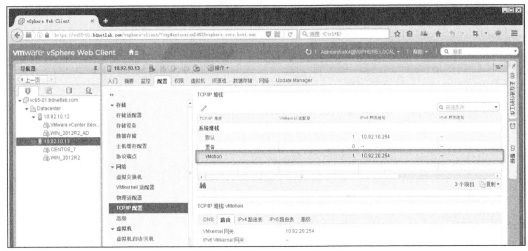

图 3-2-54　TCP/IP 协议堆栈的基本配置之八

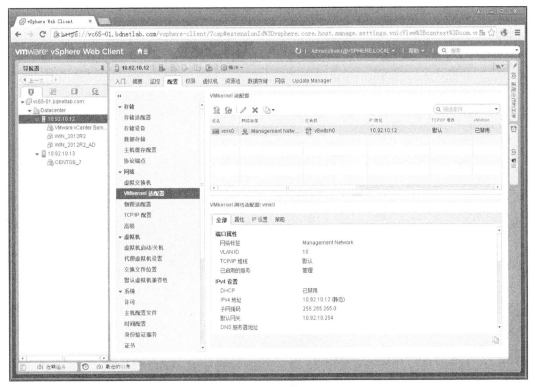

图 3-2-55　创建使用 TCP/IP 协议堆栈的 VMkernel 之一

第 2 步，选择需要配置网络的 ESXi 主机，在"管理"标签中选择"网络"，单击"添加主机网络"，选择"VMkernel 网络适配器"（如图 3-2-56 所示），单击"下一步"按钮。

图 3-2-56　创建使用 TCP/IP 协议堆栈的 VMkernel 之二

第 3 步，选择目标设备，可以使用已有的标准交换机，也可以新建。选择"现有标准交换机"（如图 3-2-57 所示），单击"下一步"按钮。

图 3-2-57　创建使用 TCP/IP 协议堆栈的 VMkernel 之三

第 4 步，输入对应的网络标签，TCP/IP 堆栈选择"vMotion"（如图 3-2-58 所示），单击"下一步"按钮。

图 3-2-58　创建使用 TCP/IP 协议堆栈的 VMkernel 之四

第 5 步，VMkernel 所需要的 IP 地址采用自动获取方式（如图 3-2-59 所示），单击"下一步"按钮。

图 3-2-59　创建使用 TCP/IP 协议堆栈的 VMkernel 之五

第 6 步，确认标准交换机相关配置正确（如图 3-2-60 所示），单击"完成"按钮。

第 3 章 部署 VMware vSphere 基本网络

图 3-2-60 创建使用 TCP/IP 协议堆栈的 VMkernel 之六

第 7 步，新创建的基于 vMotion 的 TCP/IP 堆栈完成，vMotion 处于已启用状态（如图 3-2-61 所示）。

图 3-2-61 创建使用 TCP/IP 协议堆栈的 VMkernel 之七

第 8 步，按照上述方式在 IP 地址为 10.92.10.13 的主机创建相同的基于 vMotion 的 TCP/IP 堆栈（如图 3-2-62 所示）。

3.2 配置使用标准交换机　93

图 3-2-62　创建使用 TCP/IP 协议堆栈的 VMkernel 之八

3. vMotion 测试（VMkernel 使用 DHCP）

基于 vMotion 的 TCP/IP 堆栈创建好后，需要对虚拟机进行 vMotion 操作以验证其能否正常运行。

第 1 步，选择名为 WIN_2012R2 的虚拟机。目前虚拟机位于 IP 地址为 10.92.10.12 的主机（如图 3-2-63 所示）。

图 3-2-63　vMotion 测试（VMkernel 使用 DHCP）之一

第 2 步，将名为 WIN_2012R2 的虚拟机进行迁移，选择"仅更改计算资源"（如图 3-2-64 所示），单击"下一步"按钮。

图 3-2-64　vMotion 测试（VMkernel 使用 DHCP）之二

第 3 步，选择迁移到 IP 地址为 10.92.10.13 的主机（如图 3-2-65 所示），单击"下一步"按钮。

图 3-2-65　vMotion 测试（VMkernel 使用 DHCP）之三

第 4 步，选择网络（如图 3-2-66 所示），单击"下一步"按钮。

图 3-2-66 vMotion 测试（VMkernel 使用 DHCP）之四

第 5 步，选择 vMotion 优先级，使用默认值即可（如图 3-2-67 所示），单击"下一步"按钮。

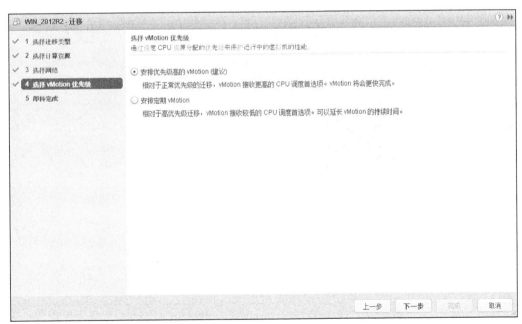

图 3-2-67 vMotion 测试（VMkernel 使用 DHCP）之五

第 6 步，确认迁移参数设置正确（如图 3-2-68 所示），单击"完成"按钮。

图 3-2-68　vMotion 测试（VMkernel 使用 DHCP）之六

第 7 步，名为 WIN_2012R2 的虚拟机成功迁移到 IP 地址为 10.92.10.13 的主机（如图 3-2-69 所示），说明基于 vMotion 的 TCP/IP 堆栈使用 DHCP 获取 IP 地址的 VMkernel 运行 vMotion 正常。

图 3-2-69　vMotion 测试（VMkernel 使用 DHCP）之七

4. vMotion 测试（VMkernel 使用静态 IP 相同网关）

测试完成 VMkernel 使用 DHCP 获取 IP 地址后，使用静态 IP 环境且相同的 VLAN 再进行测试。

第 1 步，调整 VMkernel 使用的 IP 地址，勾选"替代此适配器的默认网关"（如图 3-2-70 所示），单击"确定"按钮。

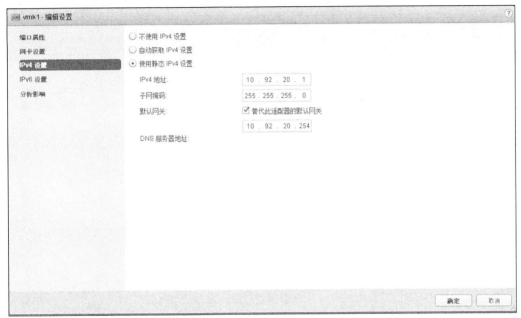

图 3-2-70　vMotion 测试（VMkernel 使用静态 IP 且 vMotion 网关相同）之一

第 2 步，IP 地址为 10.92.10.12 的主机完成，基于 vMotion 的 TCP/IP 堆栈使用静态 IP（如图 3-2-71 所示）。

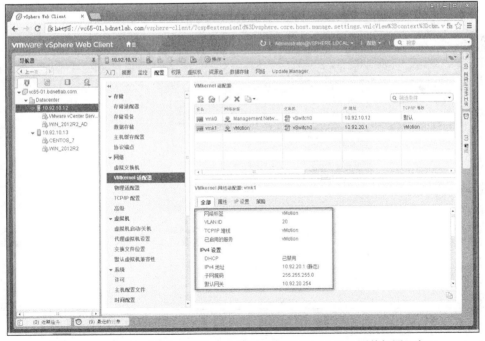

图 3-2-71　vMotion 测试（VMkernel 使用静态 IP 且 vMotion 网关相同）之二

第 3 步，按照上述方式修改 IP 地址为 10.92.10.13 的主机（如图 3-2-72 所示）。

图 3-2-72　vMotion 测试（VMkernel 使用静态 IP 且 vMotion 网关相同）之三

第 4 步，选择名为 CENTOS_7 的虚拟机。目前虚拟机位于 IP 地址为 10.92.10.13 的主机（如图 3-2-73 所示），将其迁移到 IP 地址为 10.92.10.12 的主机。

图 3-2-73　vMotion 测试（VMkernel 使用静态 IP 且 vMotion 网关相同）之四

第 5 步，虚拟机成功迁移到 IP 地址为 10.92.10.12 的主机（如图 3-2-74 所示），说明基于 vMotion 的 TCP/IP 堆栈使用静态 IP 地址的 VMkernel 运行 vMotion 正常。

图 3-2-74　vMotion 测试（VMkernel 使用静态 IP 且 vMotion 网关相同）之五

5．vMotion 测试（VMkernel 使用静态 IP 不同网关）

测试完成 VMkernel 使用静态 IP 地址后，再测试 ESXi 主机 TCP/IP 堆栈使用静态 IP 但不同网关，修改 IP 地址为 10.92.10.12 的主机即可。

第 1 步，修改 IP 地址为 10.92.10.12 主机的 TCP/IP 堆栈（如图 3-2-75 所示）。

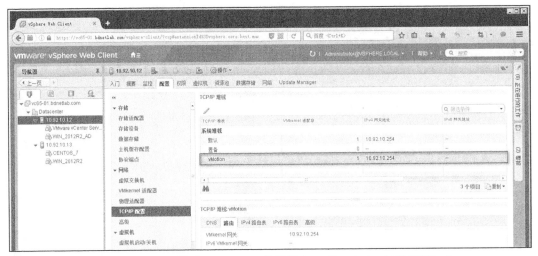

图 3-2-75　vMotion 测试（VMkernel 使用静态 IP 且 vMotion 网关不同）之一

第 2 步，修改 IP 地址为 10.92.10.12 主机 vMotion 的 VMkernel 的 IP 地址（如图 3-2-76 所示）。

100 第 3 章 部署 VMware vSphere 基本网络

图 3-2-76 vMotion 测试（VMkernel 使用静态 IP 且 vMotion 网关不同）之二

第 3 步，选择名为 CENTOS_7 的虚拟机。目前虚拟机位于 IP 地址为 10.92.10.12 的主机（如图 3-2-77 所示），将其迁移到 IP 地址为 10.92.10.13 的主机。

图 3-2-77 vMotion 测试（VMkernel 使用静态 IP 且 vMotion 网关不同）之三

第 4 步，虚拟机成功迁移到 IP 地址为 10.92.10.13 的主机（如图 3-2-78 所示），说明基于 vMotion 的 TCP/IP 堆栈使用静态 IP 地址且主机 TCP/IP 堆栈使用不同网关 vMotion 正常。

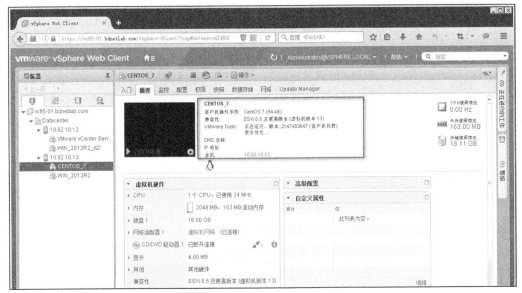

图 3-2-78　vMotion 测试（VMkernel 使用静态 IP 且 vMotion 网关不同）之四

3.3　配置使用分布式交换机

分布式交换机英文名为 Distributed Switch，VMware 一般简称为 vDS，其他厂商可能将其称为 DVS。其功能与标准交换机并没有太大的区别，可以理解为跨多台 ESXi 主机的超级交换机。它把分布在多台 ESXi 主机的标准虚拟交换机逻辑上组成一个"大"交换机。利用分布式交换机可以简化虚拟机网络连接的部署、管理和监控，为集群级别的网络连接提供一个集中控制点，使虚拟环境中的网络配置不再以主机为单位。VMware vSphere 平台允许使用第三方虚拟交换机，比如常用的是 Cisco Nexus 1000v 系列、IBM DVS 5000V 系列、HP FlexFabric 5900V 系列等。

对于中小环境来说，标准交换机可以满足其需求，但对于 ESXi 主机较多特别是有多 VLAN、网络策略等需求的中大型企业来说，如果只使用标准交换机会影响整体的管理以及网络的性能。所以使用分布式交换机是必需的选择。在生产环境中，标准交换机与分布式交换机并用，管理网络使用标准交换机上，可以把虚拟机网络、基于 VMKernel 网络迁移到分布式交换机上。本节介绍如何配置使用分布式交换机。

3.3.1　创建分布式交换机

分布式交换机的创建必须在 vCenter Server 进行操作，并且需要将 ESXi 主机加入 vCenter Server，独立的 ESXi 主机不能创建分布式交换机。创建分布式交换机之前需要至少保证 ESXi 主机有 1 个或以上未使用的以太网口。

第 1 步，使用 Web Client 登录 vCenter Server，选择"网络"，在"数据中心"上单击右键，选择"Distributed Switch"→"新建 Distributed Switch"（如图 3-3-1 所示）。

102　第 3 章　部署 VMware vSphere 基本网络

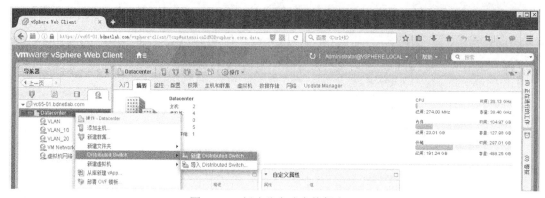

图 3-3-1　创建分布式交换机之一

第 2 步，输入新建分布式交换机的名字（如图 3-3-2 所示），单击"下一步"按钮。

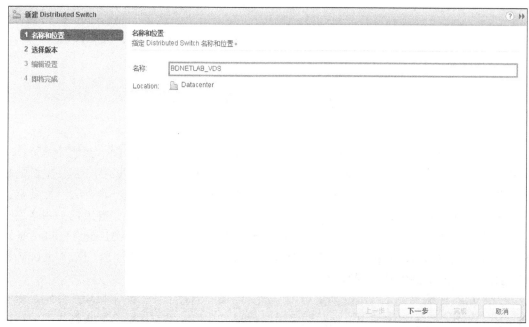

图 3-3-2　创建分布式交换机之二

第 3 步，选择分布式交换机的版本，不同的版本具有不同的功能特性，根据实际情况选择即可。此处选择 Distributed Switch :6.5.0（如图 3-3-3 所示），单击"下一步"按钮。

第 4 步，配置分布式交换机上行链路接口数量，上行链路接口数量指定的 ESXi 主机用于分布式交换机连接物理交换机的以太网口数量，一定要根据实际情况配置。例如，目前环境是两台 ESXi 主机，每台 ESXi 主机有 2 个以太网口用于分布式交换机，那么此处上行链路接口数量为 2，其他参数可以保持默认，创建好分布式交换机后可以修改（如图 3-3-4 所示），单击"下一步"按钮。

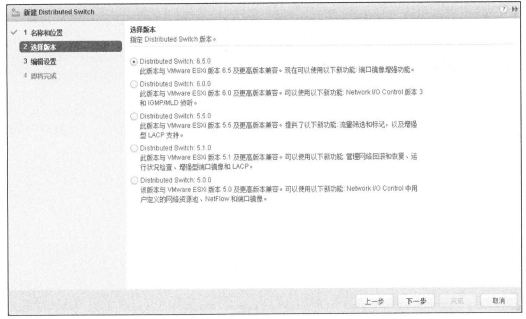

图 3-3-3　创建分布式交换机之三

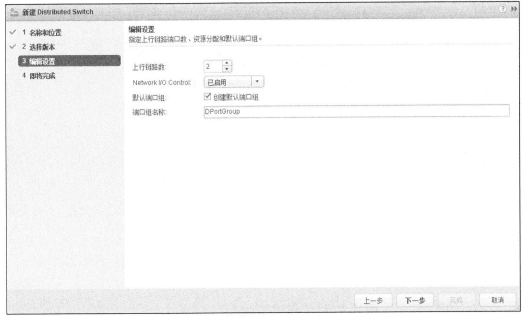

图 3-3-4　创建分布式交换机之四

第 5 步，验证分布式交换机的相关参数（如图 3-3-5 所示），单击"完成"按钮。
第 6 步，名为 BDNETLAB_VDS 的分布式交换机创建完成（如图 3-3-6 所示）。

图 3-3-5　创建分布式交换机之五

图 3-3-6　创建分布式交换机之六

分布式交换机的基本创建比较简单，但是还需要添加 ESXi 主机以及相应的端口才能正式使用分布式交换机。

3.3.2　将 ESXi 主机添加到分布式交换机

本小节介绍如何将 ESXi 主机添加到分布式交换机。

第 1 步，使用 Web Client 登录 vCenter Server，选择分布式交换机"BDNETLAB_VDS"，在"主机"处可以看到分布式交换机未添加任何 ESXi 主机（如图 3-3-7 所示）。

第 2 步，在添加管理主机窗口选择"添加主机"（如图 3-3-8 所示），单击"下一步"按钮。

第 3 步，选择"+新主机"（如图 3-3-9 所示），单击"下一步"按钮。

3.3 配置使用分布式交换机　　105

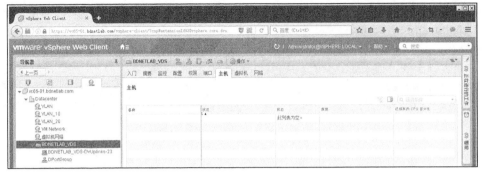

图 3-3-7　分布式交换机添加 ESXi 主机之一

图 3-3-8　分布式交换机添加 ESXi 主机之二

图 3-3-9　分布式交换机添加 ESXi 主机之三

第 4 步，勾选需要加入分布式交换机的 ESXi 主机（如图 3-3-10 所示），单击"确定"按钮。

图 3-3-10　分布式交换机添加 ESXi 主机之四

第 5 步，确认加入分布式交换机的 ESXi 主机，系统会在新加入的 ESXi 主机名前备注一个"新"的字样，提示这是新加入 ESXi 主机（如图 3-3-11 所示），单击"下一步"按钮。

图 3-3-11　分布式交换机添加 ESXi 主机之五

第 6 步，勾选"管理物理适配器"，将 ESXi 主机的适配器添加到分布式交换机（如图 3-3-12 所示），单击"下一步"按钮。

3.3 配置使用分布式交换机　　107

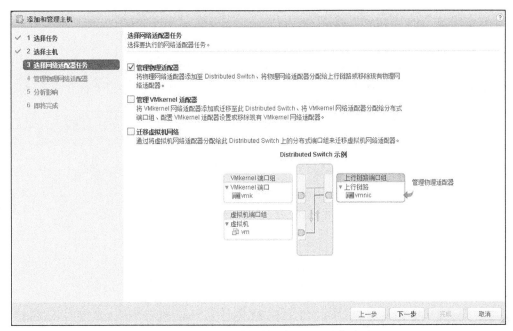

图 3-3-12　分布式交换机添加 ESXi 主机之六

第 7 步，选择 ESXi 主机需要加入分布式交换机的适配器，一般来说选择未关联其他交换机的适配器（如图 3-3-13 所示），单击"分配上行链路"。

图 3-3-13　分布式交换机添加 ESXi 主机之七

第 8 步，为 vmnic2 适配器分配上行链路编号。因为在创建分布式交换机时配置的上行链路为 2，因此此处上行链路为 Uplink1 与 Uplink2，可以手动指定，也可以选择自动分配

(如图 3-3-14 所示），单击"确定"按钮。

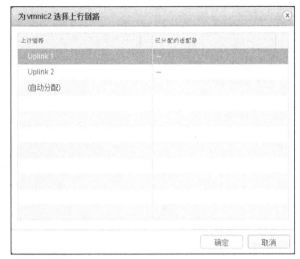

图 3-3-14 分布式交换机添加 ESXi 主机之八

第 9 步，确认将 IP 地址为 10.92.10.12 的 ESXi 主机 vmnic2 分配给上行链路 Uplink1（如图 3-3-15 所示），单击"下一步"按钮。

图 3-3-15 分布式交换机添加 ESXi 主机之九

第 10 步，使用相同的方式将 ESXi 主机其他适配器分配给分布式交换机使用（如图 3-3-16 所示），单击"下一步"按钮。

3.3 配置使用分布式交换机   109

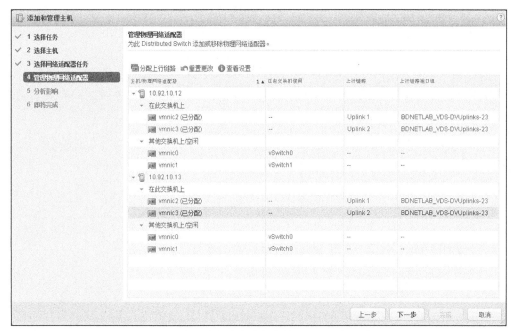

图 3-3-16 分布式交换机添加 ESXi 主机之十

第 11 步，系统会自动分析 ESXi 主机适配器加入分布式交换机是否会对已有网络造成影响（如图 3-3-17 所示），单击"下一步"按钮。

图 3-3-17 分布式交换机添加 ESXi 主机之十一

第 12 步，确认添加进入分布式交换机的 ESXi 主机参数设置正确（如图 3-3-18 所示），单击"完成"按钮。

图 3-3-18　分布式交换机添加 ESXi 主机之十二

第 13 步，在"相关对象"标签中的"主机"处可以看到 IP 地址为 10.92.10.12 以及 10.92.10.13 的两台 ESXi 主机已添加到分布式交换机（如图 3-3-19 所示）。

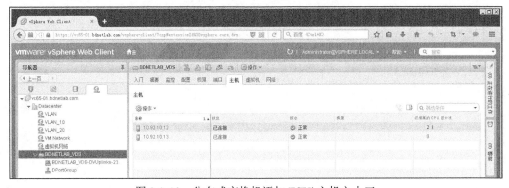

图 3-3-19　分布式交换机添加 ESXi 主机之十三

通过以上配置，成功将 ESXi 主机添加到分布式交换机。与标准交换机不同的是，标准交换机需要在每台 ESXi 主机上创建端口组，如果 ESXi 主机数量越大，工作量就越大；如果将 ESXi 主机添加到分布式交换机，分布式交换机创建的分布式端口组就可以在多台 ESXi 主机上进行调用，无需在每台 ESXi 主机进行创建，从而极大地提高工作效率，降低管理难度。

3.3.3　分布式交换机多 VLAN 配置

分布式交换机创建完成，添加 ESXi 主机以及相应的端口后就可以根据生产环境的实际情况进行使用。比较常见的是虚拟机由于各种用途，可能会划分多个 VLAN。VMware vSphere 分布式交换机不但支持 VLAN，而且支持 LACP 等，需要注意的是，分布式交换机使用的上行链路接口在物理交换机上需要配置为 TRUNK 模式。本小节介绍多 VLAN 分布式交换机的配置。

第 1 步，使用 Web Client 登录 vCenter Server，在创建好的分布式交换机"BDNETLAB_VDS"上单击右键，选择"分布式端口组"→"新建分布式端口组"（如图 3-3-20 所示）。如果不创建新的分布式端口组，可以使用默认创建的 DPortGroup 端口组。

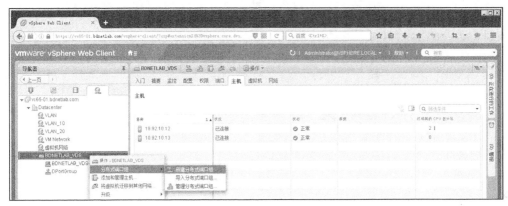

图 3-3-20　分布式交换机多 VLAN 配置之一

第 2 步，输入新建分布式端口组的名称，建议根据实际应用进行创建，以便于在日常管理中进行区分（如图 3-3-21 所示），单击"下一步"按钮。

图 3-3-21　分布式交换机多 VLAN 配置之二

第 3 步，进行端口的常规配置，一般来说只配置 VLAN 选择，其他的保持默认，需要注意的是 VLAN 类型以及 VLAN ID 建议配置生产环境实际使用的 ID（如图 3-3-22 所示），单击"下一步"按钮。

图 3-3-22 分布式交换机多 VLAN 配置之三

第 4 步，确认新建分布式端口组相关参数（如图 3-3-23 所示），单击"完成"按钮。

图 3-3-23 分布式交换机多 VLAN 配置之四

第 5 步，名为 PRO_VLAN 10 的分布式端口组创建完成（如图 3-3-24 所示）。

图 3-3-24 分布式交换机多 VLAN 配置之五

第 6 步,使用相同的方式创建名为 VLAN 20 的分布式端口组(如图 3-3-25 所示)。

图 3-3-25 分布式交换机多 VLAN 配置之六

3.3.4 迁移虚拟机到分布式交换机

当分布式交换机创建且其他后续配置完成后,就可以将虚拟机或其他应用迁移到分布式交换机上,迁移可以分为批量迁移以及虚拟机单独调整。本小节介绍如何将虚拟机迁移到分布式交换机。

1. 使用批量方式迁移虚拟机

第 1 步,使用 Web Client 登录 vCenter Server,在创建好的分布式交换机"BDNETLAB_VDS"上单击右键,选择"将虚拟机迁移到其他网络"(如图 3-3-26 所示)。

第 2 步,选择源网络和目标网络以对虚拟机的网络适配器进行迁移(如图 3-3-27 所示),单击"特定网络"的浏览。

第 3 步,选择虚拟机目前使用的"虚拟机网络"(如图 3-3-28 所示),单击"确定"按钮。

图 3-3-26　迁移虚拟机到分布式交换机之一

图 3-3-27　迁移虚拟机到分布式交换机之二

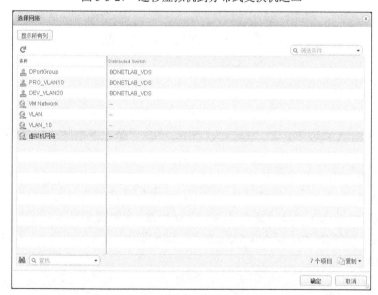

图 3-3-28　迁移虚拟机到分布式交换机之三

第4步，选择"目标网络"为分布式交换机端口组 PRO_VLAN10（如图 3-3-29 所示），单击"下一步"按钮。

图 3-3-29　迁移虚拟机到分布式交换机之四

第5步，系统自动显示出网络适配器为"虚拟机网络"的虚拟机（如图 3-3-30 所示），勾选所有交换机进行迁移，单击"下一步"按钮。

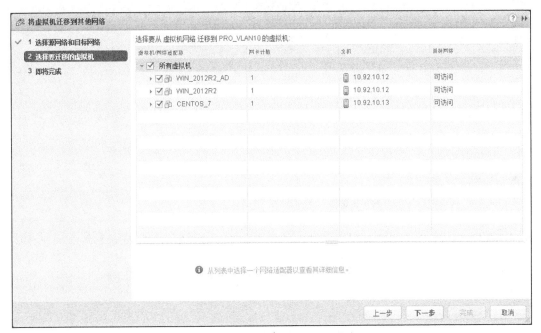

图 3-3-30　迁移虚拟机到分布式交换机之五

第 6 步，确认迁移的参数设置正确（如图 3-3-31 所示），单击"完成"按钮。

图 3-3-31　迁移虚拟机到分布式交换机之六

第 7 步，网络适配器为"虚拟机网络"的虚拟机迁移到分布式交换机 PRO_VLAN10（如图 3-3-32 所示）。

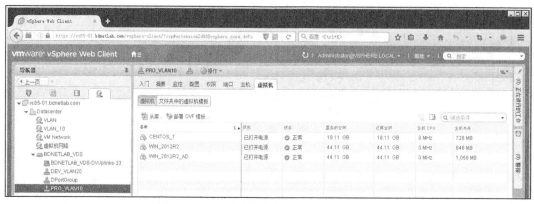

图 3-3-32　迁移虚拟机到分布式交换机之七

第 8 步，打开名为 WIN_2012R2 的虚拟机控制台，使用命令 ipconfig 查看获取 IP 地址为 10.92.10.111（如图 3-3-33 所示）。

第 9 步，使用 ping 命令测试虚拟机至网关的连通性（如图 3-3-34 所示），确认测试结果正常。

3.3 配置使用分布式交换机 117

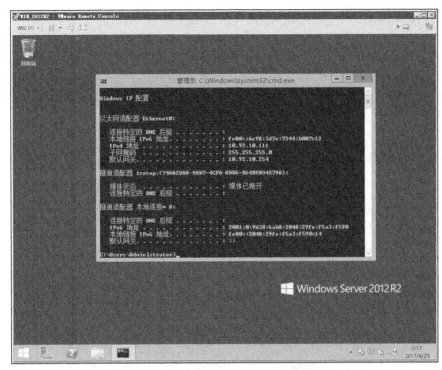

图 3-3-33 迁移虚拟机到分布式交换机之八

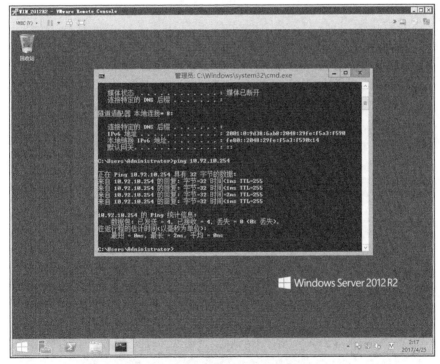

图 3-3-34 迁移虚拟机到分布式交换机之九

2. 手动调整虚拟机网络

第 1 步，编辑名为 WIN_2012R2 的虚拟机的虚拟机硬件，将网络适配器调整到分布式交换机"DEV_VLAN20"(如图 3-3-35 所示),单击"确定"按钮。

图 3-3-35　手动调整虚拟机网络之一

第 2 步,重新获取 IP 后,使用命令 ipconfig 查看获取 IP 地址为 10.92.20.4(如图 3-3-36 所示)。

图 3-3-36　手动调整虚拟机网络之二

第 3 步，使用 ping 命令测试虚拟机至网关的连通性（如图 3-3-37 所示），确认测试结果正常。

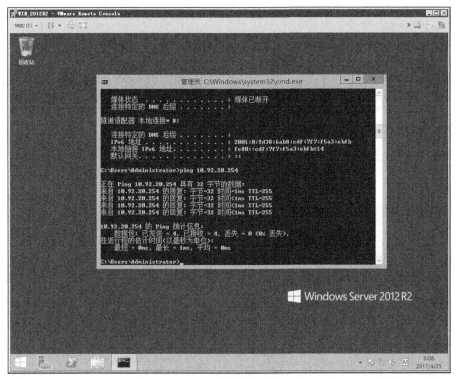

图 3-3-37　手动调整虚拟机网络之三

第 4 步，查看分布式交换机端口组 DEV_VLAN20 所在的虚拟机（如图 3-3-38 所示）。

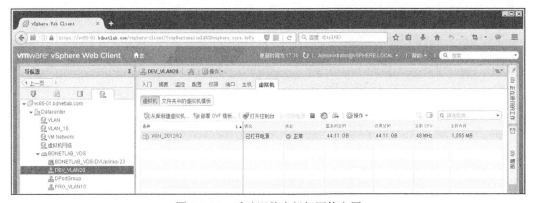

图 3-3-38　手动调整虚拟机网络之四

第 5 步，查看分布式交换机 BDNETLAB_VDS 的拓扑，虚拟机分布在新创建的 DEV_VLAN20 以及 PRO_VLAN10 两个端口组中（如图 3-3-39 所示）。

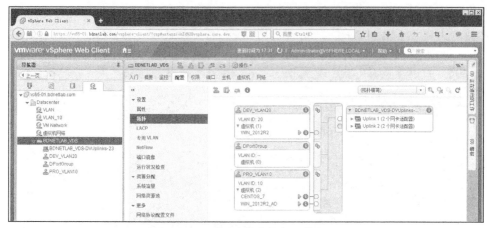

图 3-3-39　手动调整虚拟机网络之五

3.3.5　分布式交换机 LACP 配置

链路聚合在生产环境中使用非常广泛，使用基于 IP 哈希算法的负载均衡必须满足物理交换机必须支持链路聚合协议（Link Aggregation Control Protocol，简称 LACP）以及 Cisco 私有的端口聚合协议（Port Aggregation Protocol，简称 PAGP），同时要求端口必须处于同一物理交换机（如果使用 Cisco Nexus 交换机 Virtual Port Channel 功能，则不需要端口处于同一物理交换机）。大多数使用分布式交换机的环境使用开放标准的 LACP，本小节介绍分布式交换机 LACP 的配置。

第 1 步，使用 Web Client 登录 vCenter Server，选择分布式交换机"BDNETLAB_VDS"，在"管理"标签中的"设置"选择"LACP"（如图 3-3-40 所示），单击"新建链路聚合组"图标。

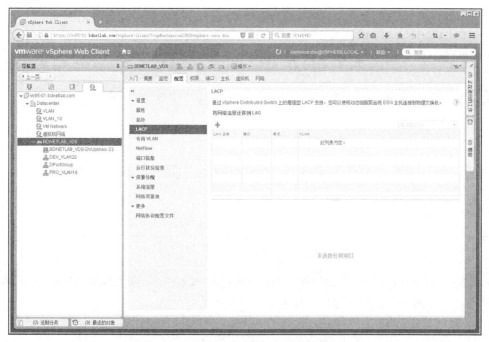

图 3-3-40　分布式交换机 LACP 配置之一

第 2 步，输入新建链路聚合组的名称以及上行链路端口数，端口数为加入到链路聚合组的数量，LACP 的模式配置为"活动"（对应物理交换机 LACP 的模式为 ACTIVE），负载平衡的方式选择"源和目标 IP 地址、TCP/UDP 端口及 VLAN"（如图 3-3-41 所示）单击"确定"按钮。

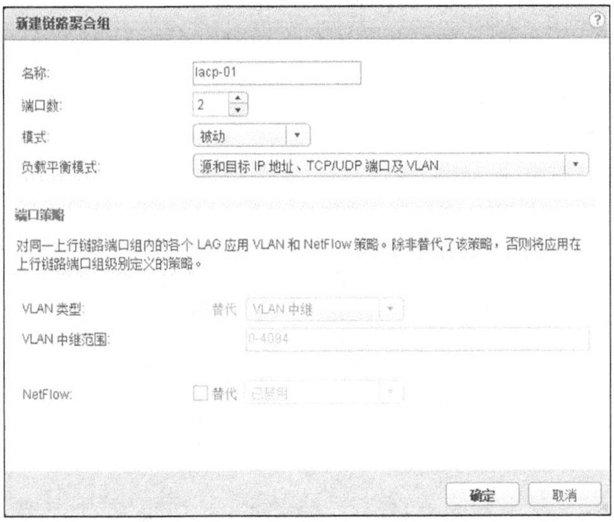

图 3-3-41　分布式交换机 LACP 配置之二

第 3 步，名为 LACP-01 的链路聚合组创建完成（如图 3-3-42 所示）。

图 3-3-42　分布式交换机 LACP 配置之三

第 4 步，不少人在完成第 3 步后即认为 LACP 配置完成，实际上只进行了基础的配置，在对应的物理交换机上可以看到适配器 line protocol 状态为 down、链路聚合组的状态为 down，其原因是新创建的链路聚合组还未添加适配器。

```
BDNETLAB_4503_01#show interfaces g3/13
GigabitEthernet3/13 is up, line protocol is down (suspended)
   Hardware is Gigabit Ethernet Port, address is 000b.fd6c.e3dc (bia 000b.fd6c.e3dc)
   Description: ESXi01-trunk
   MTU 1500 bytes, BW 1000000 Kbit, DLY 10 usec,
     reliability 255/255, txload 1/255, rxload 1/255

BDNETLAB_4503_01#show interfaces port-channel 10
Port-channel10 is down, line protocol is down (notconnect)
   Hardware is EtherChannel, address is 0000.0000.0000 (bia 0000.0000.0000)
   MTU 1500 bytes, BW 100000 Kbit, DLY 100 usec,
     reliability 255/255, txload 1/255, rxload 1/255
```

第 5 步，在分布式交换机"BDNETLAB_VDS"上单击右键，选择"添加和管理主机"（如图 3-3-43 所示）。

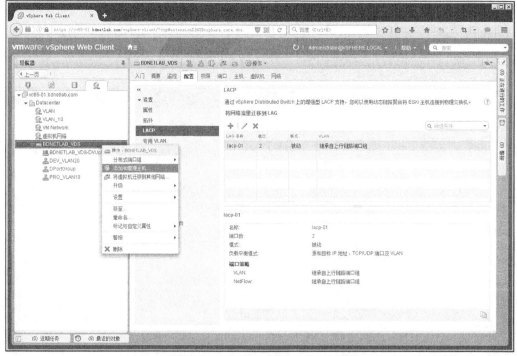

图 3-3-43　分布式交换机 LACP 配置之四

第 6 步，在添加管理主机窗口选择"管理主机网络"（如图 3-3-44 所示），单击"下一步"按钮。

图 3-3-44　分布式交换机 LACP 配置之五

第 7 步，选择 "+连接的主机"（如图 3-3-45 所示），单击 "下一步" 按钮。

图 3-3-45　分布式交换机 LACP 配置之六

第 8 步，勾选需要使用链路聚合组的 ESXi 主机（如图 3-3-46 所示），单击 "确定" 按钮。

图 3-3-46　分布式交换机 LACP 配置之七

第 9 步，确定选择的主机正确（如图 3-3-47 所示），单击"下一步"按钮。

图 3-3-47　分布式交换机 LACP 配置之八

第 10 步，选择"管理物理适配器"，将 ESXi 主机的适配器添加到分布式交换机链路聚合组（如图 3-3-48 所示），单击"下一步"按钮。

第 11 步，在 3.3.2 小节中已经将 ESXi 主机上的 vmnic2 和 vmnic3 适配器添加到了分布式交换机上行链路中，现在需要将适配器加入到链路聚合组中。选择适配器后单击"分配上行链路"（如图 3-3-49 所示）。

3.3 配置使用分布式交换机　　125

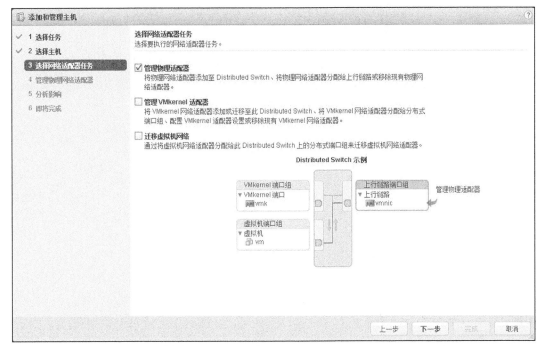

图 3-3-48　分布式交换机 LACP 配置之九

图 3-3-49　分布式交换机 LACP 配置之十

第 12 步，为适配器选择 LACP-01 端口中的 LACP-01-0 到 1 的任意一个端口（如图 3-3-50 所示），单击"确定"按钮。

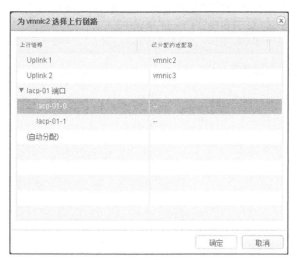

图 3-3-50　分布式交换机 LACP 配置之十一

第 13 步，对原有的适配器进行重新分配，确认适配器重新分配到上行链路 LACP-01 链路聚合组（如图 3-3-51 所示）。

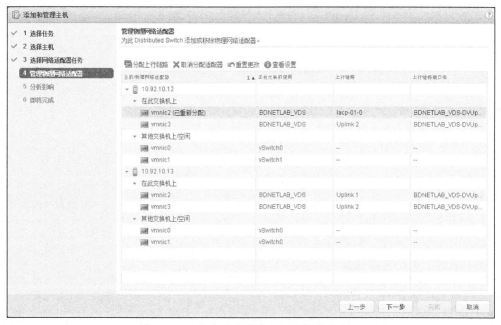

图 3-3-51　分布式交换机 LACP 配置之十二

第 14 步，使用相同的方法将 ESXi 主机其余的适配器重新分配到上行链路 LACP-01 链路聚合组（如图 3-3-52 所示），单击"下一步"按钮。

第 15 步，系统会自动分析 ESXi 主机适配器重新分配是否会对现在的网络造成影响（如图 3-3-53 所示），单击"下一步"按钮。

3.3 配置使用分布式交换机   127

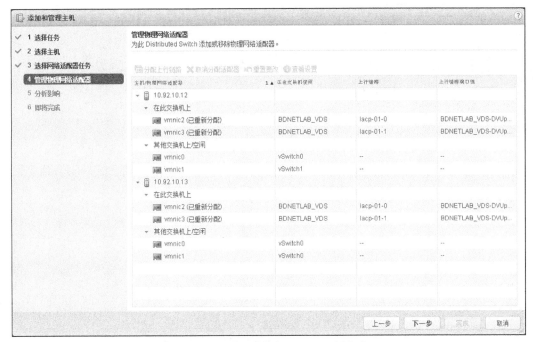

图 3-3-52　分布式交换机 LACP 配置之十三

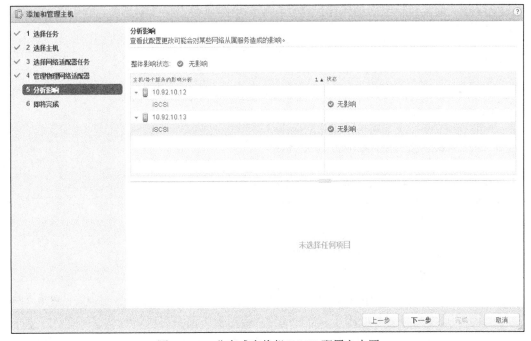

图 3-3-53　分布式交换机 LACP 配置之十四

第 16 步，确认重新分配适配器参数设置正确（如图 3-3-54 所示），单击"完成"按钮。

图 3-3-54　分布式交换机 LACP 配置之十五

第 17 步，登录对应的物理交换机查看适配器以及链路聚合组的状态，确认端口以及线协议状态均为 up。

```
BDNETLAB_4503_01#show interfaces g3/13
GigabitEthernet3/13 is up, line protocol is up (connected)
    Hardware is Gigabit Ethernet Port, address is 000b.fd6c.e3dc (bia 000b.fd6c.e3dc)
    Description: ESXi01-trunk
    MTU 1500 bytes, BW 1000000 Kbit, DLY 10 usec,
        reliability 255/255, txload 1/255, rxload 1/255

BDNETLAB_4503_01#show interfaces port-channel 10
Port-channel10 is up, line protocol is up (connected)
    Hardware is EtherChannel, address is 000b.fd6c.e3dc (bia 000b.fd6c.e3dc)
    Description: connetct to ESXi01-lacp01
    MTU 1500 bytes, BW 2000000 Kbit, DLY 10 usec,
        reliability 255/255, txload 1/255, rxload 1/255
```

第 18 步，默认情况下 PRO_VLAN10 端口组使用的是 2 个活动上行链路，LACP-01 链路聚合组处于未使用状态（如图 3-3-55 所示），使用上下箭头可以进行调整。

第 19 步，将 LACP-01 链路聚合组调整为活动上行链路，Uplink1 到 2 调整到未使用的上行链路，负载平衡类型选择"基于 IP 哈希的路由"（如图 3-3-56 所示），单击"完成"按钮。

图 3-3-55　分布式交换机 LACP 配置之十六

注意：LACP-01 链路与 Uplink 链路不能混用，否则会出现错误提示。

图 3-3-56　分布式交换机 LACP 配置之十七

第 20 步，端口组 PRO_VLAN10 配置 LACP-01 链路聚合组完成（如图 3-3-57 所示）。

图 3-3-57　分布式交换机 LACP 配置之十八

第 21 步，端口组 PRO_VLAN10 目前有两台虚拟机（如图 3-3-58 所示）。

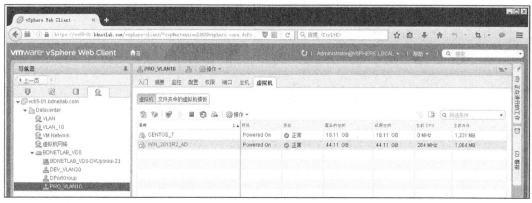

图 3-3-58　分布式交换机 LACP 配置之十九

第 22 步，重新获取 IP 后，使用命令 ipconfig 查看获取 IP 地址为 10.92.10.112（如图 3-3-59 所示）。

第 23 步，使用 ping 命令测试虚拟机至网关的连通性（如图 3-3-60 所示），确认测试结果正常。

3.3 配置使用分布式交换机 131

图 3-3-59 分布式交换机 LACP 配置之二十

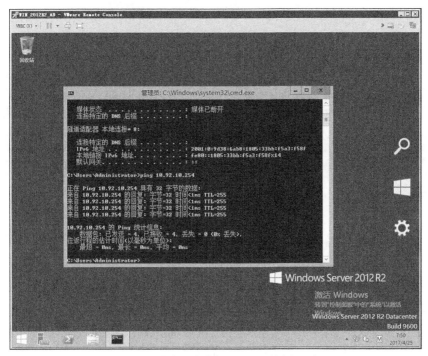

图 3-3-60 分布式交换机 LACP 配置之二十一

3.3.6 分布式交换机策略配置

对于分布式交换机来说，整体的策略配置与标准交换机相比增加了一些，最主要的差别在于分布式交换机可以对入站以及出站流量进行双向控制。与标准交换机一样，分布式交换机策略也可以基于分布式交换机全局配置以及基于端口组配置，可以根据生产环境的实际情况进行配置。

1. 分布式交换机的常规配置

常规配置主要涉及调整分布式交换机的基础信息，比如名称、上行链路数以及 Network I/O Control 等（如图 3-3-61 所示），可以根据生产环境的实际情况进行配置。

图 3-3-61　分布式交换机的常规配置

2. 分布式交换机的高级配置

高级配置主要涉及调整分布式交换机的数据交换信息，比如 MTU、多播、发现协议等（如图 3-3-62 所示）。与标准交换机一样，MTU 值可以修改为大于 1500，但需要物理交换机的支持，建议两端配置的 MTU 值一致。如果与物理交换机不匹配，可能导致网络传输问题。

图 3-3-62　分布式交换机的高级配置

3. 分布式交换机的 PVLAN 配置

分布式交换机支持 PVLAN 配置，但同时需要在物理交换机上进行配置。

第 1 步，在"BDNETLAB_VDS"分布式交换机上打开"编辑专用 VLAN 设置"窗口（如图 3-3-63 所示）。

图 3-3-63　分布式交换机的 PVLAN 配置之一

第 2 步，输入主 VLAN ID 信息以及辅助 VLAN ID 信息，注意辅助 VLAN ID 中要指定 VLAN 类型（如图 3-3-64 所示）。

图 3-3-64　分布式交换机的 PVLAN 配置之二

第 3 步，新建端口组使用 PVLAN（如图 3-3-65 所示），输入新端口组名称，单击"下一步"按钮。

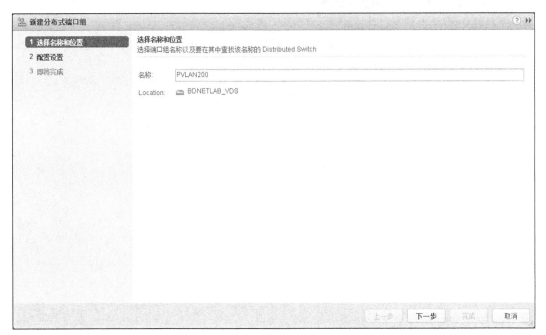

图 3-3-65　分布式交换机的 PVLAN 配置之三

第 4 步，进行端口的常规配置，注意新创建的端口组需要使用 PVLAN，VLAN 类型以及 VLAN ID 需要匹配 PVLAN 相关信息（如图 3-3-66 所示），单击"下一步"按钮。

图 3-3-66　分布式交换机的 PVLAN 配置之四

第 5 步，确定端口组参数设置正确（如图 3-3-67 所示），单击"确定"按钮。

3.3 配置使用分布式交换机　135

图 3-3-67　分布式交换机的 PVLAN 配置之五

第 6 步，查看新创建的使用 PVLAN 端口组信息（如图 3-3-68 所示）。

图 3-3-68　分布式交换机的 PVLAN 配置之六

4. 分布式交换机的 NetFlow 配置

分布式交换机支持使用 NetFlow 对流量进行分析，使用的前提是数据中心应该配置有

NetFlow 服务器用于收集流量数据。

在"BDNETLAB_VDS"分布式交换机上打开"编辑 NetFlow 设置"窗口，根据生产环境的相关服务器参数输入即可（如图 3-3-69 所示）。

图 3-3-69　分布式交换机的 NetFlow 配置

3.4　本章小结

本章对 VMware vSphere 6.5 网络部分进行了详细的介绍，包括标准交换机以及分布式交换的配置。对于标准交换机来说，配置相对简单，推荐管理网络使用标准交换机；对于分布式交换机来说，配置相对复杂，特别是 LACP 以及其他策略后的配置，一旦使用了策略，必须到物理交换机上进行验证，查看对应的接口状态是否正常。推荐虚拟机网络以及其他基于 VMkernel 网络使用分布式交换机。

第 4 章 部署 Nexus 1000V 分布式交换机

随着服务器虚拟化技术的不断发展，网络也发生了重大的变化，传统数据中心的网络设计已不太适合虚拟化数据中心的应用需求，网络虚拟化开始逐渐进入虚拟化数据中心，为虚拟化数据中心提供网络方面的控制策略。网络虚拟化可以聚合、创建或分割一个或多个数据平面、控制平台、管理平面。本章介绍基于 Cisco Nexus 系列 1000V 虚拟交换机的部署使用。

本章要点
- Cisco Nexus 1000V 介绍
- 部署 VSM
- 配置 Port-Profile
- 部署 VEM
- 虚拟机使用 Nexus 1000V
- 部署使用 VXLAN

4.1 Nexus 1000V 介绍

2009 年，VMware 公司在 vSphere 4.0 版本中引入了分布式交换机的概念。这是一个全新的领域，它被用来应对虚拟化数据中心网络应用的挑战，分布式交换机在标准交换机上增加了多个新的特性。Cisco 公司通过与 VMware 公司进行合作，开发出基于 VMware vSphere 架构的网络虚拟化产品 Cisco Nexus 1000V。

4.1.1 虚拟化架构面临的网络问题

从 VMware vSphere 虚拟化环境来看，无论是虚拟标准交换机还是分布式交换机，都将物理服务器的适配器看作对外连接的上行链路，这个上行链路承载着所有虚拟机的对外通信。然而，随着虚拟化数据中心的扩大，虚拟交换机给管理人员的日常工作提出了额外的挑战。

1. 网络的可扩展性

在虚拟化环境中，如果仅使用虚拟标准交换机，由于 vMotion 迁移特性的大规模应用，对于管理人员来说，必须在每台物理服务器上创建对应的端口组来满足 vMotion 的迁移要求，如果生产环境物理服务器数量众多，人为的错误不可避免。

2. 网络监控

虚拟化架构中网络的管理放在了物理服务器上，网络管理团队无法使用常规管理和监控工具对虚拟交换机的流量进行监控。

3. 网络管理

对于一些企业来说，物理服务器管理和网络管理分别由两个团队完成，使用虚拟化后，网络部分的配置与管理集成到物理服务器，这对于管理团队来说存在问题。有可能存在的情况是，服务器管理团队不懂网络，而网络管理团队又不懂服务器。

当然，在生产环境中不仅仅是这三大问题，可能还会有更多的问题，需要逐渐去发现并解决。

4.1.2　Nexus 1000V 基本介绍

Cisco Nexus 1000V 交换机是专门集成到 VMware vSphere 环境中基于软件的 CiscoNXOS 交换机，使用 Cisco 最新 NXOS 命令集，提供完整丰富的智能化特性。Cisco Nexus 1000V 交换机通过了 VMware 认证，与 VMware vSphere 完全兼容，其主要的功能特性如下：

- 基于 CiscoNXOS 操作系统，兼容各种交换机；
- 基于条例的虚拟机连接设置；
- 可移动虚拟网络及安全配置；
- 智能数据转向；
- 为网络管理和服务器管理提供互不干扰的运作模式；
- 对于虚拟以及物理资源有共同的管理模块。

在生产环境部署使用 Cisco Nexus 1000V 具有的优势如下：

- 服务器管理方面
  - 在虚拟机部署阶段除去网络配置工作；
  - 提高可扩展性。
- 网络管理方面
  - 统一网络配置与运作管理；
  - 改善运作安全保障，条例持久性；
  - 增强虚拟机网络功能，得到虚拟网络控制能力。

Cisco Nexus 1000V 支持以下扩展性参数：

- 2 个 VSM 虚拟机，用于高可用，分别处于活动状态与备用状态；
- 64 个 VEM 模块；
- 2048 个活动 VLAN；
- 2048 个端口；
- 2048 个端口配置文件；
- 256 个通道端口；
- 每个 VEM 支持 216 个 Veth；
- 每个 VEM 支持 32 个物理适配器；
- 每个 VEM 支持 8 个端口通道。

4.1.3 Nexus 1000V 架构介绍

Cisco Nexus 1000V 交换机主要由两大组件构成（如图 4-1-1 所示）。

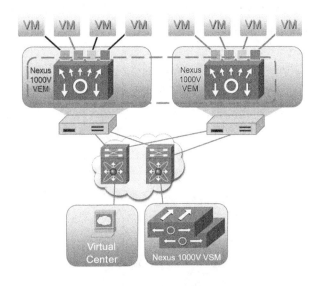

图 4-1-1　Cisco Nexus 1000V 两大组件

- VEM（Virtual Ethernet Module，虚拟以太网模块）：运行在 ESXi 主机的模块，为 ESXi 主机上运行的虚拟机提供交换端口。使用 Nexus 1000V 的 ESXi 主机都需要安装 VEM 模块。
- VSM（Virtual Supervisor Module，虚拟管理模块）：VSM 本身是虚拟机，运行在 ESXi 主机上，用于管理 ESXi 主机的 VEM 模块。

用传统交换机方式来理解 Cisco Nexus 1000V 交换机更为容易，如图 4-1-2 所示，VSM 可以理解为传统物理交换机的主备引擎，VEM 可以理解为物理交换机以太网板卡。

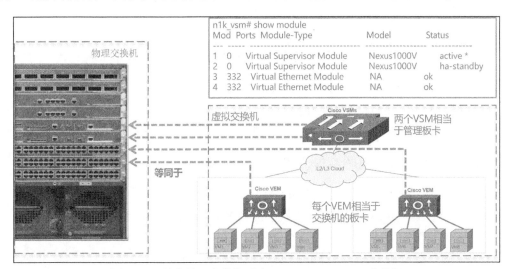

图 4-1-2　传统物理交换机对应 Cisco Nexus 1000V 交换机

Cisco Nexus 1000V 交换机的 VSM 可以控制多个 VEM，可以通过 VSM 来配置 VEM，并且可以在 VEM 之间分发配置信息，因此可以通过 VSM 统一配置所有被管理的 VEM，而不用单独配置每台 ESXi 主机的 VEM。

VSM 与 VEM 之间通信会使用到两个不同的 VLAN 接口：控制 VLAN 以及分组 VLAN，这两类 VLAN 都需要 VSM 与 VEM 之间的二层邻接关系。VSM 与 VMware vSphere 实现了紧密集成，Cisco Nexus 1000V 利用端口配置文件功能来解决了服务器虚拟化中的动态特性，利用端口配置文件功能可以为不同类型的虚拟机定义不同的虚拟机网络策略，然后将配置文件应用到具体的虚拟机，这样实现了网络资源的差异化分配，端口配置文件以及安全策略适用于虚拟机的整体生命周期，包括迁移、挂起、重启等各种状态，端口配置文件还会迁移虚拟机的网络状态，比较常见的端口计数器以及流量统计信息等，虚拟机不但可以参与流量监控操作（比如 Cisco NetFlow 或 ERSPAN），而且不会因为迁移操作而出现中断，可以说，Cisco Nexus 1000V 比较好地解决了虚拟化环境中网络问题。

Cisco Nexus 1000V 不运行 STP 生成树协议，通过丢弃 BPDU、不在物理适配器之间进行数据交换以及在入口处丢弃二层本地 MAC 地址数据包来避免环路，如图 4-1-3 所示。

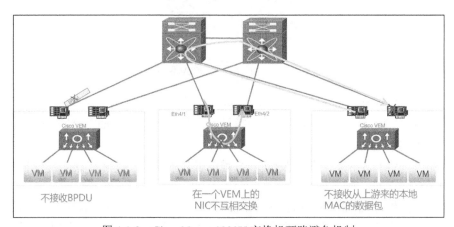

图 4-1-3　Cisco Nexus 1000V 交换机环路避免机制

4.2　部署 Nexus 1000V VSM

VSM 作为 Nexus 1000V 交换机的管理模块，是以虚拟机方式运行在 ESXi 主机上的，其稳定的运行是 Nexus 1000V 交换机正常工作的关键。本节介绍如何配置部署 Nexus 1000V VSM。

4.2.1　部署 VSM 前的准备工作

Cisco Nexus 1000V 交换机对于 VSM 提供了多种部署方式，为保证部署的成功率，推荐使用 Cisco 官方发布的 GUI 方式来部署。部署 VSM 的基础条件如下。

1. ESXi 主机

由于 VSM 部署在 ESXi 主机上，在生产环境中使用时需要考虑冗余问题，因此两个 VSM 虚拟机应该分别部署在两台 ESXi 主机上。

2. 网络地址规划

在部署过程中，VSM 需要使用到 IP 地址，因此需合理地规划 IP 地址。

3. VSM 版本与 VMware vSphere 版本

Cisco Nexus 1000V 比较常用的版本是 4.2(1)SV2 各个系列，目前广泛用于生产环境。最新发布的版本为 5.2(1)SV3 系列，根据官方文档，该系列兼容 VMware vSphere 6.0 虚拟化架构，但在测试过程中发现存在一些 BUG。

4.2.2 部署 VSM

完成 VSM 准备工作后可以开始正式部署过程，本小节详细介绍基于 GUI 的部署方式，其中 VSM 部署在 IP 地址为 10.92.10.14 以及 10.92.10.15 的两台 ESXi 主机上。

第 1 步，使用 Web Client 登录 vCenter Server，确认准备部署 VSM 的 ESXi 主机（如图 4-2-1 所示）。

图 4-2-1　部署 VSM 之一

第 2 步，解压从 Cisco 官方下载的 Cisco Nexus 1000V 软件包（如图 4-2-2 所示）。

图 4-2-2　部署 VSM 之二

第 3 步，运行 GUI 部署工具（如图 4-2-3 所示）。某些版本的 Cisco Nexus 1000V 软件包未提供 GUI 部署工具，经测试可通用。

图 4-2-3　部署 VSM 之三

第 4 步，运行 GUI 部署工具需要 JAVA 支持，应提前安装好 JAVA，单击"确定"按钮（如图 4-2-4 所示）。

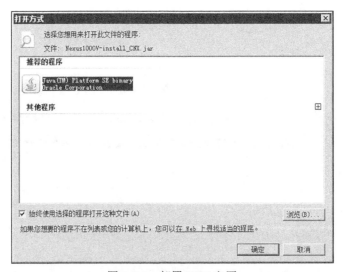

图 4-2-4　部署 VSM 之四

第 5 步，GUI 部署工具打开后，出现三个选项，选择"Cisco Nexus 1000V Complete Installation"项（如图 4-2-5 所示）。

参数解释如下：

Cisco Nexus 1000V Complete Installation:安装 VSM；

Virtual Ethernet Module Installation:安装 VEM；

vCenter Server Connection:vCenter Server 连接。

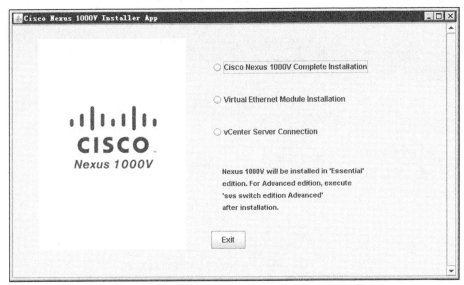

图 4-2-5　部署 VSM 之五

第 6 步，单击"Cisco Nexus 1000V Complete Installation"会有 Standard 和 Custom 两种部署方式，选择"Standard"（如图 4-2-6 所示）。

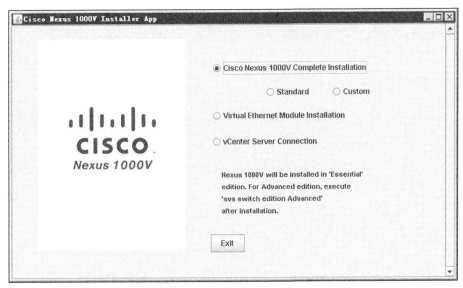

图 4-2-6　部署 VSM 之六

第 7 步，系统提示部署 Cisco Nexus 1000V 前提条件（如图 4-2-7 所示），单击"Next"按钮。

第 8 步，部署过程中会使用验证 vCenter Server，输入 vCenter Server 相关信息（如图 4-2-8 所示），单击"Next"按钮。

第 4 章 部署 Nexus 1000V 分布式交换机

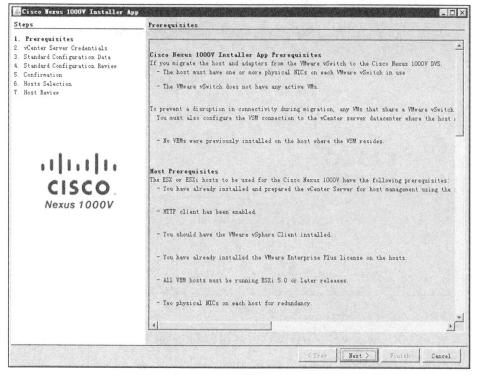

图 4-2-7 部署 VSM 之七

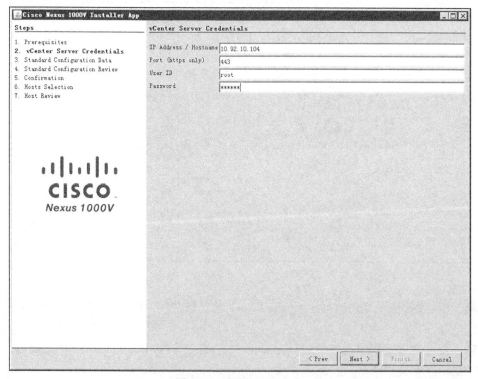

图 4-2-8 部署 VSM 之八

第 9 步，选择部署主 VSM 虚拟机所在的 ESXi 主机（如图 4-2-9 所示），单击"Browse"按钮。

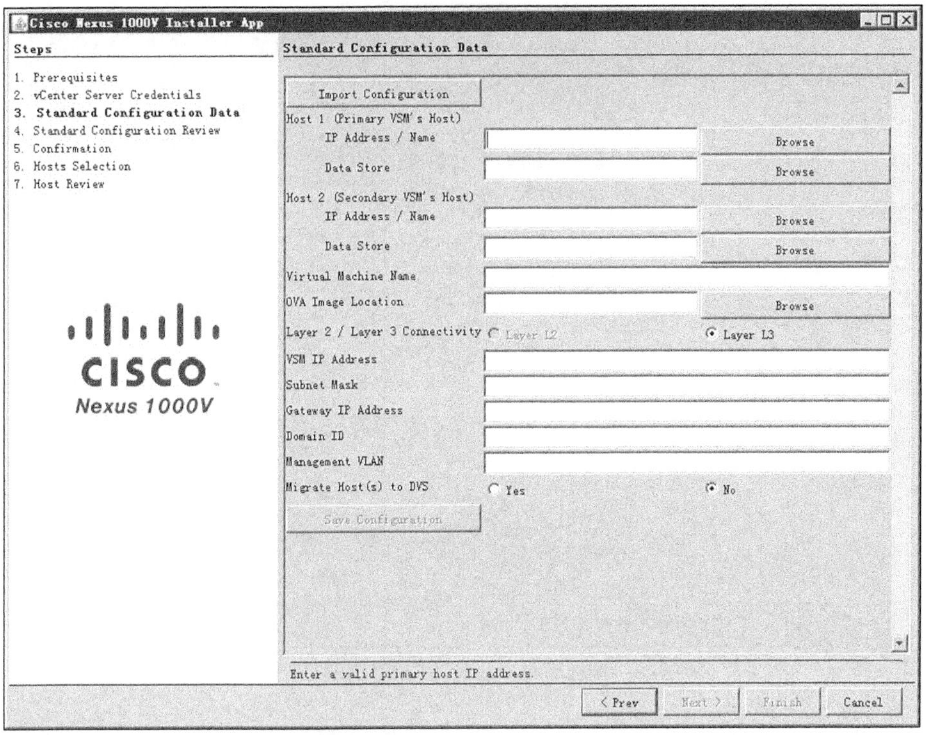

图 4-2-9 部署 VSM 之九

第 10 步，选择 IP 地址为 10.92.10.14 的 ESXi 主机（如图 4-2-10 所示），单击"Select Host"按钮。

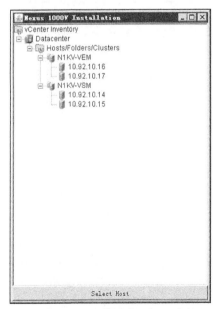

图 4-2-10 部署 VSM 之十

第 11 步，选择 ESXi 主机后会确定主 VSM 虚拟机使用的存储（如图 4-2-11 所示）。

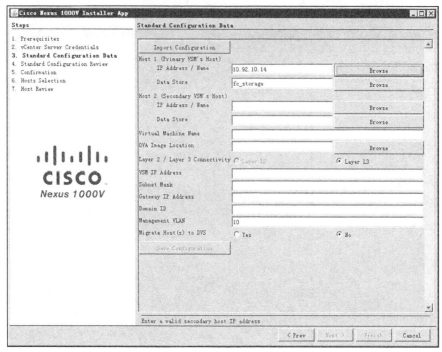

图 4-2-11　部署 VSM 之十一

第 12 步，按照相同的方式选择备 VSM 虚拟机所在的 ESXi 主机以及存储（如图 4-2-12 所示）。

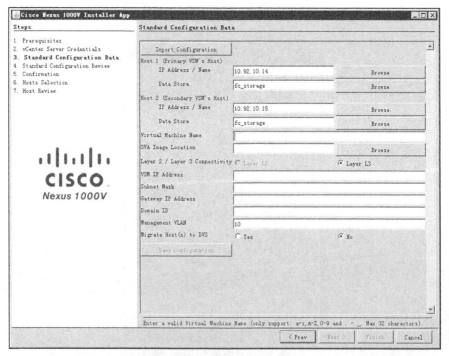

图 4-2-12　部署 VSM 之十二

第 13 步，输入 VSM 虚拟机名称、安装使用的 OVA 文件以及 VSM IP 等相关配置信息（如图 4-2-13 所示），单击 "Next" 按钮。

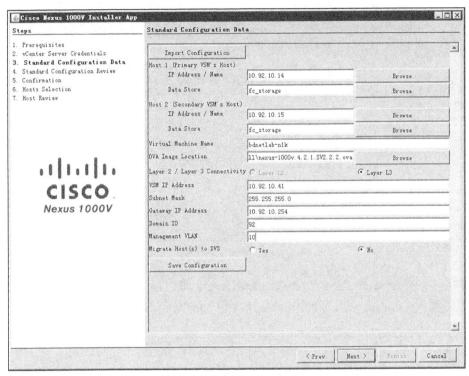

图 4-2-13　部署 VSM 之十三

第 14 步，确定主备 VSM 虚拟机参数设置正确（如图 4-2-14 所示），单击 "Next" 按钮。

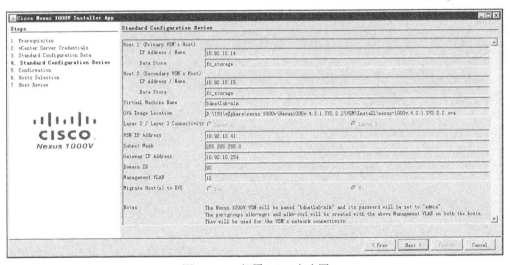

图 4-2-14　部署 VSM 之十四

第 15 步，GUI 部署工具开始部署主备 VSM 虚拟机（如图 4-2-15 所示），单击 "Next" 按钮。

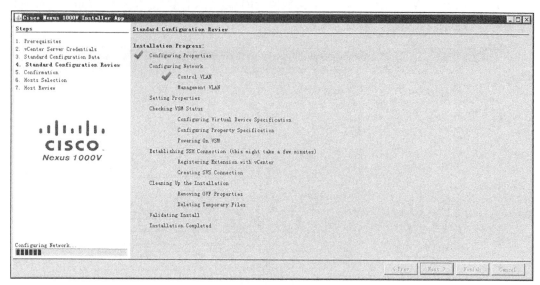

图 4-2-15　部署 VSM 之十五

第 16 步，使用 Web Client 登录 vCenter Server，通过图 4-2-16 可以看到已创建好两台 VSM 虚拟机，虚拟机名分别为 bdnetlab-n1k-1 和 bdnetlab-n1k-2。

图 4-2-16　部署 VSM 之十六

第 17 步，使用 VMware Remote Console 打开虚拟机 bdnetlab-n1k-1 控制台，通过图 4-2-17 可以看到主 VSM 虚拟机已成功启动运行。

第 18 步，系统询问是否添加 VEM 模块（按照本书的进度，VEM 模块将在后续章节添加），选择"No"（如图 4-2-18 所示），单击"Next"按钮。

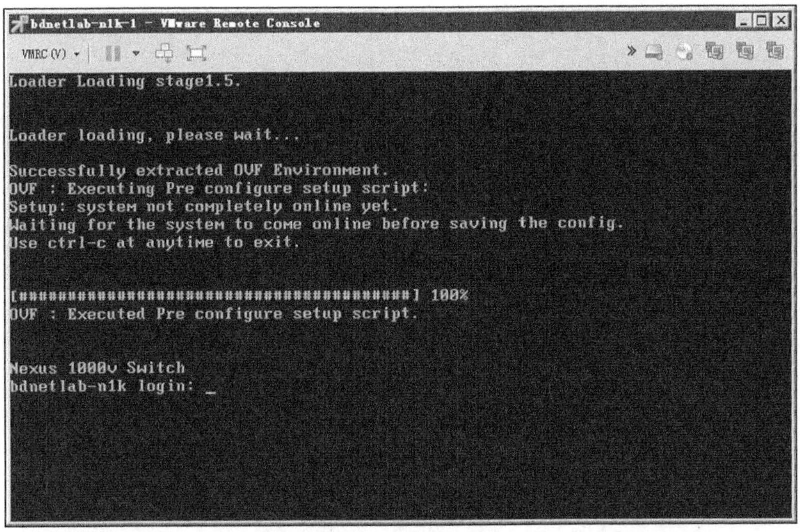

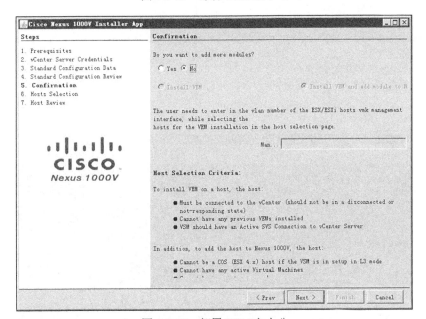

图 4-2-17　部署 VSM 之十七

图 4-2-18　部署 VSM 之十八

第 19 步，系统提示将终止安装部署（如图 4-2-19 所示），单击"是"按钮。

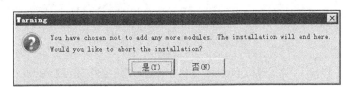

图 4-2-19　部署 VSM 之十九

第 20 步，系统启动主备 VSM 虚拟机电源（如图 4-2-20 所示）。

图 4-2-20　部署 VSM 之二十

第 21 步，输入 VSM 虚拟机账户以及密码，登录 VSM 虚拟机，实质是登录 Cisco Nexus 1000V 交换机（如图 4-2-21 所示）。

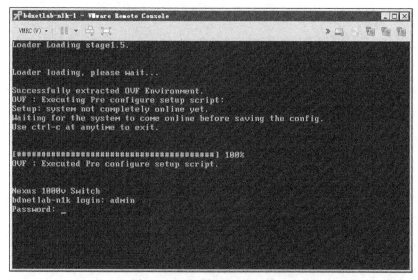

图 4-2-21　部署 VSM 之二十一

第 22 步，使用 PING 命令测试 Nexus 1000V 交换机与网关的连通性（如图 4-2-22 所示），确认测试结果正常。

第 23 步，对于生产环境来说，推荐使用第三方工具连接至 Cisco Nexus 1000V 交换机。如图 4-2-23 所示使用的是名为 SecureCRT 的工具。

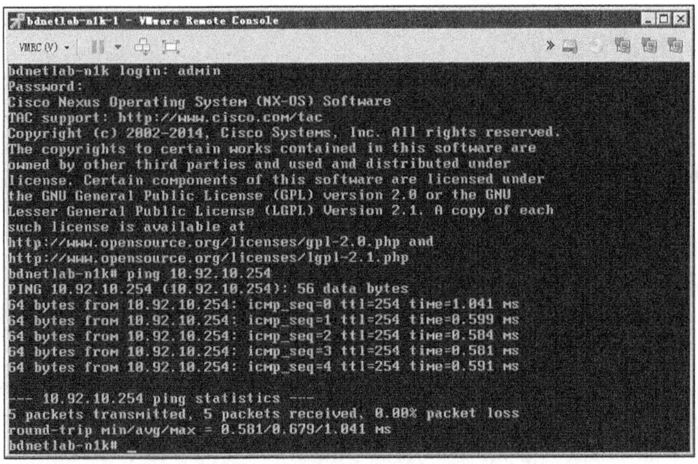

4.2 部署 Nexus 1000V VSM   151

图 4-2-22 部署 VSM 之二十二

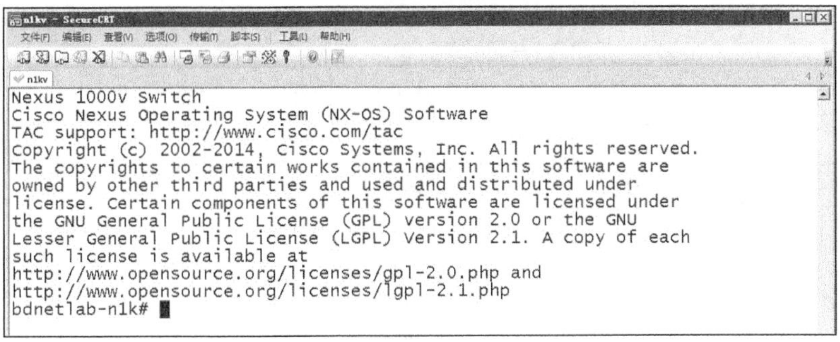

图 4-2-23 部署 VSM 之二十三

第 24 步，使用 Web Client 工具登录到 vCenter Server，在"网络"中可以看到部署好的 Cisco Nexus 1000V 交换机，也就是以虚拟机方式运行的主备 VSM 虚拟机（如图 4-2-24 所示）。

图 4-2-24 部署 VSM 之二十四

4.2.3 VSM 常用命令

1. 查看 VSM 版本信息命令

```
bdnetlab-n1k# show version    //查看 VSM 版本命令
Cisco Nexus Operating System (NX-OS) Software
TAC support: http://www.cisco.com/tac
Copyright (c) 2002-2014, Cisco Systems, Inc. All rights reserved.
The copyrights to certain works contained herein are owned by
other third parties and are used and distributed under license.
Some parts of this software are covered under the GNU Public
License. A copy of the license is available at
http://www.gnu.org/licenses/gpl.html.

Software
  loader:      version unavailable [last: loader version not available]
  kickstart: version 4.2(1)SV2(2.2)    //NXOS kickstart 文件版本
  system:      version 4.2(1)SV2(2.2)   // NXOS system 文件版本
  kickstart image file is: bootflash:/nexus-1000v-kickstart.4.2.1.SV2.2.2.bin
  kickstart compile time:   1/24/2014 15:00:00 [01/25/2014 00:20:16]
  system image file is:     bootflash:/nexus-1000v.4.2.1.SV2.2.2.bin
  system compile time:      1/24/2014 15:00:00 [01/25/2014 01:37:10]

Hardware    //Nexus 1000V 交换机硬件信息
  cisco Nexus 1000V Chassis ("Virtual Supervisor Module")
  Intel(R) Xeon(R) CPU            with 3116076 kB of memory.
  Processor Board ID T505682178F

  Device name: bdnetlab-n1k    //Nexus 1000V 交换机设备名
  bootflash:     1557496 kB

System uptime is 0 days, 0 hours, 10 minutes, 46 seconds    //Nexus 1000V 运行时间
Kernel uptime is 0 day(s), 0 hour(s), 11 minute(s), 13 second(s)

plugin
  Core Plugin, Virtualization Plugin, Ethernet Plugin
```

2. 查看 VSM 与 vCenter Server 连接情况命令

```
bdnetlab-n1k# show svs connections    //查看 VSM 注册到 vCenter Server 的相关情况
connection vcenter:
    ip address: 10.92.10.104    //vCenter Server 服务器 IP 地址
    remote port: 80    //连接 vCenter Server 使用的端口
    protocol: vmware-vim https    //连接 vCenter Server 使用的协议
    certificate: default    //vCenter Server 使用的认证方式
    datacenter name: Datacenter    //vCenter Server 数据中心名称
    admin: n1kUser(user)
```

```
        max-ports: 8192
        DVS uuid: ce dc 02 50 73 b6 a3 1f-73 ef 90 c7 8c a0 30 4d
        config status: Enabled
        operational status: Connected
        sync status: Complete
        version: VMware vCenter Server 5.5.0 build-2442329
        vc-uuid: 6B7F3CA7-653D-4847-8F93-80786FC3EE44
        ssl-cert: self-signed or not authenticated
```

3. 查看 VSM 使用 domain 以及 L2/L3 层模式命令

```
bdnetlab-n1k# show svs domain     //查看 VSM domain 相关信息
SVS domain config:
   Domain id:      92    //domain ID 信息
   Control vlan:   NA
   Packet vlan:    NA
   L2/L3 Control mode: L3    //使用 L3 控制模式
   L3 control interface: mgmt0
   Status: Config push to VC successful.
   Control type multicast: No
```

4. 查看 VSM 引擎模块以及 VEM 模块命令

```
bdnetlab-n1k# show module    //查看 Nexus 1000V 交换机模块信息
Mod  Ports  Module-Type                    Model          Status
---  -----  -----------------------------  -------------  ------------
1    0      Virtual Supervisor Module      Nexus1000V     active *      //主 VSM 引擎
2    0      Virtual Supervisor Module      Nexus1000V     ha-standby    //备 VSM 引擎

Mod  Sw                 Hw
---  -----------------  -----------------
1    4.2(1)SV2(2.2)     0.0
2    4.2(1)SV2(2.2)     0.0

Mod  Server-IP     Server-UUID                            Server-Name
---  -----------   -----------------------------------    -----------
1    10.92.10.41   NA                                     NA
2    10.92.10.41   NA                                     NA

* this terminal session
```

5. 查看 VSM 引擎冗余情况命令

```
bdnetlab-n1k# show system redundancy status     //查看 VSM 引擎冗余情况
Redundancy role
---------------
       administrative:    primary
         operational:     primary
```

```
Redundancy mode
---------------
       administrative:   HA
         operational:    HA

This supervisor (sup-1)   //sup-1 引擎处于活动状态
----------------------
    Redundancy state:    Active
    Supervisor state:    Active
     Internal state:     Active with HA standby

Other supervisor (sup-2)   //sup-2 引擎处理备份状态
----------------------
    Redundancy state:    Standby
    Supervisor state:    HA standby
     Internal state:     HA standby
```

6. 查看 VSM 接口信息命令

```
bdnetlab-n1k# show interface brief    //查看 Nexus 1000V 交换机接口信息
--------------------------------------------------------------------------------
Port      VRF        Status IP Address                    Speed      MTU
--------------------------------------------------------------------------------
mgmt0     --         up     10.92.10.41                   1000       1500
--------------------------------------------------------------------------------
Port      VRF        Status IP Address                    Speed      MTU
--------------------------------------------------------------------------------
control0 --                 up     --                     1000       1500
```

7. 查看 VSM 购买授权所需要 host-id 信息命令

```
bdnetlab-n1k# show license host-id
License hostid: VDH=9422607931493852875
```

8. 查看 VSM 全部配置命令

```
bdnetlab-n1k# show running-config
!Command: show running-config
!Time: Wed May   3 23:23:14 2017

version 4.2(1)SV2(2.2)
svs switch edition essential

no feature telnet

username admin password 5 $1$NeCjVgJ4$S/vaqyILudbfMFKYvCbim0    role network-a
dmin
```

```
banner motd #Nexus 1000v Switch#

ssh key rsa 2048
ip domain-lookup
ip host bdnetlab-n1k 10.92.10.41
hostname bdnetlab-n1k
errdisable recovery cause failed-port-state
snmp-server user admin network-admin auth md5 0xa2cb98ffa3f2bc53380d54d63b67
52db priv 0xa2cb98ffa3f2bc53380d54d63b6752db localizedkey

vrf context management
    ip route 0.0.0.0/0 10.92.10.254
vlan 1,10

port-channel load-balance ethernet source-mac
port-profile default max-ports 32
port-profile type ethernet Unused_Or_Quarantine_Uplink
    vmware port-group
    shutdown
    description Port-group created for Nexus1000V internal usage. Do not use.
    state enabled
port-profile type vethernet Unused_Or_Quarantine_Veth
    vmware port-group
    shutdown
    description Port-group created for Nexus1000V internal usage. Do not use.
    state enabled

vdc bdnetlab-n1k id 1
    limit-resource vlan minimum 16 maximum 2049
    limit-resource monitor-session minimum 0 maximum 2
    limit-resource vrf minimum 16 maximum 8192
    limit-resource port-channel minimum 0 maximum 768
    limit-resource u4route-mem minimum 1 maximum 1
    limit-resource u6route-mem minimum 1 maximum 1

interface mgmt0
    ip address 10.92.10.41/24

interface control0
line console
boot kickstart bootflash:/nexus-1000v-kickstart.4.2.1.SV2.2.2.bin sup-1
boot system bootflash:/nexus-1000v.4.2.1.SV2.2.2.bin sup-1
boot kickstart bootflash:/nexus-1000v-kickstart.4.2.1.SV2.2.2.bin sup-2
boot system bootflash:/nexus-1000v.4.2.1.SV2.2.2.bin sup-2
svs-domain
    domain id 92
```

```
        control vlan 1
        packet vlan 1
        svs mode L3 interface mgmt0
    svs connection vcenter
        protocol vmware-vim
        remote ip address 10.92.10.104 port 80
        vmware dvs uuid "ce dc 02 50 73 b6 a3 1f-73 ef 90 c7 8c a0 30 4d" datacent
    er-name Datacenter
        admin user n1kUser
        max-ports 8192
        connect
    vservice global type vsg
        tcp state-checks invalid-ack
        tcp state-checks seq-past-window
        no tcp state-checks window-variation
        no bypass asa-traffic
    vnm-policy-agent
        registration-ip 0.0.0.0
        shared-secret **********
        log-level
```

4.3 部署 Nexus Port-Profile

Nexus 1000V 中的端口配置文件（Port-Profile）用于定义一系列相同类型的端口属性，多个接口级配置命令整合在一起，创建一个完整的网络策略，使用端口配置文件，可以优化和规范交换机配置。本节介绍如何配置使用端口配置文件。

4.3.1 部署 Port-Profile 前的准备工作

在开始部署端口配置文件前，需要了解一些基础的知识。

1. 上行链路端口配置文件（ethernet Port-Profile）

用于定义 ESXi 主机连接物理交换机的适配器，多数情况下该配置文件被定义为 TRUNK 模式，在 VSM 配置中默认类别为以太网（type Ethernet）。

2. 虚拟机端口配置文件（ethernet Port-Profile）

用于定义虚拟机网络属性，多数情况下该配置文件被定义为 ACCESS 模式，在 VSM 配置中默认类别为虚拟以太网（type vEthernet）。

3. 基于 L3Control 的虚拟机端口配置文件

最初发布的 Nexus 1000V 交换机，VSM 与 VEM 之间的通信使用 2 层网络，每个 VSM 需要连接到 3 个不同的 VLAN（控制 VLAN、管理 VLAN、数据 VLAN），这样可能导致配置的复杂以及网络的不稳定，因此，后续发布的版本推荐使用三层网络来实现 VSM 与 VEM 之间的通信。

4.3 部署 Nexus Port-Profile

基于 L3Control 的虚拟机端口配置文件是为了解决在 VEM 没有被 VSM 管理配置时，确保系统 VLAN 被转发，只要在基于 L3Control 的虚拟机端口配置文件中声明系统 VLAN，它将应用到主机管理接口、VSM 接口、VEM 上行链路来承载控制流量以及 IP 存储等流量。

4.3.2 部署 Port-Profile

端口配置文件需要登录 VSM 通过命令行的模式进行配置，不能通过 vCenter Server 进行配置。

第 1 步，登录 VSM 通过命令行配置上行链路端口配置文件。

```
bdnetlab-n1k# configure terminal    //进入配置模式
Enter configuration commands, one per line.    End with CNTL/Z.
bdnetlab-n1k(config)# port-profile type ethernet uplink   //定义上行链路配置文件名为 uplink
bdnetlab-n1k(config-port-prof)# no shutdown
bdnetlab-n1k(config-port-prof)# switchport mode trunk //定义上行链路模式为 TRUNK
bdnetlab-n1k(config-port-prof)# switchport trunk allowed vlan all    //允许所有 VLAN 通过上行链路
bdnetlab-n1k(config-port-prof)# system vlan 1,10    //定义上行链路系统 VLAN 为 1，10
bdnetlab-n1k(config-port-prof)# vmware port-group    //定义为 vmware 使用的端口组
bdnetlab-n1k(config-port-prof)# state enabled    //启用配置文件

bdnetlab-n1k(config)# show running-config port-profile uplink    //查看配置文件
!Command: show running-config port-profile uplink
!Time: Wed May  3 23:29:44 2017

version 4.2(1)SV2(2.2)
port-profile type ethernet uplink
  vmware port-group
  switchport mode trunk
  switchport trunk allowed vlan 1-3967,4048-4093
  no shutdown
  system vlan 1,10
  state enabled
```

第 2 步，启用配置文件后，vCenter Server 会立即同步通过命令行创建的上行链路端口配置文件（如图 4-3-1 所示）。如果通过 VSM 命令行创建后未同步 vCenter Server，说明 VSM 与 vCenter Serve 之间的配置存在问题，则应进行检查。

图 4-3-1　查看上行链路端口配置文件

第 3 步，登录 VSM 通过命令行配置虚拟机端口配置文件。

```
bdnetlab-n1k(config)#  port-profile  type  vethernet  WEB-SERVER    //定义虚拟机端口配置文件为 WEB-SERVER
bdnetlab-n1k(config-port-prof)# no shutdown
bdnetlab-n1k(config-port-prof)# switchport mode access    //定义链路模式为 ACCESS
bdnetlab-n1k(config-port-prof)# switchport access vlan 10    //定义链路访问 VLAN 10
bdnetlab-n1k(config-port-prof)# system vlan 10    //定义链路系统 VLAN 为 10
bdnetlab-n1k(config-port-prof)# vmware port-group    //定义为 vmware 使用的端口组
bdnetlab-n1k(config-port-prof)# state enabled    //启用配置文件

bdnetlab-n1k(config)# show running-config port-profile WEB-SERVER    //查看配置文件
!Command: show running-config port-profile WEB-SERVER
!Time: Wed May   3 23:31:56 2017

version 4.2(1)SV2(2.2)
port-profile type vethernet WEB-SERVER
   vmware port-group
   switchport mode access
   switchport access vlan 10
   no shutdown
   system vlan 10
   state enabled
```

第 4 步，启用配置文件后，vCenter Server 会立即同步通过命令行创建的虚拟机端口配置文件（如图 4-3-2 所示）。

图 4-3-2　查看虚拟机端口配置文件

第 5 步，登录 VSM 通过命令行配置基于 L3Control 的虚拟机端口配置文件。

```
bdnetlab-n1k(config)# port-profile type vethernet l3-control    //定义 l3-control 端口配置文件
bdnetlab-n1k(config-port-prof)# no shutdown
bdnetlab-n1k(config-port-prof)# capability l3control    //定义 3 层控制功能，必须启用
```

```
Warning: Port-profile 'l3-control' is configured with 'capability l3control'. Also configure the corresponding
access vlan as a system vlan in:
    * Port-profile 'l3-control'.
    * Uplink port-profiles that are configured to carry the vlan
bdnetlab-n1k(config-port-prof)# system vlan 10
bdnetlab-n1k(config-port-prof)# vmware port-group
bdnetlab-n1k(config-port-prof)# state enabled

bdnetlab-n1k(config)# show running-config port-profile l3-control
!Command: show running-config port-profile l3-control
!Time: Wed May   3 23:34:20 2017

version 4.2(1)SV2(2.2)
port-profile type vethernet l3-control
    capability l3control
    vmware port-group
    switchport mode access
    switchport access vlan 10
    no shutdown
    system vlan 10
    state enabled
```

第 6 步，启用配置文件后，vCenter Server 会立即同步通过命令行创建的基于 L3Control 的虚拟机端口配置文件（如图 4-3-3 所示）。

图 4-3-3　查看基于 L3Control 的虚拟机端口配置文件

4.3.3　Port-Profile 常用命令

查看 VSM 端口配置文件命令如下。

```
bdnetlab-n1k# show port-profile

port-profile Unused_Or_Quarantine_Uplink
```

```
  type: Ethernet
  description: Port-group created for Nexus1000V internal usage. Do not use.
  status: enabled
  max-ports: 32
  min-ports: 1
  inherit:
  config attributes:
    shutdown
  evaluated config attributes:
  assigned interfaces:
  port-group: Unused_Or_Quarantine_Uplink
  system vlans: none
  capability l3control: no
  capability iscsi-multipath: no
  capability vxlan: no
  capability l3-vservice: no
  port-profile role: none
  port-binding: static

port-profile Unused_Or_Quarantine_Veth
  type: Vethernet
  description: Port-group created for Nexus1000V internal usage. Do not use.
  status: enabled
  max-ports: 32
  min-ports: 1
  inherit:
  config attributes:
    shutdown
  evaluated config attributes:
  assigned interfaces:
  port-group: Unused_Or_Quarantine_Veth
  system vlans: none
  capability l3control: no
  capability iscsi-multipath: no
  capability vxlan: no
  capability l3-vservice: no
  port-profile role: none
  port-binding: static
```

4.4 部署 Nexus 1000V VEM

VEM 作为 ESXi 主机运行的虚拟机对外通信板卡，重要性不言而喻，成功的部署以及 VEM 被 VSM 管理也非常重要。本节介绍如何部署 VEM，其中 VEM 部署在 IP 地址为 10.92.10.16 以及 10.92.10.17 的两台 ESXi 主机上。

4.4.1 部署 VEM 前的准备工作

在开始部署 VEM 前,需要了解一些基础的知识。

1. VEM 数据包转发

每个 VEM 拥有自己独立的 MAC 地址表,同一个 VEM 上的虚拟机数据本地进行交换,不同 VEM 上的虚拟机通过上游交换机进行数据交换,VEM 不接收 BPDU 以避免环路。

2. VEM 上行链路 Port-Channel

当 ESXi 主机物理适配器被放入对应的 port-profile 之后,port-channel 自动生成,每个 VEM 独立形成自己的 Port Channel,可以是传统交换机上支持的 port-channel(mode on/mode active),也可以是不要求上行交换机有任何配置的 port-channel(例如 mac pinning),只要 port-channel 中有一个物理适配器处于活动状态,这个 port-channel 就会一直工作。需要注意的是,在不同的 VEM 之间不能建立 port-channel。

3. VEM 模块上的 subgroups

每一个交换机对 VEM 来说形成一个 subgroups,也可以理解为小型 port-channel 每一个物理端口都会被分配一个 subgroup ID(如图 4-4-1 所示)。subgroups 可以自动生成,也可以手动指定。

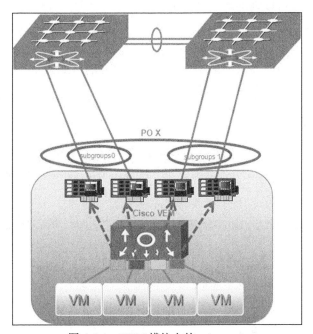

图 4-4-1　VEM 模块上的 subgroups

4. VEM 如何注册到 VSM

当 VEM 初次安装之后,它并不知道任何 port-profile 的存在,也不知道 VSM 将如何配置它,它不知道应该如何不 VSM 通信(如图 4-4-2 所示)。

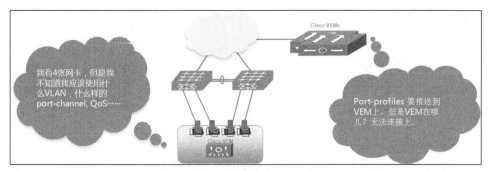

图 4-4-2　VEM 模块如何注册到 VSM

VSM 模块如何注册到 VSM 可以通过两种技术解决。
- SYSTEM VLAN
  - SYSTEM VLAN 是指在端口在被完全配置完成前就应该允许通信的 VLAN；
  - 一个 Nexus 1000V 中最多允许 32 个 SYSTEM VLAN；
  - SYSTEM VLAN 必须是上行 trunk 被允许的 VLAN 的子集；
  - 当有端口使用 port-profile 的时候，SYSTEM VLAN 不能从该 port-profile 中移除，但是可以增加新的 SYSTEM VLAN。
- Opaque Data
  - Opaque data 从 vCenter Server 复制信息到 VEM 模块，包含了 VEM 需要初始化上行端口所需要的最小信息，一旦上行端口配置完成，VEM 即可与 VSM 通信完成其他配置。

4.4.2　部署 VEM

了解 VEM 准备工作后可以开始正式部署过程，本小节详细介绍基于 GUI 的部署方式。

第 1 步，运行 GUI 部署工具，选择 "Virtual Ethernet Module Installation"（如图 4-4-3 所示）。

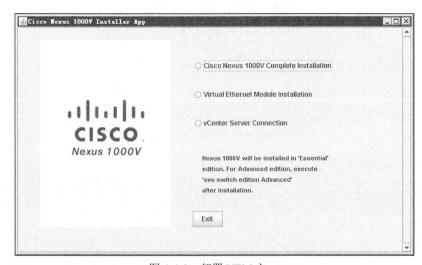

图 4-4-3　部署 VEM 之一

第 2 步，系统提示部署 VEM 需要满足的前提条件（如图 4-4-4 所示），单击"Next"按钮。

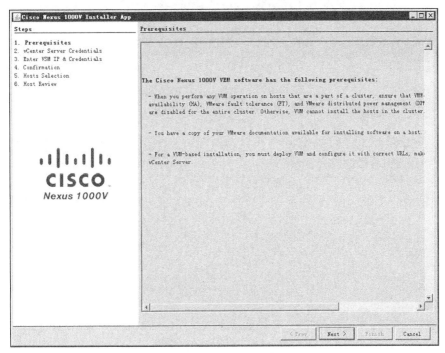

图 4-4-4　部署 VEM 之二

第 3 步，输入 vCenter Server 相关信息（如图 4-4-5 所示），单击"Next"按钮。

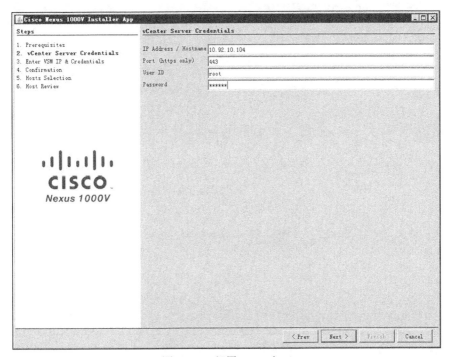

图 4-4-5　部署 VEM 之三

第 4 步，输入已部署好的 VSM 相关信息（如图 4-4-6 所示），单击"Next"按钮。

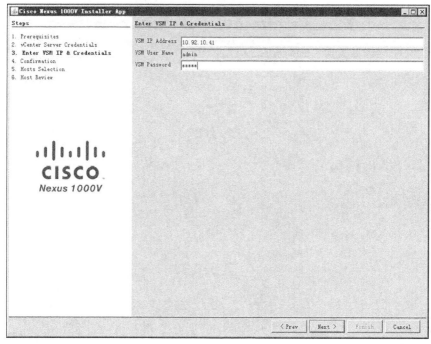

图 4-4-6　部署 VEM 之四

第 5 步，进入 VEM 模块添加界面（如图 4-4-7 所示）。

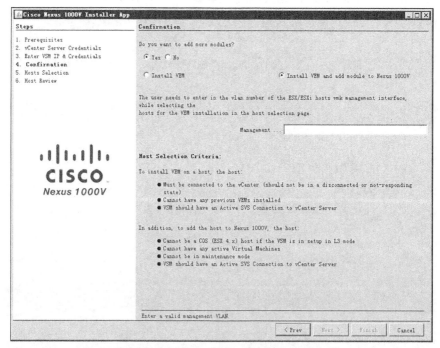

图 4-4-7　部署 VEM 之五

第 6 步，选择"Install VEM"（如图 4-4-8 所示），单击"Next"按钮。

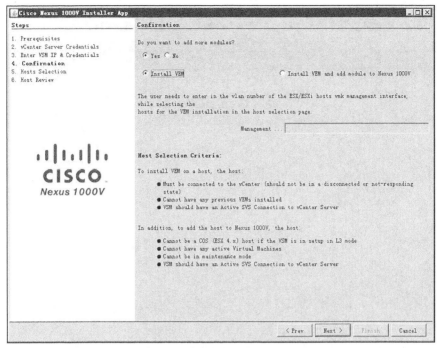

图 4-4-8　部署 VEM 之六

第 7 步，选择需要部署 VEM 模块的 ESXi 主机（如图 4-4-9 所示），单击"Next"按钮。

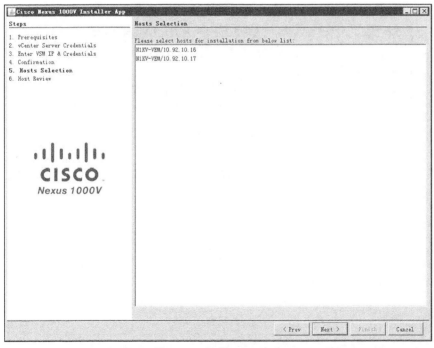

图 4-4-9　部署 VEM 之七

第8步，确认部署 VEM 模块的 ESXi 主机以及 VSM 信息（如图 4-4-10 所示），单击"Finish"按钮。

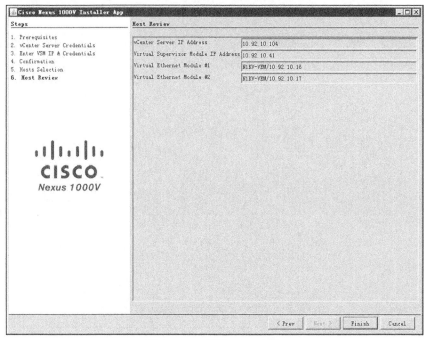

图 4-4-10　部署 VEM 之八

第9步，完成 ESXi 主机 VEM 模块的安装，安装状态为 Success（如图 4-4-11 所示），单击"Close"按钮。

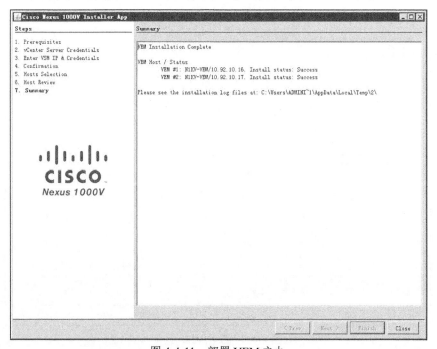

图 4-4-11　部署 VEM 之九

第 10 步，使用 Web Client 登录 vCenter Server，选择分布式交换机"bdnetlab-n1k"，在"主机"处可以看到分布式交换机未添加任何 ESXi 主机，列表为空（如图 4-4-12 所示）。

图 4-4-12　部署 VEM 之十

第 11 步，在添加管理主机窗口选择"添加主机"（如图 4-4-13 所示），单击"下一步"按钮。

图 4-4-13　部署 VEM 之十一

第 12 步，选择"+新主机"（如图 4-4-14 所示），单击"下一步"按钮。

第 13 步，勾选需要加入分布式交换机"bdnetlab-n1k"的 ESXi 主机（如图 4-4-15 所示），单击"确定"按钮。

图 4-4-14　部署 VEM 之十二

图 4-4-15　部署 VEM 之十三

第 14 步，确认加入分布式交换机"bdnetlab-n1k"的 ESXi 主机，系统会在新加入的 ESXi 主机名前备注一个"新"的字样，提示这是新加入 ESXi 主机（如图 4-4-16 所示），单击"下一步"按钮。

第 15 步，选择"管理物理适配器"，将 ESXi 主机的适配器添加到分布式交换机"bdnetlab-n1k"（如图 4-4-17 所示），单击"下一步"按钮。

4.4 部署 Nexus 1000V VEM

图 4-4-16 部署 VEM 之十四

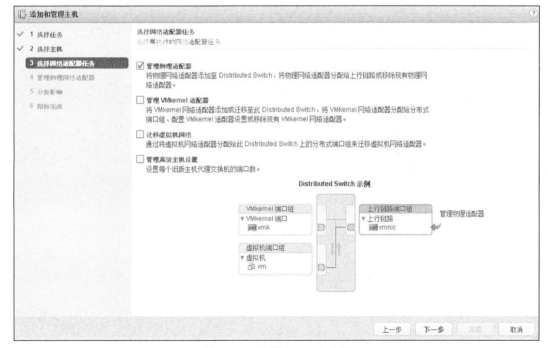

图 4-4-17 部署 VEM 之十五

第 16 步，选择 ESXi 主机需要加入分布式交换机"bdnetlab-n1k"的适配器，一般来说选择未关联其他交换机的适配器（如图 4-4-18 所示），单击"分配上行链路"。

图 4-4-18　部署 VEM 之十六

第 17 步，为 vmnic2 适配器分配上行链路编号，需要注意的是 ESXi 主机将使用 Nexus 1000V 交换机，因为上行链路端口组应选择 VSM 创建名为 uplink 的上行链路配置文件（如图 4-4-19 所示），单击"确定"按钮。

图 4-4-19　部署 VEM 之十七

第 18 步，确认将 IP 地址为 10.92.10.16 的 ESXi 主机 vmnic2 分配给上行链路端口组 uplink（如图 4-4-20 所示），单击"下一步"按钮。

图 4-4-20　部署 VEM 之十八

第 19 步，使用相同的方式将 ESXi 主机其他适配器分配给分布式交换机"bdnetlab-n1k"使用，确保上行链路端口组为 uplink（如图 4-4-21 所示），单击"下一步"按钮。

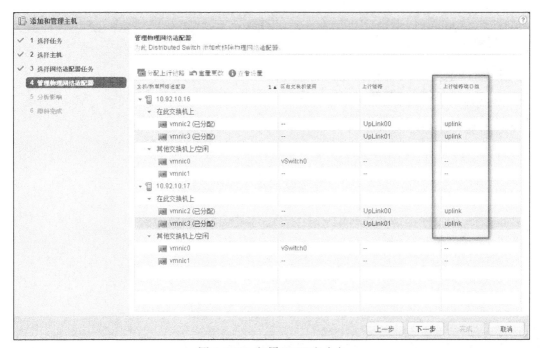

图 4-4-21　部署 VEM 之十九

第 20 步，系统自动分析 ESXi 主机适配器加入分布式交换机"bdnetlab-n1k"是否会对已有网络造成影响（如图 4-4-22 所示），单击"下一步"按钮。

图 4-4-22　部署 VEM 之二十

第 21 步，确认添加进入分布式交换机"bdnetlab-n1k"的 ESXi 主机参数设置正确（如图 4-4-23 所示），单击"完成"按钮。

图 4-4-23　部署 VEM 之二十一

第 22 步，在"相关对象"标签的"主机"处可以看到上行链路端口数量为 4（如图 4-4-24 所示）。

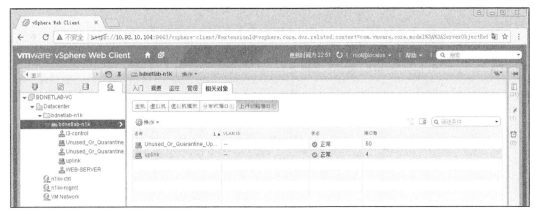

图 4-4-24　部署 VEM 之二十二

第 23 步，选择"管理 VMkernel 适配器"，调整已有 VMkernel 配置（如图 4-4-25 所示），单击"下一步"按钮。

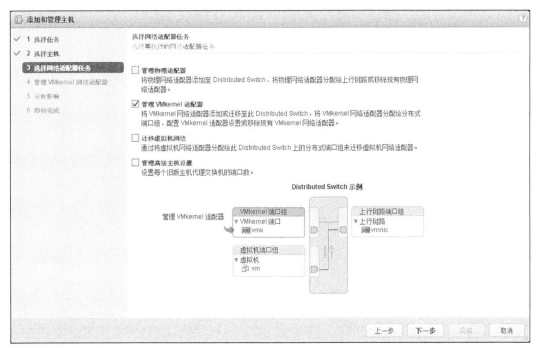

图 4-4-25　部署 VEM 之二十三

第 24 步，选择并调整 vmk0（如图 4-4-26 所示），单击"分配端口组"。

第 25 步，选择在前面章节创建好的基于 L3Control 的虚拟机端口配置文件（如图 4-4-27 所示），单击"确定"按钮。

图 4-4-26　部署 VEM 之二十四

图 4-4-27　部署 VEM 之二十五

第 26 步，确认将 IP 地址为 10.92.10.16 的 ESXi 主机 vmk0 分配给目标端口组 l3-control（如图 4-4-28 所示），单击"下一步"按钮。

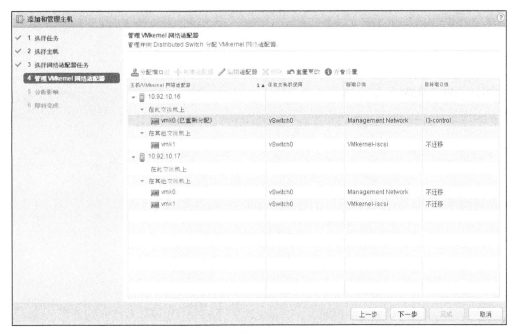

图 4-4-28 部署 VEM 之二十六

第 27 步，使用相同的方式将其他 ESXi 主机 vmk0 分配目标端口组，确保目标端口组为 l3-control（如图 4-4-29 所示），单击"下一步"按钮。

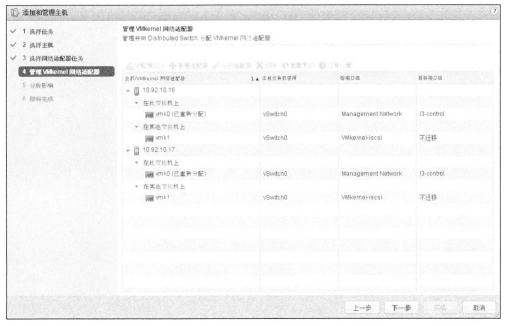

图 4-4-29 部署 VEM 之二十七

第 28 步，系统自动分析 ESXi 主机适配器加入分布式交换机"bdnetlab-n1k"是否会对已有网络造成影响（如图 4-4-30 所示），单击"下一步"按钮。

176　第 4 章　部署 Nexus 1000V 分布式交换机

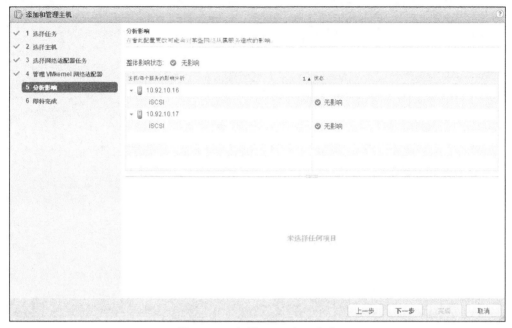

图 4-4-30　部署 VEM 之二十八

第 29 步，确认添加进入分布式交换机"bdnetlab-n1k"的 ESXi 主机参数设置正确（如图 4-4-31 所示），单击"完成"按钮。

图 4-4-31　部署 VEM 之二十九

第 30 步，在"相关对象"标签的"主机"处可以看到 IP 地址为 10.92.10.16 和 10.92.10.17 的两台 ESXi 主机已添加到分布式交换机（如图 4-4-32 所示）。

4.4 部署 Nexus 1000V VEM   177

图 4-4-32 部署 VEM 之三十

4.4.3 VEM 常用命令

1. 查看 VSM 引擎模块以及 VEM 模块命令

当 VSM 以及 VEM 部署完成后，使用该命令查看 VEM 模块是否成功注册到 VSM。如果 VEM 未注册到 VSM，则 VSM 无法对 VEM 进行配置。

```
bdnetlab-n1k# show module
Mod   Ports   Module-Type                    Model            Status
---   -----   ---------------------------    -----------      ------
1     0       Virtual Supervisor Module      Nexus1000V       active *
2     0       Virtual Supervisor Module      Nexus1000V       ha-standby
3     332     Virtual Ethernet Module        NA               ok        //VEM 注册成功信息
4     332     Virtual Ethernet Module        NA               ok        //VEM 注册成功信息

Mod   Sw                Hw
---   --------------    -----------------
1     4.2(1)SV2(2.2)    0.0
2     4.2(1)SV2(2.2)    0.0
3     4.2(1)SV2(2.2)    VMware ESXi 5.5.0 Releasebuild-2403361 (3.2)    // VEM 模块 ESXi 主机版本
4     4.2(1)SV2(2.2)    VMware ESXi 5.5.0 Releasebuild-2403361 (3.2)    // VEM 模块 ESXi 主机版本

Mod   Server-IP       Server-UUID                             Server-Name
---   -----------     ------------------------------------    -----------
1     10.92.10.41     NA                                      NA
2     10.92.10.41     NA                                      NA
3     10.92.10.16     4c4c4544-0030-5810-8032-b9c04f4d4c31    NA    //VEM 模块对应主机 IP 地址
4     10.92.10.17     4c4c4544-0054-5310-8032-b9c04f4d4c31    NA    //VEM 模块对应主机 IP 地址

* this terminal session
```

2. 查看 VSM 接口信息命令

当 VEM 模块注册到 VSM 后，ESXi 主机上行链路的物理适配器格式为模块号/接口，Eth3/3、Eth3/4 代表 IP 地址为 10.92.10.16 ESXi 主机的上行链路，Eth4/3、Eth4/4 代表 IP

地址为 10.92.10.17 ESXi 主机的上行链路。

```
bdnetlab-n1k# show interface brief
--------------------------------------------------------------------------------
Port          VRF             Status IP Address                 Speed        MTU
--------------------------------------------------------------------------------
mgmt0         --              up     10.92.10.41                1000         1500

Ethernet      VLAN    Type Mode    Status  Reason               Speed        Port
Interface                                                                    Ch #
--------------------------------------------------------------------------------
Eth3/3        1       eth  trunk   up      none                 1000
Eth3/4        1       eth  trunk   up      none                 1000
Eth4/3        1       eth  trunk   up      none                 1000
Eth4/4        1       eth  trunk   up      none                 1000

Vethernet     VLAN    Type Mode    Status  Reason               Speed
--------------------------------------------------------------------------------
Veth1         10      virt access  up      none                 auto
Veth2         10      virt access  up      none                 auto

Port          VRF             Status IP Address                 Speed        MTU
--------------------------------------------------------------------------------
control0 --                   up     --                         1000         1500
```

3. 查看 VEM 模块对应 ESXi 主机 vmnic 信息

查看 VEM 模块号为 3、IP 地址为 10.92.10.16 ESXi 主机对应 vmnic 信息。

```
bdnetlab-n1k# module vem 3 execute vemcmd show port
  LTL    VSM Port   Admin Link   State   PC-LTL   SGID   Vem Port   Type
   19    Eth3/3     UP    UP     FWD     0               vmnic2     //Eth3/3 对应 vmnic2
   20    Eth3/4     UP    UP     FWD     0               vmnic3     //Eth3/4 对应 vmnic3
   49    Veth1      UP    UP     FWD     0               vmk0

* F/B: Port is BLOCKED on some of the vlans.
       One or more vlans are either not created or
       not in the list of allowed vlans for this port.
  Please run "vemcmd show port vlans" to see the details.
```

查看 VEM 模块号为 4、IP 地址为 10.92.10.17 ESXi 主机对应 vmnic 信息。

```
bdnetlab-n1k# module vem 4 execute vemcmd show port
  LTL    VSM Port   Admin Link   State   PC-LTL   SGID   Vem Port   Type
   19    Eth4/3     UP    UP     FWD     0               vmnic2     //Eth4/3 对应 vmnic2
   20    Eth4/4     UP    UP     FWD     0               vmnic3     //Eth4/4 对应 vmnic3
   49    Veth2      UP    UP     FWD     0               vmk0
```

```
  * F/B: Port is BLOCKED on some of the vlans.
           One or more vlans are either not created or
           not in the list of allowed vlans for this port.
    Please run "vemcmd show port vlans" to see the details.
```

4. 查看 Nexus 1000V 交换机详细信息

```
bdnetlab-n1k# show vms internal info dvs
  DVS INFO:
------------
  DVS name: [bdnetlab-n1k]
       UUID: [ce dc 02 50 73 b6 a3 1f-73 ef 90 c7 8c a0 30 4d]
       Description: [(null)]
       Config version: [1]
       Max ports: [8192]
       DC name: [Datacenter]
       OPQ data: size [636], data: [data-version 1.0
switch-domain 92
switch-name bdnetlab-n1k
cp-version 4.2(1)SV2(2.2)
control-vlan 1
system-primary-mac 00:50:56:82:cf:c5
active-vsm packet mac 00:50:56:82:d0:de
active-vsm mgmt mac 00:50:56:82:17:8f
standby-vsm ctrl mac 0050-5682-8795
inband-vlan 1
svs-mode L3
l3control-ipaddr 10.92.10.41
upgrade state 0 mac 0050-5682-8795 l3control-ipv4 null
cntl-type-mcast 0
profile dvportgroup-49 trunk 1,10
profile dvportgroup-49 mtu 1500
profile dvportgroup-50 access 10
profile dvportgroup-50 mtu 1500
profile dvportgroup-51 access 10
profile dvportgroup-51 mtu 1500
profile dvportgroup-51 capability l3control
  end-version 1.0
]
       push_opq_data flag: [1]
```

5. 查看 Nexus 1000V 交换机 VSM 与 VEM 连接信息

```
bdnetlab-n1k# show svs neighbors
Active Domain ID: 92


AIPC Interface MAC: 0050-5682-cfc5
```

```
Inband Interface MAC: 0050-5682-d0de

Src MAC            Type    Domain-id   Node-id    Last learnt (Sec. ago)
-------------------------------------------------------------------------
0050-5682-8795     VSM     92          0201       5267.92
0002-3d40-5c02     VEM     92          0302       192.50
0002-3d40-5c03     VEM     92          0402       191.80
```

6. 从 Nexus 1000V 交换机 PING ESXi 主机测试连通性

```
bdnetlab-n1k# ping 10.92.10.16
PING 10.92.10.16 (10.92.10.16): 56 data bytes
64 bytes from 10.92.10.16: icmp_seq=0 ttl=63 time=1.3 ms
64 bytes from 10.92.10.16: icmp_seq=1 ttl=63 time=1.026 ms
64 bytes from 10.92.10.16: icmp_seq=2 ttl=63 time=0.924 ms
64 bytes from 10.92.10.16: icmp_seq=3 ttl=63 time=0.909 ms
64 bytes from 10.92.10.16: icmp_seq=4 ttl=63 time=0.938 ms

--- 10.92.10.16 ping statistics ---
5 packets transmitted, 5 packets received, 0.00% packet loss
round-trip min/avg/max = 0.909/1.019/1.3 ms
bdnetlab-n1k# ping 10.92.10.17
PING 10.92.10.17 (10.92.10.17): 56 data bytes
64 bytes from 10.92.10.17: icmp_seq=0 ttl=63 time=1.213 ms
64 bytes from 10.92.10.17: icmp_seq=1 ttl=63 time=0.986 ms
64 bytes from 10.92.10.17: icmp_seq=2 ttl=63 time=0.832 ms
64 bytes from 10.92.10.17: icmp_seq=3 ttl=63 time=0.855 ms
64 bytes from 10.92.10.17: icmp_seq=4 ttl=63 time=0.865 ms

--- 10.92.10.17 ping statistics ---
5 packets transmitted, 5 packets received, 0.00% packet loss
round-trip min/avg/max = 0.832/0.95/1.213 ms
```

7. 查看完成基本部署后的完整配置信息

```
bdnetlab-n1k# show running-config
!Command: show running-config
!Time: Thu May   4 00:40:03 2017

version 4.2(1)SV2(2.2)
svs switch edition essential

no feature telnet

username admin password 5 $1$NeCjVgJ4$S/vaqylLudbfMFKYvCbim0    role network-a
dmin
```

```
banner motd #Nexus 1000v Switch#

ssh key rsa 2048
ip domain-lookup
ip host bdnetlab-n1k 10.92.10.41
hostname bdnetlab-n1k
errdisable recovery cause failed-port-state
vem 3
    host id 4c4c4544-0030-5810-8032-b9c04f4d4c31
vem 4
    host id 4c4c4544-0054-5310-8032-b9c04f4d4c31
snmp-server user admin network-admin auth md5 0xa2cb98ffa3f2bc53380d54d63b67
52db priv 0xa2cb98ffa3f2bc53380d54d63b6752db localizedkey


vrf context management
    ip route 0.0.0.0/0 10.92.10.254
vlan 1,10


port-channel load-balance ethernet source-mac
port-profile default max-ports 32
port-profile type ethernet Unused_Or_Quarantine_Uplink
    vmware port-group
    shutdown
    description Port-group created for Nexus1000V internal usage. Do not use.
    state enabled
port-profile type vethernet Unused_Or_Quarantine_Veth
    vmware port-group
    shutdown
    description Port-group created for Nexus1000V internal usage. Do not use.
    state enabled
port-profile type ethernet uplink
    vmware port-group
    switchport mode trunk
    switchport trunk allowed vlan 1-3967,4048-4093
    no shutdown
    system vlan 1,10
    state enabled
port-profile type vethernet WEB-SERVER
    vmware port-group
    switchport mode access
    switchport access vlan 10
    no shutdown
    system vlan 10
    state enabled
```

```
port-profile type vethernet l3-control
    capability l3control
    vmware port-group
    switchport mode access
    switchport access vlan 10
    no shutdown
    system vlan 10
    state enabled

vdc bdnetlab-n1k id 1
    limit-resource vlan minimum 16 maximum 2049
    limit-resource monitor-session minimum 0 maximum 2
    limit-resource vrf minimum 16 maximum 8192
    limit-resource port-channel minimum 0 maximum 768
    limit-resource u4route-mem minimum 1 maximum 1
    limit-resource u6route-mem minimum 1 maximum 1

interface mgmt0
    ip address 10.92.10.41/24

interface Vethernet1
    inherit port-profile l3-control
    description VMware VMkernel, vmk0
    vmware dvport 65 dvswitch uuid "ce dc 02 50 73 b6 a3 1f-73 ef 90 c7 8c a0 30 4d"
    vmware vm mac 0026.6CFB.612C

interface Vethernet2
    inherit port-profile l3-control
    description VMware VMkernel, vmk0
    vmware dvport 64 dvswitch uuid "ce dc 02 50 73 b6 a3 1f-73 ef 90 c7 8c a0 30 4d"
    vmware vm mac 0026.6CFA.FCA8

interface Ethernet3/3
    inherit port-profile uplink

interface Ethernet3/4
    inherit port-profile uplink

interface Ethernet4/3
    inherit port-profile uplink

interface Ethernet4/4
    inherit port-profile uplink
```

```
interface control0
line console
boot kickstart bootflash:/nexus-1000v-kickstart.4.2.1.SV2.2.2.bin sup-1
boot system bootflash:/nexus-1000v.4.2.1.SV2.2.2.bin sup-1
boot kickstart bootflash:/nexus-1000v-kickstart.4.2.1.SV2.2.2.bin sup-2
boot system bootflash:/nexus-1000v.4.2.1.SV2.2.2.bin sup-2
svs-domain
  domain id 92
  control vlan 1
  packet vlan 1
  svs mode L3 interface mgmt0
svs connection vcenter
  protocol vmware-vim
  remote ip address 10.92.10.104 port 80
  vmware dvs uuid "ce dc 02 50 73 b6 a3 1f-73 ef 90 c7 8c a0 30 4d" datacent
er-name Datacenter
  admin user n1kUser
  max-ports 8192
  connect
vservice global type vsg
  tcp state-checks invalid-ack
  tcp state-checks seq-past-window
  no tcp state-checks window-variation
  no bypass asa-traffic
vnm-policy-agent
  registration-ip 0.0.0.0
  shared-secret **********
  log-level
```

4.4.4 VEM 常见故障排除

VEM 在部署过程中比较常见的错误就是无法注册到 VSM，导致 VSM 无法配置 VEM。下面讲述如何处理这样的故障。

第 1 步，确认 Nexus 1000V 的固件版本与 ESXi 主机匹配。ESXi 5.X 之前，ESXi 的每个版本都对应有 VEM 固件版本，可以访问 Cisco 官方网站进行下载。

第 2 步，登录 ESXi 主机查看 VEM 版本信息以及连接状态。

```
~ # vem version    //查看 ESXi 主机 VEM 模块版本信息
Running esx version -2403361 x86_64
VEM Version: 4.2.1.2.2.2.0-3.2.1
VSM Version: 4.2(1)SV2(2.2)
System Version: VMware ESXi 5.5.0 Releasebuild-2403361

~ # vem status    //查看 ESXi 主机 VEM 模块状态
```

```
VEM modules are loaded
Switch Name       Num Ports    Used Ports   Configured Ports   MTU      Uplinks
vSwitch0          2352         4            128                1500     vmnic0
DVS Name          Num Ports    Used Ports   Configured Ports   MTU      Uplinks
bdnetlab-n1kv     1024         14           1024               1500     vmnic3,vmnic2
VEM Agent (vemdpa) is running
```

第 3 步，确认 VSM 与 vCenter Server 连接正常。

```
bdnetlab-n1k# show svs connections    //查看 VSM 注册到 vCenter Server 相关情况
connection vcenter:
    ip address: 10.92.10.104    //vCenter Server 服务器 IP 地址
    remote port: 80    //连接 vCenter Server 使用的端口
    protocol: vmware-vim https    //连接 vCenter Server 使用的协议
    certificate: default    //vCenter Server 使用的认证方式
    datacenter name: Datacenter    //vCenter Server 数据中心名称
    admin: n1kUser(user)
    max-ports: 8192
    DVS uuid: ce dc 02 50 73 b6 a3 1f-73 ef 90 c7 8c a0 30 4d
    config status: Enabled
    operational status: Connected
    sync status: Complete
    version: VMware vCenter Server 5.5.0 build-2442329
    vc-uuid: 6B7F3CA7-653D-4847-8F93-80786FC3EE44
    ssl-cert: self-signed or not authenticated
```

第 4 步，确认 VSM 与 VEM 连接正常。

```
bdnetlab-n1k# show svs neighbors
Active Domain ID: 92

AIPC Interface MAC: 0050-5682-cfc5
Inband Interface MAC: 0050-5682-d0de

Src MAC            Type     Domain-id    Node-id    Last learnt (Sec. ago)
-----------------------------------------------------------------------
0050-5682-8795     VSM      92           0201       5267.92
0002-3d40-5c02     VEM      92           0302       192.50
0002-3d40-5c03     VEM      92           0402       191.80
```

第 5 步，在 ESXi 主机上查看 VEM 模块注册信息。

```
~ # vemcmd show card
Card UUID type   2: 4c4c4544-0030-5810-8032-b9c04f4d4c31
Card name:
Switch name: bdnetlab-n1kv
Switch alias: DvsPortset-0
```

```
Switch uuid: 2f 8f 05 50 12 7a 28 85-8b b4 d9 d9 79 34 d2 e6
Card domain: 92
Card slot: 4
VEM Tunnel Mode: L3 Mode
L3 Ctrl Index: 49
L3 Ctrl VLAN: 10
VEM Control (AIPC) MAC: 00:02:3d:10:5c:03
VEM Packet (Inband) MAC: 00:02:3d:20:5c:03
VEM Control Agent (DPA) MAC: 00:02:3d:40:5c:03
VEM SPAN MAC: 00:02:3d:30:5c:03
Primary VSM MAC : 00:50:56:85:99:7f
Primary VSM PKT MAC : 00:50:56:85:45:25
Primary VSM MGMT MAC : 00:50:56:85:c8:38
Standby VSM CTRL MAC : 00:50:56:85:80:21
Management IPv4 address: 10.92.10.3
Management IPv6 address: 0000:0000:0000:0000:0000:0000:0000:0000
Primary L3 Control IPv4 address: 10.92.10.41
Secondary VSM MAC : 00:00:00:00:00:00
Secondary L3 Control IPv4 address: 0.0.0.0
Upgrade : Default
Max physical ports: 32
Max virtual ports: 300
Card control VLAN: 1
Card packet VLAN: 1
Control type multicast: No
Card Headless Mode : No
         Processors: 16
    Processor Cores: 8
Processor Sockets: 2
    Kernel Memory:    33544932
Port link-up delay: 5s
Global UUFB: DISABLED
Heartbeat Set: True
PC LB Algo: source-mac
Datapath portset event in progress : no
Licensed: Yes
```

第 6 步，在 ESXi 主机 PING VSM 地址，确认 L3 路由连通性。

```
~ # vmkping 10.92.10.41
PING 10.92.10.41 (10.92.10.41): 56 data bytes
64 bytes from 10.92.10.41: icmp_seq=0 ttl=255 time=0.684 ms
64 bytes from 10.92.10.41: icmp_seq=1 ttl=255 time=0.701 ms
64 bytes from 10.92.10.41: icmp_seq=2 ttl=255 time=0.665 ms
--- 10.92.10.41 ping statistics ---
3 packets transmitted, 3 packets received, 0% packet loss
```

round-trip min/avg/max = 0.665/0.683/0.701 ms

4.5 虚拟机使用 Nexus 1000V

当完成 VSM、Port-Profile 以及 VEM 部署后，虚拟机已经可以使用 Nexus 1000V 交换机对外进行通信，首先要做的是将虚拟机网络迁移到 Nexus 1000V 交换机，然后根据生产环境的需求进行各种策略控制。本节介绍如何迁移虚拟机到 Nexus 1000V 交换机以及基本策略的配置。

4.5.1 迁移虚拟机到 Nexus 1000V 交换机

在开始迁移虚拟机前应该准备好虚拟机，且保证虚拟机网络是可用的状态。

第 1 步，使用 VMware Remote Console 工具打开名为 WEB-SERVER 的虚拟机控制台，通过图 4-5-1 可以看到目前的 IP 地址以及到网关是可达状态。

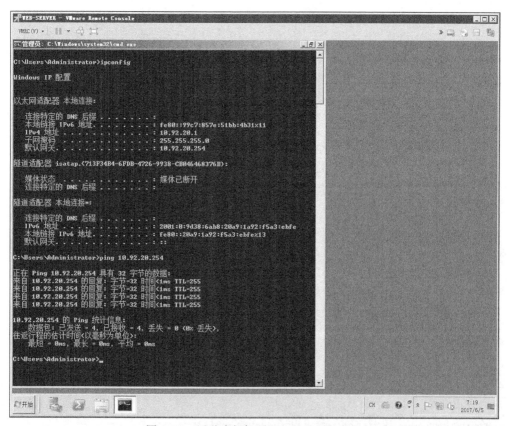

图 4-5-1　迁移虚拟机至 Nexus 1000V 之一

第 2 步，使用 VMware Remote Console 工具打开名为 MS_SQL_SERVER 的虚拟机控制台，通过图 4-5-2 可以看到目前的 IP 地址以及到网关是可达状态。

4.5 虚拟机使用 Nexus 1000V   187

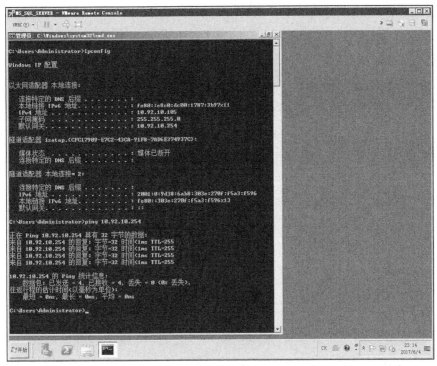

图 4-5-2   迁移虚拟机至 Nexus 1000V 之二

第 3 步，在 bdnetlab-n1k 分布式交换机上单击右键，选择"将虚拟机迁移到其他网络"（如图 4-5-3 所示）。

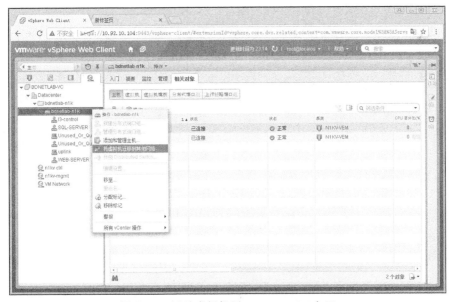

图 4-5-3   迁移虚拟机至 Nexus 1000V 之三

第 4 步，进入迁移虚拟机网络向导，选择"特定网络"→"浏览"的按钮（如图 4-5-4 所示）。

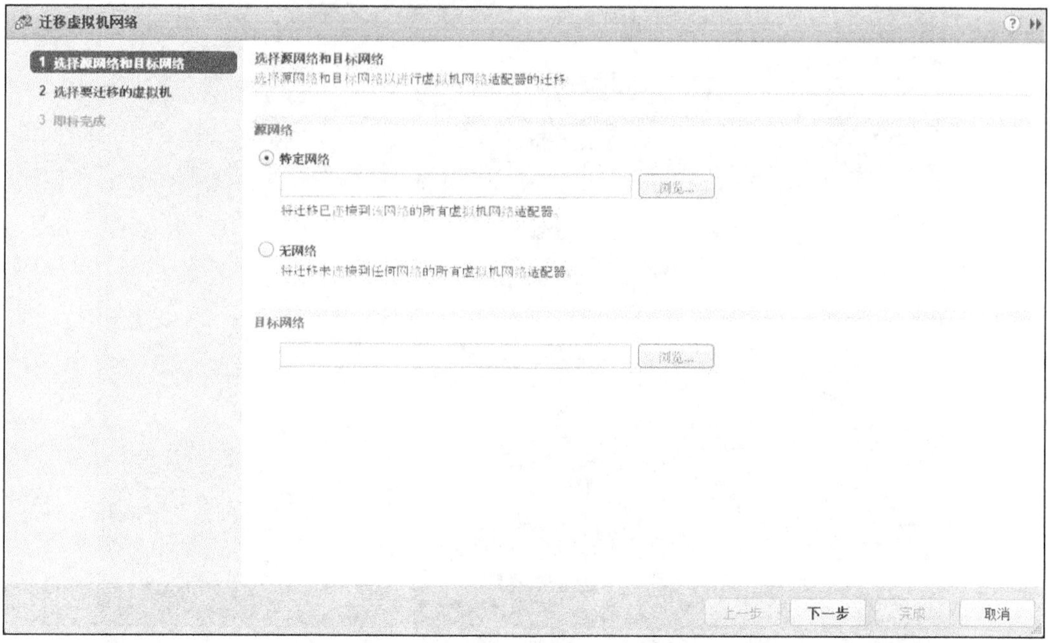

图 4-5-4　迁移虚拟机至 Nexus 1000V 之四

第 5 步，选择虚拟机默认使用的"VM Network"（如图 4-5-5 所示），单击"确定"按钮。

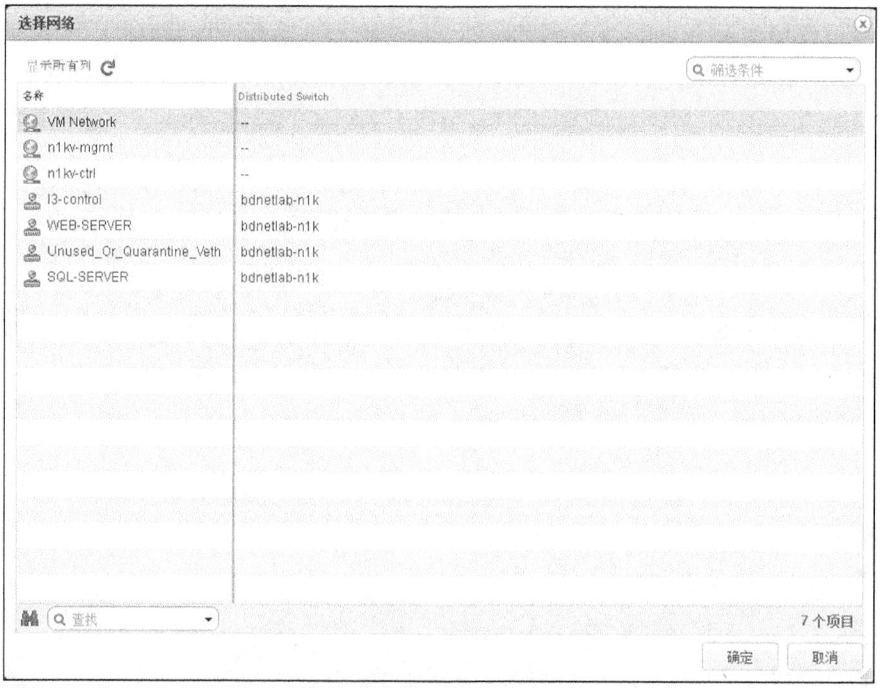

图 4-5-5　迁移虚拟机至 Nexus 1000V 之五

第 6 步，选择"目标网络"→"浏览"的按钮（如图 4-5-6 所示）。

4.5 虚拟机使用 Nexus 1000V   189

图 4-5-6　迁移虚拟机至 Nexus 1000V 之六

第 7 步，选择虚拟机迁移到的目标网络"WEB-SERVER"（如图 4-5-7 所示），单击"确定"按钮。

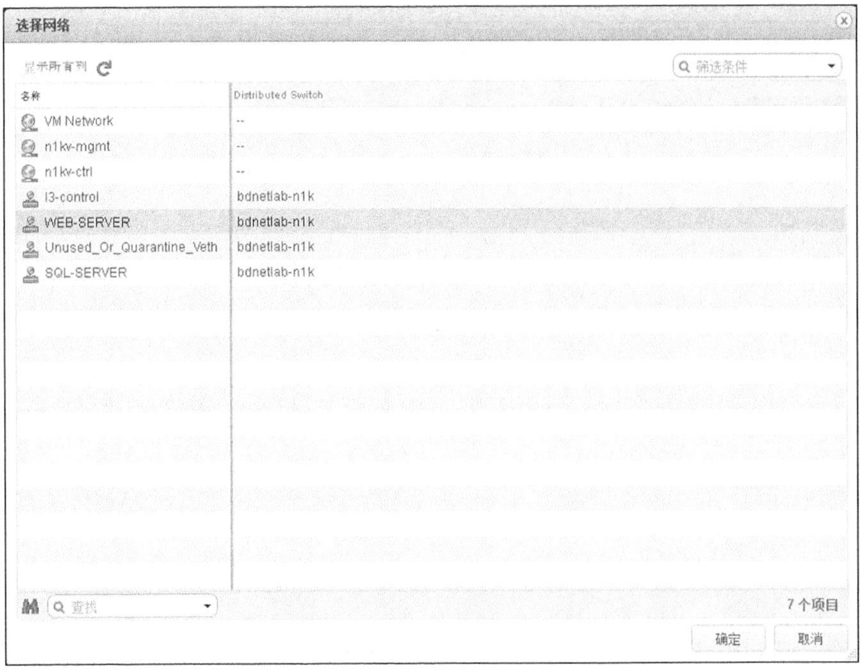

图 4-5-7　迁移虚拟机至 Nexus 1000V 之七

第 8 步，确认虚拟机源网络以及目标网络正确（如图 4-5-8 所示），单击"下一步"按钮。

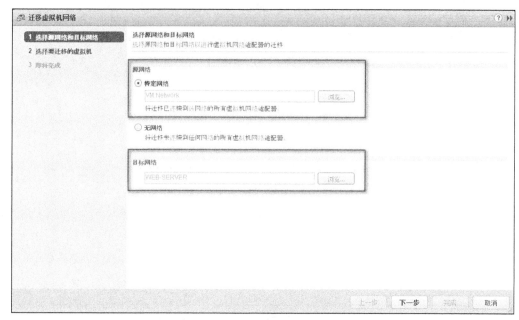

图 4-5-8　迁移虚拟机至 Nexus 1000V 之八

第 9 步，系统列出满足条件的虚拟机，勾选需要迁移的虚拟机（如图 4-5-9 所示），单击"下一步"按钮。

图 4-5-9　迁移虚拟机至 Nexus 1000V 之九

第 10 步，确认准备迁移的虚拟机参数设置正确（如图 4-5-10 所示），单击"完成"按钮。

图 4-5-10　迁移虚拟机至 Nexus 1000V 之十

第 11 步，WEB-SERVER 虚拟机网络迁移完成，网络从标准交换机迁移到 Nexus 1000V 交换机（如图 4-5-11 所示）。

图 4-5-11　迁移虚拟机至 Nexus 1000V 之十一

第 12 步，使用 VMware Remote Console 工具打开名为 WEB-SERVER 的虚拟机控制台，通过图 4-5-12 可以看到目前的 IP 地址已经发生变化，PING 新的网关是可达状态。

第 13 步，除使用批量方式迁移虚拟机网络外，也可以通过修改虚拟机配置进行。编辑名为 MS_SQL_SERVER 虚拟机设置，通过图 4-5-13 可以看到虚拟机网络默认为"VM Network"。

192 第 4 章 部署 Nexus 1000V 分布式交换机

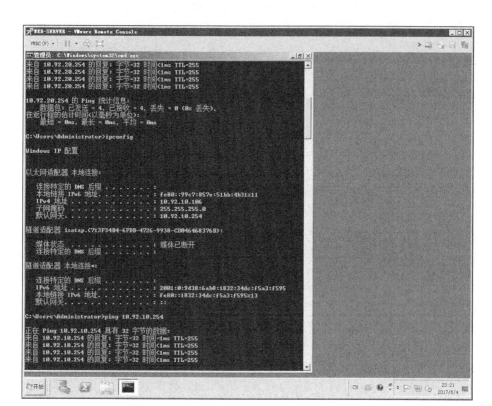

图 4-5-12 迁移虚拟机至 Nexus 1000V 之十二

图 4-5-13 迁移虚拟机至 Nexus 1000V 之十三

第 14 步，将名为 MS_SQL_SERVER 的虚拟机网络调整为 Nexus 1000V 交换机上的"SQL-SERVER"（如图 4-5-14 所示），单击"确定"按钮。

图 4-5-14　迁移虚拟机至 Nexus 1000V 之十四

第 15 步，MS_SQL_SERVER 虚拟机网络迁移完成，网络从标准交换机迁移到 Nexus 1000V 交换机（如图 4-5-15 所示）。

图 4-5-15　迁移虚拟机至 Nexus 1000V 之十五

第 16 步，使用 VMware Remote Console 工具打开名为 MS_SQL_SERVER 的虚拟机控制台，通过图 4-5-16 可以看到目前的 IP 地址已经发生变化，PING 新的网关是可达状态。

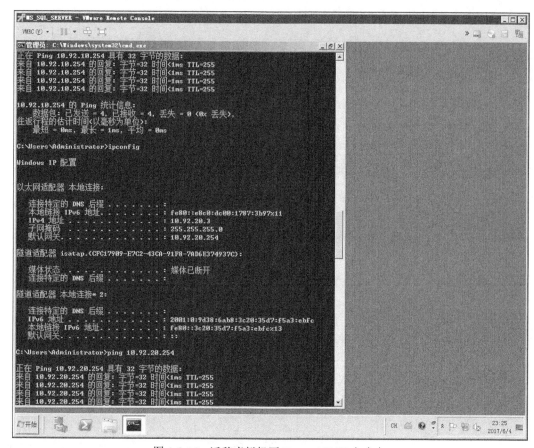

图 4-5-16 迁移虚拟机至 Nexus 1000V 之十六

4.5.2 Nexus 1000V 安全策略配置

Cisco Nexus 1000V 交换机继承了 NXOS 系列交换机的安全特性，可以通过命令行的方式对虚拟机网络进行安全策略配置。本小节介绍比较常用的安全策略配置。

1. TELNET 安全策略配置

在名为 MS_SQL_SERVER 的虚拟机上启用 TELNET 服务器，从名为 WEB-SERVER 的虚拟机上进行登录，在不进行安全策略情况下，登录正常，通过对 Nexus 1000V 相应的端口进行配置，禁用 TELNET 登录。

第 1 步，在名为 WEB-SERVER 的虚拟机上使用 TELNET 访问 MS_SQL_SERVER 的虚拟机 TELNET 服务（如图 4-5-17 所示），成功登录 TELNET 服务器，通过命令 hostname 查看到 TELNET 服务器名。

第 2 步，使用命令查看虚拟机网络在 Nexus 1000V 交换机的信息。

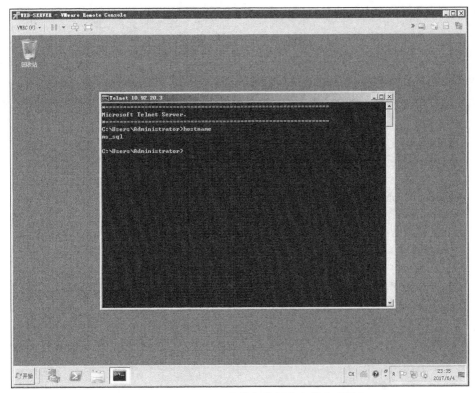

图 4-5-17　Nexus 1000V 交换机常用安全策略配置之一

查看 Nexus 1000V 整体接口信息。

```
bdnetlab-n1k# show interface brief
--------------------------------------------------------------------------
Port        VRF             Status IP Address                Speed       MTU

mgmt0       --              up     10.92.10.41               1000        1500
--------------------------------------------------------------------------
Ethernet    VLAN    Type Mode   Status  Reason               Speed       Port
Interface                                                                Ch #
--------------------------------------------------------------------------
Eth3/3      1       eth  trunk  up      none                 1000
Eth3/4      1       eth  trunk  up      none                 1000
Eth4/3      1       eth  trunk  up      none                 1000
Eth4/4      1       eth  trunk  up      none                 1000
--------------------------------------------------------------------------
Vethernet   VLAN    Type Mode   Status  Reason               Speed
--------------------------------------------------------------------------
Veth1       10      virt access up      none                 auto
Veth2       10      virt access up      none                 auto
Veth3       10      virt access up      none                 auto
Veth4       20      virt access up      none                 auto
```

Port	VRF	Status	IP Address	Speed	MTU
control0	--	up	--	1000	1500

查看虚拟机 WEB-SERVER 以及 MS_SQL_SERVER 对应的 vethernet 信息。

```
bdnetlab-n1k# show interface vethernet 3
Vethernet3 is up
  Port description is WEB-SERVER, Network Adapter 1    //WEB-SERVER 虚拟机对应 vethernet 3 接口
  Hardware: Virtual, address: 0050.5682.6052 (bia 0050.5682.6052)
  Owner is VM "WEB-SERVER", adapter is Network Adapter 1
  Active on module 3
  VMware DVS port 160
  Port-Profile is WEB-SERVER
  Port mode is access
  5 minute input rate 88 bits/second, 0 packets/second
  5 minute output rate 4488 bits/second, 1 packets/second
  Rx
    189 Input Packets 18 Unicast Packets
    97 Multicast Packets 74 Broadcast Packets
    17560 Bytes
  Tx
    1162 Output Packets 32 Unicast Packets
    0 Multicast Packets 1130 Broadcast Packets 1130 Flood Packets
    333771 Bytes
    0 Input Packet Drops 0 Output Packet Drops

bdnetlab-n1k# show interface vethernet 4
Vethernet4 is up
  Port description is MS_SQL_SERVER, Network Adapter 1 //MS_SQL_SERVER 虚拟机对应 vethernet 4 接口
  Hardware: Virtual, address: 0050.5682.726b (bia 0050.5682.726b)
  Owner is VM "MS_SQL_SERVER", adapter is Network Adapter 1
  Active on module 4
  VMware DVS port 194
  Port-Profile is SQL-SERVER
  Port mode is access
  5 minute input rate 488 bits/second, 0 packets/second
  5 minute output rate 48 bits/second, 0 packets/second
  Rx
    351 Input Packets 14 Unicast Packets
    148 Multicast Packets 189 Broadcast Packets
    31980 Bytes
  Tx
    32 Output Packets 28 Unicast Packets
    0 Multicast Packets 4 Broadcast Packets 4 Flood Packets
    3323 Bytes
    0 Input Packet Drops 0 Output Packet Drops
```

第 3 步，配置访问控制列表。

```
bdnetlab-n1k# conf t
Enter configuration commands, one per line.   End with CNTL/Z.
bdnetlab-n1k(config)# ip access-list deny_telnet        //定义访问控制列表
bdnetlab-n1k(config-acl)# deny tcp any any eq 23    //拒绝 TELNET 使用的 23 端口
bdnetlab-n1k(config-acl)# permit ip any any         //允许其他端口通过
bdnetlab-n1k(config-acl)# statistics per-entry      //打开计数器
bdnetlab-n1k(config-acl)# exit
bdnetlab-n1k(config)# interface vethernet 4         //进入虚拟机所使用接口
bdnetlab-n1k(config-if)# ip port access-group deny_telnet in    //调用访问控制列表
bdnetlab-n1k(config-if)# ip port access-group deny_telnet out   //调用访问控制列表

bdnetlab-n1k(config)# show run interface vethernet 4    //查看虚拟机接口配置
!Command: show running-config interface Vethernet4
!Time: Mon Jun   5 00:48:21 2017
version 4.2(1)SV2(2.2)
interface Vethernet4
   inherit port-profile SQL-SERVER
   description MS_SQL_SERVER, Network Adapter 1
   vmware dvport 194 dvswitch uuid "ce dc 02 50 73 b6 a3 1f-73 ef 90 c7 8c a0 30 4d"
   vmware vm mac 0050.5682.726B
   ip port access-group deny_telnet in
   ip port access-group deny_telnet out
```

第 4 步，再次访问 MS_SQL_SERVER 的虚拟机 TELNET 服务（如图 4-5-18 所示），无法连接到 TELNET 服务器，说明安全策略生效。

图 4-5-18   Nexus 1000V 交换机常用安全策略配置之二

第 5 步，通过命令查看安全策略计数器相关信息。

```
bdnetlab-n1k(config)# show ip access-lists
IP access list deny_telnet
        statistics per-entry
        10 deny tcp any any eq telnet [match=3]      //匹配 3 次拒绝
        20 permit ip any any [match=78]
```

2. 虚拟机端口安全策略配置

在生产环境中，为了保证服务器安全，可能会配置端口安全策略来绑定 MAC 地址。当接口匹配 MAC 地址时，数据包进行转发；当接口不匹配 MAC 地址时，数据包进行丢弃并将端口处于 shutdown 状态。

第 1 步，查看名为 MS_SQL_SERVER 虚拟机的 MAC 地址为 00-50-56-82-72-6B（如图 4-5-19 所示）。

图 4-5-19　Nexus 1000V 交换机常用安全策略配置之三

第 2 步，配置启用端口安全策略。

```
bdnetlab-n1k(config)# interface vethernet 4
bdnetlab-n1k(config-if)# switchport port-security            //配置端口安全策略
bdnetlab-n1k(config-if)# switchport port-security mac-address 11-11-11-11-11-11   //绑定虚拟机 MAC 地址
bdnetlab-n1k(config-if)# switchport port-security violation shutdown   //定义不满足条件关闭端口
bdnetlab-n1k(config-if)# exit
bdnetlab-n1k(config)# show running-config interface vethernet 4
```

```
!Command: show running-config interface Vethernet4
!Time: Mon Jun  5 00:49:33 2017
version 4.2(1)SV2(2.2)
interface Vethernet4
  inherit port-profile SQL-SERVER
  description MS_SQL_SERVER, Network Adapter 1
  vmware dvport 194 dvswitch uuid "ce dc 02 50 73 b6 a3 1f-73 ef 90 c7 8c a0 30 4d"
  vmware vm mac 0050.5682.726B
  switchport port-security
  switchport port-security mac-address 1111.1111.1111
  switchport port-security violation shutdown
  ip port access-group deny_telnet in
  ip port access-group deny_telnet out
```

第 3 步，验证端口安全策略应用情况。

```
bdnetlab-n1k(config)# show interface vethernet 4
Vethernet4 is down (Error disabled)    //端口已处于 down 状态
  Port description is MS_SQL_SERVER, Network Adapter 1
  Hardware: Virtual, address: 0050.5682.726b (bia 0050.5682.726b)
  Owner is VM "MS_SQL_SERVER", adapter is Network Adapter 1
  Active on module 4
  VMware DVS port 194
  Port-Profile is SQL-SERVER
  Port mode is access
  5 minute input rate 128 bits/second, 0 packets/second
  5 minute output rate 48 bits/second, 0 packets/second
  Rx
    732 Input Packets 222 Unicast Packets
    244 Multicast Packets 266 Broadcast Packets
    78598 Bytes
  Tx
    280 Output Packets 275 Unicast Packets
    0 Multicast Packets 5 Broadcast Packets 5 Flood Packets
    25758 Bytes
  11 Input Packet Drops 0 Output Packet Drops

bdnetlab-n1k(config)# show port-security interface vethernet 4
Port Security              : Enabled
Port Status                : Secure Down
Violation Mode             : Shutdown
Aging Time                 : 0 mins
Aging Type                 : Absolute
Maximum MAC Addresses      : 1
Total MAC Addresses        : 1
Configured MAC Addresses   : 1
Sticky MAC Addresses       : 0
Security violation count   : 1
```

第 4 步，使用 VMware Remote Console 打开虚拟机控制台验证其网络连接性，通过图 4-5-20 可以看到网关为不可达状态，说明安全策略配置生效。

图 4-5-20　Nexus 1000V 交换机常用安全策略配置之四

3. 虚拟机端口安全默认启用 igmp snooping

IGMP 是多播网络中非常重要的组件，主机通过 IGMP 协议来表示其希望加入特定的多播组，路由器看到 IGMP 加入消息后，即开始将主机所请求的多播流量发送给接收端。由于交换机一般运行在二层模式，因此开发了 IGMP Snooping 技术，可以不采取广播或泛洪机制来实现多播流量的智能转发，通过 IGMP Snooping，二层交换机可以检查 IGMP 成员关系报告消息并仅向特定端口发送多播放流量。如果没有 IGMP Snooping，二层交换机通常会所有多播流量泛洪到所有端口，从而产生巨大的网络流量给网络带来严重的负担。

Cisco Nexus 1000V 完全支持 IGMP Snooping 功能，并且默认为启动状态，因此无需配置就可以使用。Nexus 1000V 中的 vetherent 端口也支持 IGMP Snooping 功能，因此 VMware vSphere 虚拟化架构中的虚拟机可以使用 IP 多播，可以更好地改善网络性能。

```
bdnetlab-n1k# show ip igmp snooping vlan 10    //查看 VLAN 10 多播配置情况
IGMP Snooping information for vlan 10
  IGMP snooping enabled        //IGMP Snooping 处于启用状态
  IGMP querier none
  Switch-querier disabled
  IGMPv3 Explicit tracking enabled
  IGMPv2 Fast leave disabled
```

```
        IGMPv1/v2 Report suppression disabled
        IGMPv3 Report suppression disabled
        Link Local Groups suppression enabled
        Router port detection using PIM Hellos, IGMP Queries
        Number of router-ports: 0
        Number of groups: 0
        Active ports:        //IGMP Snooping 激活使用的端口
            Veth1        Eth3/3    Eth3/4   Veth2
    Eth4/3        Eth4/4    Veth3

    bdnetlab-n1k# show ip igmp snooping vlan 20    //查看 VLAN 20 多播配置情况
    IGMP Snooping information for vlan 20
        IGMP snooping enabled        //IGMP Snooping 处于启用状态
        IGMP querier none
        Switch-querier disabled
        IGMPv3 Explicit tracking enabled
        IGMPv2 Fast leave disabled
        IGMPv1/v2 Report suppression disabled
        IGMPv3 Report suppression disabled
        Link Local Groups suppression enabled
        Router port detection using PIM Hellos, IGMP Queries
        Number of router-ports: 0
        Number of groups: 0
        Active ports:        //IGMP Snooping 激活使用的端口
            Eth3/3        Eth3/4    Eth4/3    Eth4/4
            Veth4
```

4. 交换式端口分析器 SPAN（Switch Port Analyzer）配置

Cisco Nexus 1000V 支持以太网流量配置的 SPAN，SPAN 使用源端口和目的端口概念，源端口就是通过 SPAN 向目的端口发送流量的源，目的端口就是 SPAN 源复制的流量所要发送的目标位置。Nexus 1000V 最多可以配置 64 条 SPAN。

```
    bdnetlab-n1k(config)# interface ethernet 4/4     //选择端口
    bdnetlab-n1k(config-if)# switchport monitor      //配置 SPAN 监控端口
    bdnetlab-n1k(config-if)# exit
    bdnetlab-n1k(config)# monitor session 1          //配置监控会话
    bdnetlab-n1k(config-monitor)# source interface vethernet 4       //指定源端口
    bdnetlab-n1k(config-monitor)# destination interface ethernet 4/4  //指定目的端口
    bdnetlab-n1k(config-monitor)# no shutdown
    bdnetlab-n1k(config-monitor)# exit
    bdnetlab-n1k(config)# show monitor session 1     //查看监控会话状态
        session 1
    ---------------
    type              : local
    state             : up
    source intf       :
```

```
    rx              : Veth4
    tx              : Veth4
    both            : Veth4
source VLANs        :
    rx              :
    tx              :
    both            :
source port-profile :
    rx              :
    tx              :
    both            :
filter VLANs        : filter not specified
destination ports   : Eth4/4
destination port-profile :
//SPAN 监控默认方向是双向监控，使用 tx 和 rx 可以限定所要监控流量的方向
```

5. 封闭的远程交换式端口分析器 ERSPAN（Encapsulated Remote Switch Port Analyzer）配置

Cisco Nexus 1000V 支持以太网流量配置的 ERSPAN，ERSPAN 也使用源端口和目的端口概念，源端口就是通过 ERSPAN 向目的端口发送流量的源，目的端口就是 ERSPAN 源复制的流量所要发送的目标位置。ERSPAN 目的端口是由 IP 地址指定的，Nexus 1000V 最多可以配置 64 条 ERSPAN。

```
bdnetlab-n1k(config)# monitor session 2 type erspan-source    //配置 ERSPAN 监控会话
bdnetlab-n1k(config-erspan-src)# source interface vethernet 3 //指定源端口
bdnetlab-n1k(config-erspan-src)# destination ip 10.92.30.248  //指定目的 IP 地址
bdnetlab-n1k(config-erspan-src)# erspan-id 23                 //指定 ERSPAN ID 端口
bdnetlab-n1k(config-erspan-src)# no shutdown
bdnetlab-n1k(config-erspan-src)# exit
bdnetlab-n1k(config)# show monitor session 2                  //查看监控会话状态
    session 2
---------------
type                : erspan-source
state               : up
source intf         :
    rx              : Veth3
    tx              : Veth3
    both            : Veth3
source VLANs        :
    rx              :
    tx              :
    both            :
source port-profile :
    rx              :
    tx              :
    both            :
filter VLANs        : filter not specified
```

```
destination IP      : 10.92.30.248
ERSPAN ID           : 23
ERSPAN TTL          : 64
ERSPAN IP Prec.     : 0
ERSPAN DSCP         : 0
ERSPAN MTU          : 1500
ERSPAN Header Type: 2
```

6. NetFlow 配置

Cisco Nexus 1000V 提供了强大的网络统计信息采集工具 NetFlow，利用 NetFlow 可以为网络监控、规划等提供统计数据。NetFlow 将流定义为到达某个源接口或 VLAN 的数据包，且这些数据包拥有相同的关键属性值，如果需要导出并分析 NetFlow 数据，可以使用 exporter 来完成，NetFlow 使用 UDP 导出数据流，支持版本 5 以及版本 9 两种格式。

第 1 步，启用 NetFlow 并创建流记录。

```
bdnetlab-n1k(config)# feature NetFlow         //启用 NetFlow 特性，默认为禁用状态
bdnetlab-n1k(config)# flow record inbound     //创建流记录
bdnetlab-n1k(config-flow-record)# match ipv4 source address          //匹配 IPv4 源地址
bdnetlab-n1k(config-flow-record)# match ipv4 destination address     //匹配 IPv4 目的地址
bdnetlab-n1k(config-flow-record)# match transport source-port        //匹配流端口
bdnetlab-n1k(config-flow-record)# match transport destination-port   //匹配流目的端口
bdnetlab-n1k(config-flow-record)# collect counter bytes              //定义采集流字节数
bdnetlab-n1k(config-flow-record)# collect counter packets            //定义采集流数据包
bdnetlab-n1k# show flow record inbound        //查看创建流记录
Flow record inbound:
    No. of users: 0
    Template ID: 0
    Fields:
        match ipv4 source address
        match ipv4 destination address
        match transport source-port
        match transport destination-port
        match interface input
        match interface output
        match flow direction
        collect counter bytes
        collect counter packets
```

第 2 步，配置流导出器。

```
bdnetlab-n1k(config)# flow exporter netflow               //配置流导出器
bdnetlab-n1k(config-flow-exporter)# destination 10.92.30.248     //配置流导出器目的端口
bdnetlab-n1k(config-flow-exporter)# source mgmt 0         //配置流导出器源端口
bdnetlab-n1k(config-flow-exporter)# version 9             //配置流导出器使用版本 9
bdnetlab-n1k(config-flow-exporter-version-9)# exit
bdnetlab-n1k# show flow exporter netflow        //查看流导出器配置
```

```
Flow exporter netflow:
    Destination: 10.92.30.248
    VRF: default (1)
    Source Interface mgmt0 (10.92.10.41)
    Export Version 9
        Data template timeout 0 seconds
    Exporter Statistics
        Number of Flow Records Exported 0
        Number of Templates Exported 0
        Number of Export Packets Sent 0
        Number of Export Bytes Sent 0
        Number of No Buffer Events 0
        Number of Packets Dropped (other) 0
        Number of Packets Dropped (LC to RP Error) 0
        Number of Packets Dropped (Output Drops) 0
        Time statistics were last cleared: Never
```

第 3 步，定义并应用流监控器

```
bdnetlab-n1k(config)# flow monitor netflow          //定义流监控器
bdnetlab-n1k(config-flow-monitor)# record inbound          //调用流记录
bdnetlab-n1k(config-flow-monitor)# exporter netflow          //配置流导出
bdnetlab-n1k(config-flow-monitor)# exit
bdnetlab-n1k(config)# port-profile BDNETLAB-N1K-VLAN10          //创建应用
bdnetlab-n1k(config-port-prof)# ip flow monitor netflow in          //应用流监控器
bdnetlab-n1k(config-port-prof)# exit
bdnetlab-n1k(config)# show flow monitor          //查看流监控器
Flow Monitor netflow:
    Use count: 0
    Flow Record: inbound
    Flow Exporter: netflow
    Inactive timeout: 300
    Active timeout: 1800
Cache Size: 4096
```

4.6　部署使用 VXLAN

VXLAN，全称 Virtual extensible Local Area Network，是近年来软件定义网络比较常见的技术，由于传统的 VLAN 具有一定的局限性，VXLAN 主要解决在虚拟化数据中心遇到的网络扩展性问题。Nexus 1000V 支持基本的 VXLAN，本节对如何在 VMware vSphere 虚拟化环境中使用基于 Nexus 1000V 的 VXLAN 进行简要介绍。

4.6.1　VXLAN 基础知识介绍

在传统的 802.1Q 中，VLAN 使用 12 位标识，也就是说 VLAN 最大可分配数量为 2^{12}，这个数量对于传统的数据中心来说可能够用，但对于虚拟化数据中心就可能存在不足。随之而来诞生了 VXLAN。目前，VXLAN 使用 24 位标识，也就是说 VLAN 最大可分配数量为 2^{24}，VXLAN 在一定程度上解决了大型虚拟化数据中心网络扩展问题。

1. 为什么使用 VXLAN 技术

（1）服务器虚拟化后需要大量的 MAC 地址

数据中心的服务器，如果不使用虚拟化技术，一台服务器配置 4 个物理适配器，正常情况下就是 4 个 MAC 地址；如果使用服务器虚拟化技术，一台服务器上运行 20 个虚拟机，每个虚拟机 1 个适配器，则至少需要 20 个 MAC 地址。也就是说，使用服务器虚拟化技术后，需要提供大量的 MAC 地址。

（2）802.1Q VLAN 数量限制

对于传统数据中心来说，多数在转向虚拟化云计算数据中心，802.1Q 技术限制了 VLAN 的数量最大为 2^{12}，已经不能满足大型云计算数据中心的需求。

（3）物理网络区域限制

对于一个大型云计算数据中心来说，可能是由异地多个数据中心组成，多个数据中心都有 VLAN 以及 IP 段，在 VMware vSphere 6.0 发布之前，虚拟机的迁移需要使用二层网络，显然多数据中心的环境会影响虚拟机的迁移，这样可能限制某些需在二层网络应用的部署。如何通过三层网络实现多个数据中心的大二层网络也是一个需要解决的重要问题。

2. VXLAN 常用参数解释

（1）VTEP

VTEP，全称 VXLAN Tunnel End Point，VXLAN 隧道终结点，与终端设备连接设备，可以是虚拟机，负责原始以太报文的 VXLAN 封装和解封装，形态可以是虚拟交换机，也可以是物理交换机。如果在虚拟化环境中使用 VTEP，那么所有虚拟机流量在进入物理交换机之前，可以在 ESXi 主机上打上 VXLAN 标签和 UDP 包头，相当于在任意两点之间建立了隧道；如果在物理交换机上使用 VTEP，那么物理交换机与虚拟机交换机通信时，需要进行 VXLAN 至 VLAN 的转换。

对于 Nexus 1000V 交换机来说，VEM 模块就是 VTEP，Nexus 1000V 使用一个 VMkernel 来终结 VETP 流量，这个 VMkernel 连接到一个 VLAN，用来传输 VXLAN 封装的流量。由于虚拟机 VLAN 对外不可见，VXLAN 会使用 VNI（VXLAN Network Identifier，VXLAN 标识），VNI 取代 VLAN 用来标识 VXLAN 网段，虚拟机可以在拥有相同 VNI 的二层网络中进行通信。

（2）VXLAN 二层网关

VXLAN 二层网关用于终结 VXLAN 网络，将 VXLAN 报文转换成对应的传统二层网络送到传统以太网络，适用于 VXLAN 网络内服务器与远端终端或远端服务器的二层互联。如在不同网络中进行虚拟机迁移，当业务需要传统网络中服务器与 VXLAN 网络中服务器在同一个二层中时，需要使用 VXLAN 二层网关打通 VXLAN 网络和二层网络。可见，它除了具备 VTEP 的功能外，还负责 VLAN 报文与 VXLAN 报文之间的映射和转发，主要以物理交换机为主。

Cisco Nexus N5600、N7000（F3）、N9000、N3100 系列交换机支持 VXLAN 二层网关。

（3）VXLAN 三层网关

VXLAN 二层网关用于终结 VXLAN 网络，将 VXLAN 报文转换成传统三层报文送至

IP 网络，适用于 VXLAN 网络内服务器与远端终端之间的三层互访；同时也用作不同 VXLAN 网络互通。当服务器访问外部网络时，VXLAN 三层网关剥离对应 VXLAN 报文封装，送入 IP 网络；当外部终端访问 VXLAN 内的服务器时，VXLAN 根据目的 IP 地址确定所属 VXLAN 及所属的 VTEP，加上对应的 VXLAN 报文头封装进入 VXLAN 网络。VXLAN 之间的互访流量与此类似，VXLAN 网关剥离 VXLAN 报文头，并基于目的 IP 地址确定所属 VXLAN 及所属的 VTEP，重新封装后送入另外的 VXLAN 网络。VXLAN 三层网关还负责处理不同 VXLAN 之间的报文通信，同时也是数据中心内部服务向往发布业务的出口，主要以高性能物理交换机为主。

Cisco Nexus N5600、N7000（F3）、N9000 系列交换机支持 VXLAN 三层网关。

（4）IP 组播

对于二层网络来说，如果所有主机在一个子网，IP 组播路由不是必需；如果承载 VXLAN VMkernel 位于不同的子网，则必须在上游交换机或路由器上启用 IP 组播路由。

4.6.2 配置 VXLAN

本节介绍在 Nexus 1000V 上配置使用 VXLAN 以及如何在虚拟机上使用 VXLAN。

第 1 步，在 Nexus 1000V 交换机上启用 VXLAN。

```
bdnetlab-n1k(config)# feature segmentation    //启用 VXLAN
bdnetlab-n1k(config)#  2017  Jun  15  00:10:29  bdnetlab-n1k  %SEG_BD-2-SEG_BD_ENABLED:  Feature
Segmentation enabled

bdnetlab-n1k(config)# show feature    //查看启用 feature
Feature Name              Instance   State
--------------------      --------   --------
cts                       1          disabled
dhcp-snooping             1          disabled
evb                       1          disabled
http-server               1          enabled
lacp                      1          disabled
netflow                   1          enabled
network-segmentation      1          disabled
port-profile-roles        1          disabled
private-vlan              1          disabled
segmentation              1          enabled
sshServer                 1          enabled
tacacs                    1          disabled
telnetServer              1          disabled
vff                       1          disabled
vtracker                  1          disabled
vxlan-gateway             1          disabled
```

第 2 步，创建 VXLAN，在 Nexus 1000V 交换机上定义桥接域以及组播地址。

```
bdnetlab-n1k(config)# bridge-domain vxlan       //创建 VXLAN
bdnetlab-n1k(config-bd)# segment id 222222      //定义分层 ID
```

```
bdnetlab-n1k(config-bd)# group 239.1.1.1        //定义组播地址
bdnetlab-n1k(config-bd)# exit

bdnetlab-n1k(config)# show bridge-domain        //查看 VXLAN
Global Configuration:
Mode: Unicast-only          //使用单播模式
MAC Distribution: Disable
Bridge-domain vxlan (0 ports in all)     //因为还没有配置完成，所以没有端口
Segment ID: 222222 (Manual/Active)
Mode: Unicast-only
MAC Distribution: Disable
Group IP: 239.1.1.1
State: UP                   Mac learning: Enabled
```

第 3 步，创建用于封装 VXLAN 流量传输的 Port-Profile 文件。

```
bdnetlab-n1k(config)# port-profile type vethernet vxlan
bdnetlab-n1k(config-port-prof)# switchport mode access
bdnetlab-n1k(config-port-prof)# switchport access vlan 20
bdnetlab-n1k(config-port-prof)# capability vxlan        //定义使用 VXLAN
bdnetlab-n1k(config-port-prof)# system vlan 20
bdnetlab-n1k(config-port-prof)# vmware port-group
bdnetlab-n1k(config-port-prof)# state enabled
bdnetlab-n1k(config-port-prof)# no shutdown
bdnetlab-n1k(config-port-prof)# exit

bdnetlab-n1k(config)# show run port-profile vxlan
!Command: show running-config port-profile vxlan
!Time: Thu Jun 15 00:12:35 2017

version 4.2(1)SV2(2.2)
port-profile type vethernet vxlan
    vmware port-group
    switchport mode access
    switchport access vlan 20
    capability vxlan
    no shutdown
    state enabled
```

第 4 步，创建用于虚拟机使用的 VXLAN Port-Profile 文件。

```
bdnetlab-n1k(config)# port-profile type vethernet vxlan-vm
bdnetlab-n1k(config-port-prof)# switchport mode access
bdnetlab-n1k(config-port-prof)# switchport access bridge-domain vxlan    //定义端口访问 VXLAN
bdnetlab-n1k(config-port-prof)# vmware port-group
bdnetlab-n1k(config-port-prof)# state enabled
```

```
bdnetlab-n1k(config-port-prof)# no shutdown
bdnetlab-n1k(config-port-prof)# exit

bdnetlab-n1k(config)# show run port-profile vxlan-vm
!Command: show running-config port-profile vxlan-vm
!Time: Thu Jun 15 00:13:44 2017

version 4.2(1)SV2(2.2)
port-profile type vethernet vxlan-vm
  vmware port-group
  switchport mode access
  switchport access bridge-domain vxlan
  no shutdown
  state enabled
```

第 5 步，在 IP 地址为 10.92.10.16 的 ESXi 主机上创建用于传输 VXLAN 的 VMKernel，需要关联到刚创建的 vxlan（如图 4-6-1 所示）。

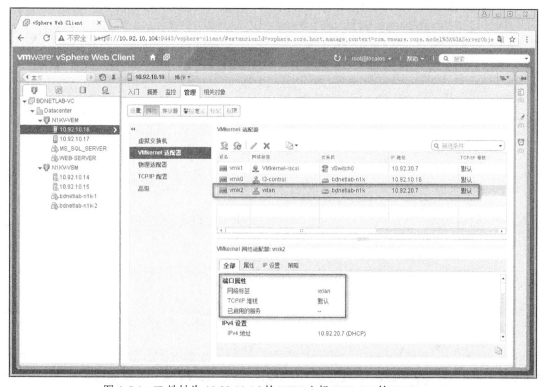

图 4-6-1　IP 地址为 10.92.10.16 的 ESXi 主机 VXLAN 的 VMKernel

第 6 步，在 IP 地址为 10.92.10.17 的 ESXi 主机上创建用于传输 VXLAN 的 VMKernel，需要关联到刚创建的 vxlan（如图 4-6-2 所示）。

图 4-6-2 IP 地址为 10.92.10.17 的 ESXi 主机 VXLAN 的 VMKernel

第 7 步，编辑名为 MS_SQL_SERVER 虚拟机网络适配器，关联到刚创建的 vxlan-vm（如图 4-6-3 所示），单击"确定"按钮。

图 4-6-3 MS_SQL_SERVER 虚拟机网络适配器

第 8 步，编辑名为 WEB-SERVER 虚拟机网络适配器，关联到刚创建的 vxlan-vm（如图 4-6-4 所示），单击"确定"按钮。

图 4-6-4　WEB-SERVER 虚拟机网络适配器

第 9 步，在名为 MS_SQL_SERVER 虚拟机进行网络连通性测试（如图 4-6-5 所示），PING 名为 WEB-SERVER 虚拟机网络正常。

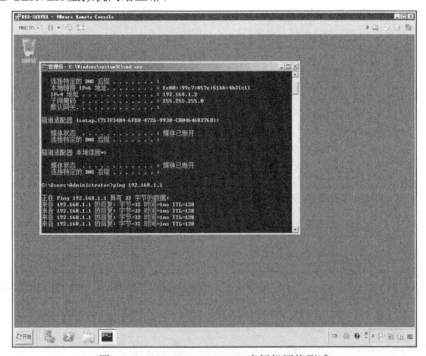

图 4-6-5　MS_SQL_SERVER 虚拟机网络测试

第 10 步，在名为 WEB-SERVER 的虚拟机进行网络连通性测试（如图 4-6-6 所示），PING 名为 MS_SQL_SERVER 的虚拟机网络正常。

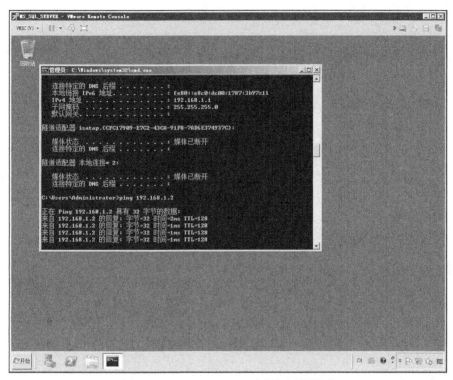

图 4-6-6　WEB-SERVER 虚拟机网络测试

第 11 步，再次查看 VXLAN 信息。

```
bdnetlab-n1k(config)# show bridge-domain
Global Configuration:
Mode: Unicast-only
MAC Distribution: Disable

Bridge-domain vxlan (2 ports in all)
Segment ID: 222222 (Manual/Active)
Mode: Unicast-only
MAC Distribution: Disable
Group IP: 239.1.1.1
State: UP                  Mac learning: Enabled
Veth3, Veth4      //使用 VXLAN 接口为 veth3 和 veth4
```

第 12 步，查看 veth3 和 veth4 接口信息。

```
bdnetlab-n1k(config)# show interface vethernet 3
Vethernet3 is up
  Port description is WEB-SERVER, Network Adapter 1     //对应虚拟机为 WEB-SERVER
  Hardware: Virtual, address: 0050.5682.6052 (bia 0050.5682.6052)
```

```
Owner is VM "WEB-SERVER", adapter is Network Adapter 1
Active on module 3
VMware DVS port 323
Port-Profile is vxlan-vm      //端口配置文件使用 vxlan-vm
Port mode is access
5 minute input rate 600 bits/second, 0 packets/second
5 minute output rate 8 bits/second, 0 packets/second
Rx
    229 Input Packets 802 Unicast Packets
    226 Multicast Packets 785 Broadcast Packets
    20436 Bytes
Tx
    798 Output Packets 0 Unicast Packets
    40 Multicast Packets 73 Broadcast Packets 5 Flood Packets
    58604 Bytes
    0 Input Packet Drops 0 Output Packet Drops

bdnetlab-n1k(config)# show interface vethernet 4
Vethernet4 is up
Port description is MS_SQL_SERVER, Network Adapter 1    //对应虚拟机为 MS_SQL_SERVER
Hardware: Virtual, address: 0050.5682.726b (bia 0050.5682.726b)
Owner is VM "MS_SQL_SERVER", adapter is Network Adapter 1
Active on module 4
VMware DVS port 322
Port-Profile is vxlan-vm      //端口配置文件使用 vxlan-vm
Port mode is access
5 minute input rate 464 bits/second, 0 packets/second
5 minute output rate 8 bits/second, 0 packets/second
Rx
    1952 Input Packets 793 Unicast Packets
    270 Multicast Packets 889 Broadcast Packets
    141984 Bytes
Tx
    881 Output Packets 801 Unicast Packets
    36 Multicast Packets 44 Broadcast Packets 80 Flood Packets
    66020 Bytes
    0 Input Packet Drops 0 Output Packet Drops
```

第 13 步，查看 VXLAN MAC 地址分配情况。

```
bdnetlab-n1k(config)# show mac address-table bridge-domain vxlan
Bridge-domain: vxlan
         MAC Address       Type       Age        Port              IP Address      Mod
---------------------+--------+-----------+----------------+----------------+---
         0050.5682.6052    static    0          Veth3             0.0.0.0         3
         0050.5682.726b    dynamic   52         Eth3/3            10.92.20.7      3
```

		0050.5682.726b	static 0	Veth4	0.0.0.0	4
		0050.5682.6052	dynamic 57	Eth4/3	10.92.20.8	4

Total MAC Addresses: 4

第 14 步，登录 IP 地址为 10.92.10.16 的 ESXi 主机查看端口信息。

```
~ # vemcmd show port
  LTL   VSM Port   Admin Link   State   PC-LTL   SGID   Vem Port   Type
   19    Eth3/3     UP    UP    FWD       0              vmnic2
   20    Eth3/4     UP    UP    FWD       0              vmnic3
   49    Veth1      UP    UP    FWD       0              vmk0
   50    Veth3      UP    UP    FWD       0              WEB-SERVER.eth0
   51    Veth6      UP    UP    FWD       0              vmk2      VXLAN

~ # vemcmd show vxlan-stats
  LTL   Ucast   Mcast/Repl   Ucast    Mcast    Total
        Encaps  Encaps       Decaps   Decaps   Drops
   50    12                   22       34       0        0
   51    12                   22       34       0        0

~ # vemcmd show vxlan interfaces
  LTL   VSM Port      IP          Seconds since Last   Vem Port
                                  IGMP Query Received
(* = IGMP Join Interface/Designated VTEP)
-----------------------------------------------------------
   51    Veth6    10.92.20.7       1245515              vmk2        *
```

第 15 步，登录 IP 地址为 10.92.10.17 的 ESXi 主机查看端口信息。

```
~ # vemcmd show port
  LTL   VSM Port   Admin Link   State   PC-LTL   SGID   Vem Port   Type
   19    Eth4/3     UP    UP    FWD       0              vmnic2
   20    Eth4/4     UP    UP    F/B*      0              vmnic3
   49    Veth2      UP    UP    FWD       0              vmk0
   50    Veth4      UP    UP    FWD       0              MS_SQL_SERVER.eth0
   51    Veth5      UP    UP    FWD       0              vmk2      VXLAN

~ # vemcmd show vxlan-stats
  LTL   Ucast   Mcast/Repl   Ucast    Mcast    Total
        Encaps  Encaps       Decaps   Decaps   Drops
   50    11                   23       33       0        0
   51    11                   23       33       0        0

~ # vemcmd show interfaces
~ # vemcmd show vxlan interfaces
  LTL   VSM Port      IP          Seconds since Last   Vem Port
                                  IGMP Query Received
```

```
(* = IGMP Join Interface/Designated VTEP)
-----------------------------------------------------------
 51         Veth5      10.92.20.8     1245627            vmk2                *
```

第 16 步，虚拟机基本网络连通没有问题后，可以测试虚拟机在 VXLAN 环境下迁移是否正常。通过图 4-6-7 可以看到，名为 WEB-SERVER 的虚拟机运行在 IP 地址为 10.92.0.17 的 ESXi 主机。

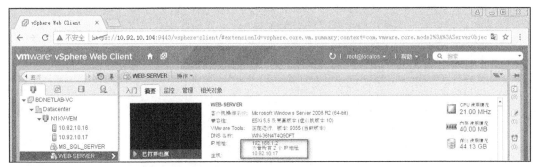

图 4-6-7　虚拟机在 VXLAN 环境 vMotion 测试之一

第 17 步，将名为 WEB-SERVER 的虚拟机进行迁移，选择"更改主机"（如图 4-6-8 所示），单击"下一步"按钮 。

图 4-6-8　虚拟机在 VXLAN 环境 vMotion 测试之二

第 18 步，选择迁移的集群（如图 4-6-9 所示），单击"下一步"按钮。
第 19 步，选择迁移的 ESXi 主机（如图 4-6-10 所示），单击"下一步"按钮。

图 4-6-9　虚拟机在 VXLAN 环境 vMotion 测试之三

图 4-6-10　虚拟机在 VXLAN 环境 vMotion 测试之四

第 20 步，选择"为最优 vMotion 性能预留 CPU（建议）"选项（如图 4-6-11 所示），单击"下一步"按钮。

图 4-6-11 虚拟机在 VXLAN 环境 vMotion 测试之五

第 21 步,确认迁移参数正确(如图 4-6-12 所示),单击"完成"按钮。

图 4-6-12 虚拟机在 VXLAN 环境 vMotion 测试之六

第 22 步,名为 WEB-SERVER 的虚拟机迁移到 IP 地址为 10.92.0.16 的 ESXi 主机(如图 4-6-13 所示)。

图 4-6-13　虚拟机在 VXLAN 环境 vMotion 测试之七

第 23 步，使用 VMware Remote Console 打开虚拟机控制台，通过图 4-6-14 可以看到，虚拟机网络出现 1 个包的丢失即恢复正常，说明虚拟机 VXLAN 环境下 vMotion 操作正常。

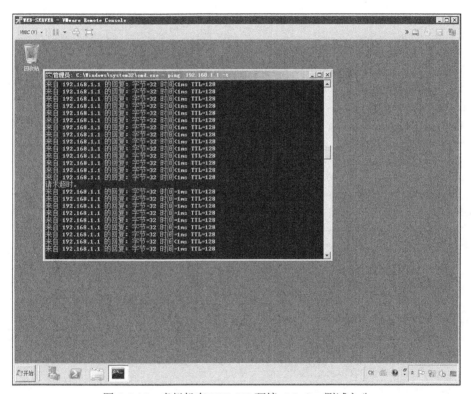

图 4-6-14　虚拟机在 VXLAN 环境 vMotion 测试之八

第 24 步，在本章的结束部分，查看 Nexus 1000V 交换机的所有配置信息。

```
bdnetlab-n1k(config)# show running-config

!Command: show running-config
!Time: Thu Jun 15 00:43:54 2017

version 4.2(1)SV2(2.2)
svs switch edition essential
```

```
no feature telnet
feature netflow
feature segmentation
segment mode unicast-only

username admin password 5 $1$NeCjVgJ4$S/vaqylLudbfMFKYvCbim0  role network-admin

banner motd #Nexus 1000v Switch#

ssh key rsa 2048
ip domain-lookup
ip host bdnetlab-n1k 10.92.10.41
hostname bdnetlab-n1k
errdisable recovery cause failed-port-state
ip access-list deny_telnet
    statistics per-entry
    10 deny tcp any any eq telnet
    20 permit ip any any
class-map type qos match-any WEB-SERVER
    match precedence 0
policy-map type qos SERVER
    class WEB-SERVER
vem 3
    host id 4c4c4544-0030-5810-8032-b9c04f4d4c31
vem 4
    host id 4c4c4544-0054-5310-8032-b9c04f4d4c31
bridge-domain vxlan
    segment id 222222
    group 239.1.1.1
snmp-server user admin network-admin auth md5 0xa2cb98ffa3f2bc53380d54d63b6752db priv 0xa2cb98f
fa3f2bc53380d54d63b6752db localizedkey

vrf context management
    ip route 0.0.0.0/0 10.92.10.254
flow exporter netflow
    destination 10.92.30.248
    source mgmt0
    version 9
flow record inbound
    match ipv4 source address
    match ipv4 destination address
    match transport source-port
    match transport destination-port
    collect counter bytes
    collect counter packets
```

```
flow monitor netflow
  record inbound
  exporter netflow
vlan 1,10,20

port-channel load-balance ethernet source-mac
port-profile default max-ports 32
port-profile type ethernet Unused_Or_Quarantine_Uplink
  vmware port-group
  shutdown
  description Port-group created for Nexus1000V internal usage. Do not use.
  state enabled
port-profile type vethernet Unused_Or_Quarantine_Veth
  vmware port-group
  shutdown
  description Port-group created for Nexus1000V internal usage. Do not use.
  state enabled
port-profile type ethernet uplink
  vmware port-group
  switchport mode trunk
  switchport trunk allowed vlan 1-3967,4048-4093
  no shutdown
  system vlan 10
  state enabled
port-profile type vethernet WEB-SERVER
  vmware port-group
  switchport mode access
  switchport access vlan 10
  no shutdown
  system vlan 10
  state enabled
port-profile type vethernet l3-control
  capability l3control
  vmware port-group
  switchport mode access
  switchport access vlan 10
  no shutdown
  system vlan 10
  state enabled
port-profile type vethernet SQL-SERVER
  vmware port-group
  switchport mode access
  switchport access vlan 20
  no shutdown
  state enabled
port-profile type vethernet BDNETLAB-N1K-VLAN10
```

```
    ip flow monitor netflow input
port-profile type vethernet vxlan
    vmware port-group
    switchport mode access
    switchport access vlan 20
    capability vxlan
    no shutdown
    system vlan 20
    state enabled
port-profile type vethernet vxlan-vm
    vmware port-group
    switchport mode access
    switchport access bridge-domain vxlan
    no shutdown
    state enabled

vdc bdnetlab-n1k id 1
    limit-resource vlan minimum 16 maximum 2049
    limit-resource monitor-session minimum 0 maximum 2
    limit-resource vrf minimum 16 maximum 8192
    limit-resource port-channel minimum 0 maximum 768
    limit-resource u4route-mem minimum 1 maximum 1
    limit-resource u6route-mem minimum 1 maximum 1

interface mgmt0
    ip address 10.92.10.41/24

interface Vethernet1
    inherit port-profile l3-control
    description VMware VMkernel, vmk0
    vmware dvport 66 dvswitch uuid "ce dc 02 50 73 b6 a3 1f-73 ef 90 c7 8c a0 30 4d"
    vmware vm mac 0026.6CFB.612C

interface Vethernet2
    inherit port-profile l3-control
    description VMware VMkernel, vmk0
    vmware dvport 67 dvswitch uuid "ce dc 02 50 73 b6 a3 1f-73 ef 90 c7 8c a0 30 4d"
    vmware vm mac 0026.6CFA.FCA8

interface Vethernet3
    inherit port-profile vxlan-vm
    description WEB-SERVER, Network Adapter 1
    vmware dvport 323 dvswitch uuid "ce dc 02 50 73 b6 a3 1f-73 ef 90 c7 8c a0 30 4d"
    vmware vm mac 0050.5682.6052

interface Vethernet4
```

```
    inherit port-profile vxlan-vm
    description MS_SQL_SERVER, Network Adapter 1
    vmware dvport 322 dvswitch uuid "ce dc 02 50 73 b6 a3 1f-73 ef 90 c7 8c a0 30 4d"
    vmware vm mac 0050.5682.726B

interface Vethernet5
    inherit port-profile vxlan
    description VMware VMkernel, vmk2
    vmware dvport 290 dvswitch uuid "ce dc 02 50 73 b6 a3 1f-73 ef 90 c7 8c a0 30 4d"
    vmware vm mac 0050.566D.846E

interface Vethernet6
    inherit port-profile vxlan
    description VMware VMkernel, vmk2
    vmware dvport 291 dvswitch uuid "ce dc 02 50 73 b6 a3 1f-73 ef 90 c7 8c a0 30 4d"
    vmware vm mac 0050.5666.BC28

interface Ethernet3/3
    inherit port-profile uplink

interface Ethernet3/4
    inherit port-profile uplink

interface Ethernet4/3
    inherit port-profile uplink

interface Ethernet4/4
    inherit port-profile uplink
    no shutdown
    switchport monitor

interface control0
line console
boot kickstart bootflash:/nexus-1000v-kickstart.4.2.1.SV2.2.2.bin sup-1
boot system bootflash:/nexus-1000v.4.2.1.SV2.2.2.bin sup-1
boot kickstart bootflash:/nexus-1000v-kickstart.4.2.1.SV2.2.2.bin sup-2
boot system bootflash:/nexus-1000v.4.2.1.SV2.2.2.bin sup-2
monitor session 1
    destination interface Ethernet4/4
    no shut
monitor session 2 type erspan-source
    destination ip 10.92.30.248
    erspan-id 23
    ip ttl 64
    ip prec 0
    ip dscp 0
```

```
    mtu 1500
    header-type 2
    no shut
svs-domain
    domain id 92
    control vlan 1
    packet vlan 1
    svs mode L3 interface mgmt0
svs connection vcenter
    protocol vmware-vim
    remote ip address 10.92.10.104 port 80
    vmware dvs uuid "ce dc 02 50 73 b6 a3 1f-73 ef 90 c7 8c a0 30 4d" datacenter-name Datacenter
    admin user n1kUser
    max-ports 8192
    connect
vservice global type vsg
    tcp state-checks invalid-ack
    tcp state-checks seq-past-window
    no tcp state-checks window-variation
    no bypass asa-traffic
vnm-policy-agent
    registration-ip 0.0.0.0
    shared-secret **********
    log-level
```

4.7 本章小结

本章对 Cisco Nexus 1000V 进行了详细介绍，包括 VSM 安装部署、Port-Profile 配置、VEM 安装部署、Nexus 1000V 交换机常用安全策略使用，最后还增加了基于 Nexus 1000V 交换机 VXLAN 基本配置。

对于生产环境来说，VMware vSphere 集成的分布式交换机无法通过命令行进行细节控制以及功能受限，如果网络环境需要使用传统命令行管理以及需要较细的安全策略控制，可以考虑使用 Cisco Nexus 1000V 交换机，在本章的结尾部分，列出了 Cisco Nexus 1000V 交换机在本章实验的所有配置，读者可以进行参考。

当然，除了 Nexus 1000V 交换机外，VMware 也提供 NSX 软件定义网络解决方案，在写作本书的时候刚发布，由于其发布时候比较短，还未在生产环境中大规模使用，因此读者有兴趣可以对其进行测试。

第 5 章 部署 Nexus N5K&N2K 交换机

1993 年，Cisco 公司推出了一个交换机的品牌——Catalyst，这个品牌在市场上处于绝对的领导地位。经过 15 年的时间，到 2008 年 1 月，Cisco 公司又对外宣布了另一个交换机的品牌——Nexus。Nexus 交换机是适应未来云计算、虚拟化、整合化数据中心的新一代交换机产品，对以前的 Catalyst 交换机有重大改进和扩展，是业界最先进的产品。本章介绍 Cisco Nexus N5K&N2K 交换机的基本配置以及在 VMware vSphere 环境的应用。

本章要点
- Nexus N5K&N2K 交换机介绍
- 配置使用 Nexus FEX
- 配置使用 Nexus Vpc
- 虚拟化架构使用 N5K&N2K

5.1 Nexus N5K&N2K 交换机介绍

Cisco Nexus 交换机是适应未来数据中心的高密度交换机，具有比 Catalyst 交换机高得多的极高的性能，端口交换延迟很低，同时支持国际标准不丢包的以太网技术。

5.1.1 Nexus 系列交换机介绍

作为全新的品牌，Nexus 系列交换机适应未来数据中心的先进架构，与 Catalyst 交换机及其他厂家等交换机有本质上不同。Nexus 交换机不简单是传输数据，而是一种全新的架构设计——计算机总线的延伸，目前只有 Brocade 交换机能部分支持，其他厂商等无法支持这些特性。Nexus 交换机具有以下特点。

1. 高性能/低延迟/不丢包的以太网

高性能/低延迟/不丢包的结合使得 Nexus 交换机具有一个独特的优势——计算机总线的延伸，相当于把彼此通信的众多计算机的总线直接连接起来。

- 高性能相当于修了一条非常宽的路，宽到有多少车都可以直接上来；
- 低延迟相当于每辆车的速度都很快；
- 不丢包相当于车上的货物在道路上不丢失，以前的所有交换机都做不到这点，相当于目的地有个货物检查员来检查车上的货物，一旦发现丢了，再重发货，效率会降低。

2. 统一的架构

Nexus 交换机能支持以太网和存储网的统一架构。使用 Nexus 交换机的统一架构 FCOE 后，布线大大减少，考虑到冗余，只需要 2 个接口卡（CNA）即可，使得原来机房杂乱的布线变得非常简单、整洁，维护管理异常方便（如图 5-1-1 所示）。

图 5-1-1　统一架构

3. Nexus VDC

Nexus 交换机支持完全虚拟化的架构。从大的方面讲，这种虚拟化的架构与 Vmware 结合后，构成一个完全适应未来数据中心即云计算数据中心的架构。从小的方面讲，Nexus N7000 支持 VDC 虚拟交换机技术，支持将一台 Nexus N7000 交换机划分为 4 台或 8 台完全软件和端口独立的 Nexus N7000 交换机，软件进程和端口完全隔离，购买一台 Nexus7000 交换机，可以当作 4 台或 8 台独立的 Nexus N7000 使用，给不同的业务系统隔离使用，降低了需要购买设备的数量和经费，而且只安装一次，非常灵活（如图 5-1-2 所示）。

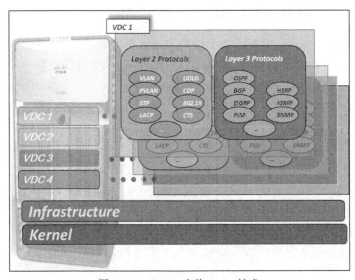

图 5-1-2　Nexus 交换 VDC 技术

4. 简化管理

Nexux N5000 交换机通过 10GE 接口与 N2K 连接，每个 N2K 相当于 N5K 的一块 48 口板卡，配置管理完全在 N5K 上进行，把 N2K 放在每个机柜上，与机柜上服务器连接，然后再与 N5K 连接。通过这种独特的 N5K+N2K 的虚拟板卡架构，管理简化，只管理一个 N5K 即可，不像以前那样要管理多个连接服务器的交换机（如图 5-1-3 所示）。

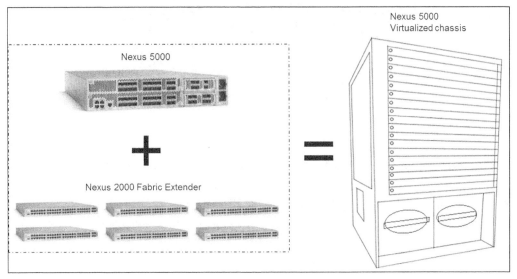

图 5-1-3 Nexus N5K+N2K 组合

5. NV-LINK 技术以及 NXOS

（1）NV-LINK 技术

在 VMware 环境中使用的 Nexus 1000V 虚拟交换机能软件支持 NV-Link 功能，Nexus N5000 交换机也将会硬件支持这个功能，这样在交换机上就能识别出各个虚拟机 VM 的特征（QOS、安全等），从而能够为每个 VM 制定一些策略，有效地与 VM 环境结合。NV-Link 在 VMware 的虚拟环境中对用户很有吸引力，如果有这种环境或对虚拟机很感兴趣，可以有针对性地告诉用户，只有 Cisco 的交换机上能支持。

（2）NX-OS

Nexus 交换机集中了 Cisco 存储交换机 SAN OS 及 IP 网络设备 IOS 中最精华的部分，形成了最先进的操作系统 NX-OS。NX-OS 使用 Linux 内核，具有极高的稳定性，可以达到 99.999%。

5.1.2 Nexus N5K 介绍

统一、融合是 Nexus N5K 系列交换机的最大特点，Nexus N5K 系列交换机提供统一端口\FEX\FC\FCoE\10G\40GB 以太网端口，Nexus N5K 系列包括 5000\5500\5600 系列。

如图 5-1-4 所示为一款 Cisco 主流 Nexus 5548UP 交换机，提供 32 个固定统一端口（可通过命令的方式将端口修改为以太网或 FC 接口），同时提供一个扩展插槽用于增加接口数量。要了解更多 Nexus N5K 型号的相关情况可以访问 Cisco 官方网站。

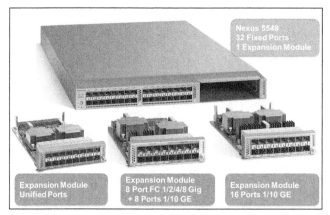

图 5-1-4　Cisco Nexus 5548UP 交换机

5.1.3　Nexus N2K 介绍

从某种意义上说，Nexus N2K 不是一款完整功能的交换机，它作为 N7K 或 N5K 交换机的远程板卡使用，因为它无法独立工作，也无法独立配置，必须依赖于上游交换机才能进行配置以及数据转发。要了解更多 Nexus N2K 型号的相关情况可以访问 Cisco 官方网站。

如图 5-1-5 所示为一款 Cisco 主流 Nexus 2248TP 交换机。该机提供 48 个固定 1GE 以太网接口用于连接主机、4 个 SFP+接口用于连接 N7K 或 N5K。

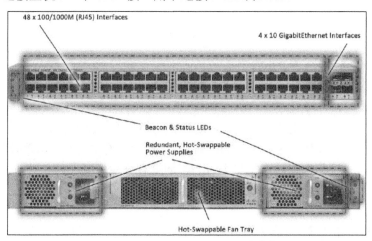

图 5-1-5　Cisco Nexus 2248TP 交换机

如图 5-1-6 所示为一款 Cisco 主流 Nexus 2232PP 交换机。该机提供 32 个固定 SFP+接口用于连接主机、8 个 SFP+用于连接 N7K 或 N5K。

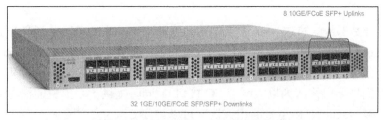

图 5-1-6　Cisco Nexus 2232PP 交换机

5.1.4 Nexus NXOS 基本命令行介绍

1. 查看 Nexus 交换机软/硬件信息

```
DC2-N5K-01# show version
Cisco Nexus Operating System (NX-OS) Software     //运行 NXOS 软件
TAC support: http://www.cisco.com/tac
Documents: http://www.cisco.com/en/US/products/ps9372/tsd_products_support_series_home.html
Copyright (c) 2002-2016, Cisco Systems, Inc. All rights reserved.
The copyrights to certain works contained herein are owned by
other third parties and are used and distributed under license.
Some parts of this software are covered under the GNU Public
License. A copy of the license is available at
http://www.gnu.org/licenses/gpl.html.

Software
  BIOS:            version 3.5.0
  Power Sequencer Firmware:
              Module 1: v1.0
              Module 2: v2.0
  Microcontroller Firmware:       version v1.2.0.1
  QSFP Microcontroller Firmware:
              Module not detected
  CXP Microcontroller Firmware:
              Module not detected
  kickstart: version 7.3(0)N1(1)     //kickstart 包括 linux 内核、基本驱动以及初始化文件系统
  system: version 7.3(0)N1(1)        //system 包括系统软件、基础以及四到七层功能
  BIOS compile time:       02/03/2011
  kickstart image file is: bootflash:///n5000-uk9-kickstart.7.3.0.N1.1.bin
  kickstart compile time:  2/17/2016 22:00:00 [02/18/2016 06:55:14]
  system image file is:    bootflash:///n5000-uk9.7.3.0.N1.1.bin
  system compile time:     2/17/2016 22:00:00 [02/18/2016 09:34:14]


Hardware
  cisco Nexus5548 Chassis ("O2 32X10GE/Modular Universal Platform Supervisor")   //Nexus 交换机型号
  Intel(R) Xeon(R) CPU        with 8253812 kB of memory.      //Nexus 交换机 CPU、内存信息
  Processor Board ID FOC16166M6S

  Device name: DC2-N5K-01
  bootflash:     2007040 kB

Kernel uptime is 0 day(s), 0 hour(s), 4 minute(s), 49 second(s)

Last reset
  Reason: Unknown
  System version: 7.3(0)N1(1)
```

Service:

plugin
    Core Plugin, Ethernet Plugin

Active Package(s)

2. 查看 Nexus 交换机 NXOS 软件

```
DC2-N5K-01# show boot
Current Boot Variables:

kickstart variable = bootflash:/n5000-uk9-kickstart.7.3.0.N1.1.bin
system variable = bootflash:/n5000-uk9.7.3.0.N1.1.bin
Boot POAP Disabled

Boot Variables on next reload:

kickstart variable = bootflash:/n5000-uk9-kickstart.7.3.0.N1.1.bin
system variable = bootflash:/n5000-uk9.7.3.0.N1.1.bin
Boot POAP Disabled
DC2-N5K-01#
```

3. 查看 Nexus 交换机模块信息

```
DC2-N5K-01# show module
Mod Ports Module-Type                              Model              Status
--- ----- -------------------------------------- -------------------- ----------
1   32    O2 32X10GE/Modular Universal Platfo    N5K-C5548UP-SUP      active *
3   0     O2 Non L3 Daughter Card                N55-DL2              ok

Mod  Sw           Hw     World-Wide-Name(s) (WWN)
---  -----------  -----  --------------------------------------------
1    7.3(0)N1(1)  1.0    22:61:77:62:2f:6c:69:62 to 70:6c:61:74:66:6f:72:6d
3    7.3(0)N1(1)  1.0    --

Mod   MAC-Address(es)                         Serial-Num
---   --------------------------------------  ----------
1     547f.ee99.8c48 to 547f.ee99.8c67        FOC16166M6S
3     0000.0000.0000 to 0000.0000.000f        FOC16151ZR1
```

4. 查看 Nexus 交换机电源状态

```
DC2-N5K-01# show environment power

Power Supply:
Voltage: 12 Volts
-----------------------------------------------------
```

PS	Model	Input Power Type	Power (Watts)	Current (Amps)	Status
1	N55-PAC-750W	AC	780.00	65.00	ok
2	N55-PAC-750W	AC	780.00	65.00	ok

Mod	Model	Power Requested (Watts)	Current Requested (Amps)	Power Allocated (Watts)	Current Allocated (Amps)	Status
1	N5K-C5548UP-SUP	492.00	41.00	492.00	41.00	powered-up
3	N55-DL2	24.00	2.00	24.00	2.00	powered-up

Power Usage Summary:

Power Supply redundancy mode: Redundant
Power Supply redundancy operational mode: Redundant

Total Power Capacity   1560.00 W

Power reserved for Supervisor(s)   492.00 W
Power currently used by Modules   24.00 W

Total Power Available   1044.00 W
                                             ----

5. 查看 Nexus 交换机 feature 信息

```
DC2-N5K-01# show feature
Feature Name          Instance    State
--------------------  --------    --------
Flexlink              1           disabled
amt                   1           disabled
bfd                   1           disabled
bfd_app               1           disabled
bgp                   1           disabled
bulkstat              1           disabled
cable-management      1           disabled
cts                   1           disabled
dhcp                  1           disabled
dot1x                 1           disabled
……
```

6. 查看 Nexus 交换机 host-id 信息

```
DC2-N5K-01# show license host-id
License hostid: VDH=SSI16070H4F
```

7. 查看 Nexus 交换机许可文件

```
DC2-N5K-01# show license file license_SSI16070H4F_10.lic
SERVER this_host ANY
VENDOR cisco
INCREMENT FC_FEATURES_PKG cisco 1.0 permanent uncounted \
        VENDOR_STRING=MDS HOSTID=VDH=SSI16070H4F \

NOTICE=<LicFileID>20120524122358000</LicFileID><LicLineID>1</LicLineID><PAK>N5K-C5548UP
    -B-S32SSI16070H4F</PAK> \
        SIGN=CAF8EC3E1ED8
INCREMENT ENTERPRISE_PKG cisco 1.0 permanent uncounted \
        VENDOR_STRING=MDS HOSTID=VDH=SSI16070H4F \

NOTICE=<LicFileID>20120524122358000</LicFileID><LicLineID>1</LicLineID><PAK>N5K-C5548UP
    -B-S32SSI16070H4F</PAK> \
        SIGN=253ADB94F086
```

8. 查看 Nexus 交换机安装的许可授权

```
DC2-N5K-01# show license usage
Feature                        Ins  Lic   Status  Expiry Date  Comments
                                    Count
--------------------------------------------------------------------------------
FCOE_NPV_PKG                   No   -     Unused                -
FM_SERVER_PKG                  No   -     Unused                -
ENTERPRISE_PKG                 Yes  -     Unused  never         -
FC_FEATURES_PKG                Yes  -     Unused  never         -
VMFEX_FEATURE_PKG              No   -     Unused                -
ENHANCED_LAYER2_PKG            No   -     Unused                -
NETWORK_SERVICES_PKG           No   -     Unused                -
LAN_BASE_SERVICES_PKG          No   -     Unused                -
LAN_ENTERPRISE_SERVICES_PKG    No   -     Unused                -
--------------------------------------------------------------------------------
```

9. 查看 Nexus 交换机全部配置

```
DC2-N5K-01# show running-config
!Command: show running-config
!Time: Thu Mar  5 12:22:17 2009

version 7.3(0)N1(1)
hostname DC2-N5K-01

feature telnet
feature lldp
feature fex
```

```
username admin password 5 $1$XZkOxUET$S5djE6W2TUbJ86gcRIeLM1   role network-admin
no password strength-check

banner motd #
Welcome to BDNETLAB DataCenter Lab
You are using now is "DC2-N5K-01"
#
ip domain-lookup
control-plane
   service-policy input copp-system-policy-customized
policy-map type control-plane copp-system-policy-customized
   class copp-system-class-default
      police cir 2048 kbps bc 6400000 bytes
fex 105
   pinning max-links 1
   description "FEX0105"
slot 1
   port 1-24 type ethernet
   port 25-32 type fc
snmp-server user admin network-admin auth md5 0x2e7a7da98ecd5a860d4510e8a094ec66 priv 0x2e7a7da
   98ecd5a860d4510e8a094ec66 localizedkey
rmon event 1 log description FATAL(1) owner PMON@FATAL
rmon event 2 log description CRITICAL(2) owner PMON@CRITICAL
rmon event 3 log description ERROR(3) owner PMON@ERROR
rmon event 4 log description WARNING(4) owner PMON@WARNING
rmon event 5 log description INFORMATION(5) owner PMON@INFO
system qos
   service-policy type queuing input fcoe-default-in-policy
   service-policy type queuing output fcoe-default-out-policy
   service-policy type qos input fcoe-default-in-policy
   service-policy type network-qos fcoe-default-nq-policy

vlan 1
vrf context management
   ip route 0.0.0.0/0 10.92.30.254

interface Ethernet1/1
……
interface Ethernet1/24

interface mgmt0
   vrf member management
   ip address 10.92.30.244/24
line console
line vty
```

```
boot kickstart bootflash:/n5000-uk9-kickstart.7.3.0.N1.1.bin
boot system bootflash:/n5000-uk9.7.3.0.N1.1.bin
```

5.2 配置使用 Nexus FEX

FEX，全称为 Fabric Extenders，由 Cisco 公司提出，简单来说就是用于 N7K 或 N5K 连接和管理 N2K 平台的技术，通过 FEX，N7K 或 N5K 交换机可以配置 N2K 交换机。

5.2.1 Nexus FEX 技术介绍

在配置 N2K 交换机之前，需要对 N2K 所使用的 FEX 技术进行一些了解。

1. 什么是 FEX

N2K 交换机作为 N7K 和/或 N5K 的一个远程线卡，被 N7K 或 N5K 交换机管理和配置，N2K 交换机本身不能配置。

N2K 交换机使用 FEX 技术与 N7K 和/或 N5K 合并在一起，实现了最好的 Top-of-rack 的接线和管理。

FEX 协议为 VN-Tag/VN-LINK，公有化标准为 802.1BR，由 Cisco 提供。

VN-Tag 技术是 Cisco Nexus 交换架构的基础，用于交换机之间或交换机与服务器之前的通信。

图 5-2-1 完整地显示了 FEX 技术的创新以及 FEX 架构的使用。

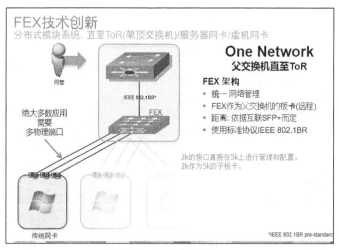

图 5-2-1　FEX 技术

2. FEX 连接模式

对于使用 FEX 连接的 N2K 连接，有 Static Pinning 以及 Port Channel 两种模式。

（1）Static Pinning 模式

Static Pinning 是将 N2K 交换机端口进行分组，一条上行链路映射一组端口，Static Pinning 模式不支持负载均衡，如果某条上行链路出现故障，这条链路映射的端口将全部无法访问，无法切换到其他正常上行链路。图 5-2-2 显示了使用 Static Pinning 模式端口分组进行数据转发的情况。N2K 交换机具有 4 条上行链路，每条上行链路对应 8 个连接主机或服务器使用的以太网端口进行数据转发，当某条上行链路接口出现故障，则所对应的 8 个以太端口将无法转发数据。

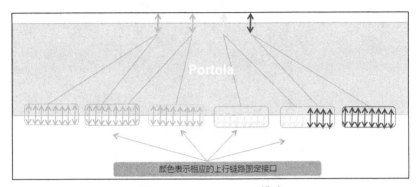

图 5-2-2　Static Pinning 模式

（2）Port Channel

Port Channel 模式是将 N2K 交换机上行链路进行捆绑形成上行链路，这条上行链路映射所有端口进行数据转发，如果某条上行链路端口出现故障，端口立即切换到这条链路其他正常上行链路端口进行数据转发。图 5-2-3 显示了使用 Port Channel 模式端口分组进行数据转发的情况。使用 Port Channel 后，4 条上行链路捆绑为一个整体提供数据转发以及负载均衡，以 4 条上行链路环境为例，最多可以允许 3 条链路出现故障而不影响数据转发。

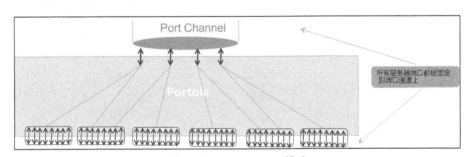

图 5-2-3　Port Channel 模式

3．FEX 使用规则

（1）Port Channel

N5K 连接 N2K 接口如果使用 Port Channel，不支持 LACP 协议。

（2）FEX 端口命名规则

N2K 交换机接口命名为 chassis/slot/port（E100/1/1），范围为 100～199。

（3）N2K 交换机接口

N2K 交换机不能连接交换机，这种接口设计只用于提供主机和服务器的连接。N2K 交换机接口默认激活 BPDU Guard，如果该接口连接交换机，一旦从这个接口收到 BPDU，这个接口将被置为"error-disable"状态而不能使用，同时不能在 N2K 交换机禁用 BPDU Guard 功能。

4. FEX 支持的连接模式

图 5-2-4 显示了 FEX 支持的多种方式的连接模式。

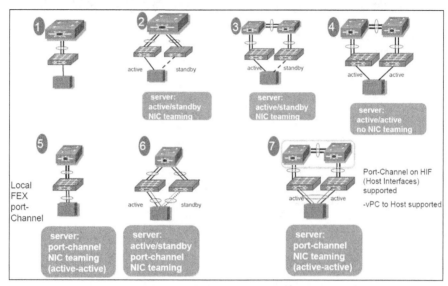

图 5-2-4　FEX 支持的连接模式

图 5-2-5 显示了 FEX 支持的针对 N5K 交换机的连接模式。

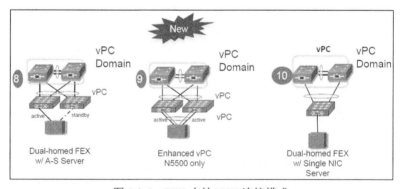

图 5-2-5　FEX 支持 N5K 连接模式

5. FEX 不支持的拓扑

图 5-2-6 显示了 FEX 不支持的连接模式。

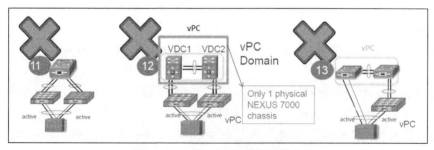

图 5-2-6　FEX 不支持的连接模式

5.2.2 配置 Nexus FEX

对 FEX 进行了解后即可进行 FEX 配置。本小节介绍基本的 FEX 配置。

1. Static Pinning 配置

第 1 步，在 N5K 上启用 FEX 特性，并配置 E1/1 为 Fabric Port。

```
BDNETLAB-N5K-01# configure terminal       //进入配置模式
Enter configuration commands, one per line.   End with CNTL/Z.
BDNETLAB-N5K-01(config)# feature fex      //启用 FEX 特性，默认为禁用状态
BDNETLAB-N5K-01(config)# interface ethernet 1/1
BDNETLAB-N5K-01(config-if)# switchport mode fex-fabric    //定义端口模式为 fex-fabric
BDNETLAB-N5K-01(config-if)# fex associate 103             //定义 fex 接口 ID
BDNETLAB-N5K-01(config-if)# no shutdown
```

第 2 步，当 N5K 接口配置好，N2K 电源打开后，N2K 开始注册到 N5K 交换机。通过 N5K 交换机可以看到 N2K 注册到 N5K 的日志信息。

```
BDNETLAB-N5K-01#show fex detail
08/08/2016 03:13:33.62356: Module register received
08/08/2016 03:13:33.63661: Image Version Mismatch    //N2K 出厂有原始的 NXOS 版本，可能与 N5K 交
换机 NXOS 不匹配，如果不匹配就需要重新下载
08/08/2016 03:13:33.64315: Registration response sent    //发送注册信息
08/08/2016 03:13:33.64634: Requesting satellite to download image    //请求下载 NXOS
08/08/2016 03:17:09.673301: Image preload successful.    //下载成功
08/08/2016 03:17:10.856586: Deleting route to FEX    //更新 NXOS 需要删除原路径信息
08/08/2016 03:17:10.898752: Module disconnected    //N2K 交换机作为模块断开
08/08/2016 03:17:10.899738: Module Offline    //N2K 交换机作为模块处于离线状态
08/08/2016 03:17:10.904162: Deleting route to FEX
08/08/2016 03:17:10.915101: Module disconnected
08/08/2016 03:17:10.968614: Offlining Module
08/08/2016 03:18:24.310323: Module register received
08/08/2016 03:18:24.312119: Registration response sent    //更新后再次发送注册信息
08/08/2016 03:18:25.463578: Module Online Sequence    //N2K 模块在线注册成功
08/08/2016 03:18:31.749348: Module Online    //N2K 模块在线
```

第 3 步，查看 N2K 通过 FEX 注册到 N5K 的信息。

```
BDNETLAB-N5K-01# show fex    //查看注册到 N5K 的 N2K 信息
  FEX         FEX          FEX                              FEX
  Number      Description  State           Model            Serial
  ------------------------------------------------------------------
  103         FEX0103      Online          N2K-C2248TP-1GE  SSI153916DA

BDNETLAB-N5K-01# show interface ethernet 1/1 fex-intf    //查看 N5K 交换机 E1/1 映射的 N2K 端口
```

Fabric Interface	FEX Interfaces			
Eth1/1	Eth103/1/1	Eth103/1/2	Eth103/1/3	Eth103/1/4
	Eth103/1/5	Eth103/1/6	Eth103/1/7	Eth103/1/8
	Eth103/1/9	Eth103/1/10	Eth103/1/11	Eth103/1/12
	Eth103/1/13	Eth103/1/14	Eth103/1/15	Eth103/1/16
	Eth103/1/17	Eth103/1/18	Eth103/1/19	Eth103/1/20
	Eth103/1/21	Eth103/1/22	Eth103/1/23	Eth103/1/24
	Eth103/1/25	Eth103/1/26	Eth103/1/27	Eth103/1/28
	Eth103/1/29	Eth103/1/30	Eth103/1/31	Eth103/1/32
	Eth103/1/33	Eth103/1/34	Eth103/1/35	Eth103/1/36
	Eth103/1/37	Eth103/1/38	Eth103/1/39	Eth103/1/40
	Eth103/1/41	Eth103/1/42	Eth103/1/43	Eth103/1/44
	Eth103/1/45	Eth103/1/46	Eth103/1/47	Eth103/1/48

第 4 步，配置 Static Pinning 双上行，也就是 N5K 与 N2K 之间两条上行链路，增加 E1/2 为 Fabric Port。

```
BDNETLAB-N5K-01(config)# interface ethernet 1/2
BDNETLAB-N5K-01(config-if)# switchport mode fex-fabric
BDNETLAB-N5K-01(config-if)# fex associate 103
BDNETLAB-N5K-01(config-if)# no shutdown
BDNETLAB-N5K-01# show interface ethernet 1/2 fex-intf
```

Fabric Interface	FEX Interfaces
Eth1/2	

第 5 步，由于 Static Pinning 上行链路使用 1 个端口，使用多个端口后，需要调整 max-links。当增加 1 个端口后，自动对 N2K 交换机端口进行了分组，其中 E103/1/1 至 E103/1/24 使用 E1/1 端口进行数据转发，E103/1/25 至 E103/1/48 使用 E1/2 端口进行数据转发。

```
BDNETLAB-N5K-01(config)# fex 103
BDNETLAB-N5K-01(config-fex)# pinning max-links 2
Change in Max-links will cause traffic disruption.
BDNETLAB-N5K-01(config)# show interface ethernet 1/1 fex-intf
```

Fabric Interface	FEX Interfaces			
Eth1/1	Eth103/1/24	Eth103/1/23	Eth103/1/22	Eth103/1/21
	Eth103/1/20	Eth103/1/19	Eth103/1/18	Eth103/1/17
	Eth103/1/16	Eth103/1/15	Eth103/1/14	Eth103/1/13
	Eth103/1/12	Eth103/1/11	Eth103/1/10	Eth103/1/9

```
                     Eth103/1/8    Eth103/1/7    Eth103/1/6    Eth103/1/5
                     Eth103/1/4    Eth103/1/3    Eth103/1/2    Eth103/1/1

BDNETLAB-N5K-01(config)# show interface ethernet 1/2 fex-intf
Fabric               FEX
Interface            Interfaces
---------------------------------------------------------------
Eth1/2               Eth103/1/48   Eth103/1/47   Eth103/1/46   Eth103/1/45
                     Eth103/1/44   Eth103/1/43   Eth103/1/42   Eth103/1/41
                     Eth103/1/40   Eth103/1/39   Eth103/1/38   Eth103/1/37
                     Eth103/1/36   Eth103/1/35   Eth103/1/34   Eth103/1/33
                     Eth103/1/32   Eth103/1/31   Eth103/1/30   Eth103/1/29
                     Eth103/1/28   Eth103/1/27   Eth103/1/26   Eth103/1/25
```

第 6 步，如果对注册到 N5K 的 N2K 的交换机有安全性要求，可以 FEX 扩展特性，通过 N2K 交换机唯一序列号以及型号来加入。如果不匹配，N2K 交换机无法注册成功。

```
BDNETLAB-N5K-01(config)# fex 103
BDNETLAB-N5K-01(config-fex)# description BDNETLAB-N2K-01
BDNETLAB-N5K-01(config-fex)# serial SSI153916DA      //定义注册 N2K 交换机唯一序列号
Changing serial will offline fex.
BDNETLAB-N5K-01(config-fex)# type ?          //定义注册 N2K 交换机型号
  N2148T      Fabric Extender 48x1G 4x10G Module
  N2224TP     Fabric Extender 24x1G 2x10G SFP+ Module
  N2232P      Fabric Extender 32x10G 8x10G Module
  N2232TM     Fabric Extender 32x10GBase-T 8x10G SFP+ Module
  N2232TM-E   Fabric Extender 32x10GBase-T 8x10G SFP+ Module
  N2232TP     Fabric Extender 32x10GBase-T 8x10G SFP+ Module
  N2232TT     Fabric Extender 32x10GBase-T 8x10GBase-T Module
  N2248T      Fabric Extender 48x1G 4x10G Module
  N2248TP-E   Fabric Extender 48x1G 4x10G Module
  NB22DELL    Fabric Extender 16x10G SFP+ 8x10G SFP+ Module
  NB22FJ      Fabric Extender 16x10G SFP+ 8x10G SFP+ Module
  NB22HP      Fabric Extender 16x10G SFP+ 8x10G SFP+ Module
```

第 7 步，测试在 Static Pinning 模式双上行链路正常的情况下，虚拟机 PING 都是正常的（如图 5-2-7 所示）。

第 8 步，把 E1/1 接口 shutdown 后，虚拟机之间的通信断开（如图 5-2-8 所示）。因为 Static Pinning 模式不支持负载均衡，所以如果某条上行链路端口出现故障，这条链路映射的端口将全部无法访问，无法切换到其他正常上行链路。

238　第 5 章　部署 Nexus N5K&N2K 交换机

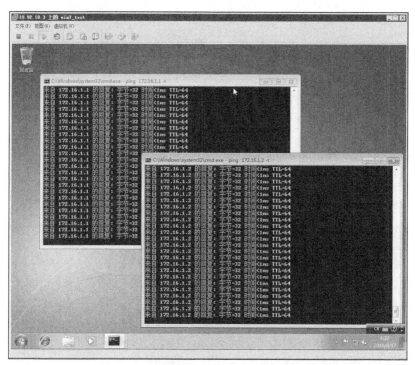

图 5-2-7　Static Pinning 模式下双上行链路正常状态

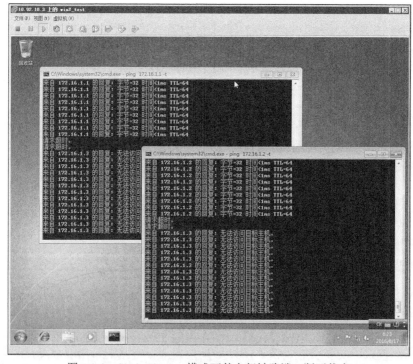

图 5-2-8　Static Pinning 模式下某上行链路端口断开状态

第 9 步，查看端口信息，可以看到当 E1/1 接口断开后，所映射的 E103/1/1 至 E103/1/24 端口全部无法映射。

```
BDNETLAB-N5K-01# show interface ethernet 1/1 fex-intf
Fabric          FEX
Interface       Interfaces
---------------------------------------------
 Eth1/1
BDNETLAB-N5K-01# show interface ethernet 1/2 fex-intf
Fabric          FEX
Interface       Interfaces
---------------------------------------------
 Eth1/2         Eth103/1/25    Eth103/1/26    Eth103/1/27    Eth103/1/28
                Eth103/1/29    Eth103/1/30    Eth103/1/31    Eth103/1/32
                Eth103/1/33    Eth103/1/34    Eth103/1/35    Eth103/1/36
                Eth103/1/37    Eth103/1/38    Eth103/1/39    Eth103/1/40
                Eth103/1/41    Eth103/1/42    Eth103/1/43    Eth103/1/44
                Eth103/1/45    Eth103/1/46    Eth103/1/47    Eth103/1/48
```

2. Port Channel 配置

第 1 步，在 N5K 上启用 FEX 特性，并配置 E1/1-2 为 port-channel。

```
BDNETLAB-N5K-01# configure terminal
Enter configuration commands, one per line.   End with CNTL/Z.
BDNETLAB-N5K-01(config)# feature fex
BDNETLAB-N5K-01(config)# interface ethernet 1/1-2    //配置 E1/1-2 接口
BDNETLAB-N5K-01(config-if)# channel-group 103        //配置 port channel 为 103
BDNETLAB-N5K-01(config-if)# no shutdown
BDNETLAB-N5K-01(config-if)# exit
BDNETLAB-N5K-01(config-if)# interface port-channel 103
BDNETLAB-N5K-01(config-if)# switchport mode fex-fabric  //定义端口模式为 fex-fabric
BDNETLAB-N5K-01(config-if)# fex associate 103    //定义 fex 接口 ID
BDNETLAB-N5K-01(config-if)# no shutdown
```

第 2 步，N2K 注册到 N5K 交换机，使用 Port Channel 后，N2K 下属的 48 个端口不再映射到某一个具体上行链路端口，而是直接映射到 port channel。

```
BDNETLAB-N5K-01# show fex
  FEX         FEX            FEX                         FEX
  Number      Description    State          Model        Serial
---------------------------------------------------------------------
  103         FEX0103        Online         N2K-C2248TP-1GE   SSI153916DA
BDNETLAB-N5K-01# show interface port-channel 103 fex-intf
Fabric          FEX
Interface       Interfaces
---------------------------------------------
 Po103          Eth103/1/1     Eth103/1/2     Eth103/1/3     Eth103/1/4
                Eth103/1/5     Eth103/1/6     Eth103/1/7     Eth103/1/8
```

Eth103/1/9	Eth103/1/10	Eth103/1/11	Eth103/1/12
Eth103/1/13	Eth103/1/14	Eth103/1/15	Eth103/1/16
Eth103/1/17	Eth103/1/18	Eth103/1/19	Eth103/1/20
Eth103/1/21	Eth103/1/22	Eth103/1/23	Eth103/1/24
Eth103/1/25	Eth103/1/26	Eth103/1/27	Eth103/1/28
Eth103/1/29	Eth103/1/30	Eth103/1/31	Eth103/1/32
Eth103/1/33	Eth103/1/34	Eth103/1/35	Eth103/1/36
Eth103/1/37	Eth103/1/38	Eth103/1/39	Eth103/1/40
Eth103/1/41	Eth103/1/42	Eth103/1/43	Eth103/1/44
Eth103/1/45	Eth103/1/46	Eth103/1/47	Eth103/1/48

第 3 步，测试在 Port Channel 模式下上行链路正常的情况下，虚拟机 PING 都是正常的（如图 5-2-9 所示）。

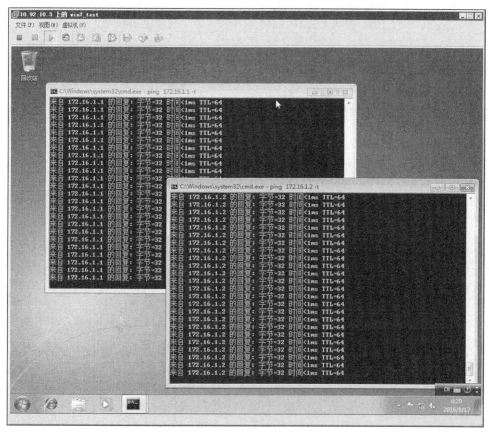

图 5-2-9　Port Channel 模式上行链路正常状态

第 4 步，把 E1/1 接口 shutdown 后，虚拟机之间的通信依然正常（如图 5-2-10 所示）。因为 Port Channel 模式本身支持负载均衡，所以如果某条上行链路端口出现故障，会自动切换到其他正常上行链路端口。

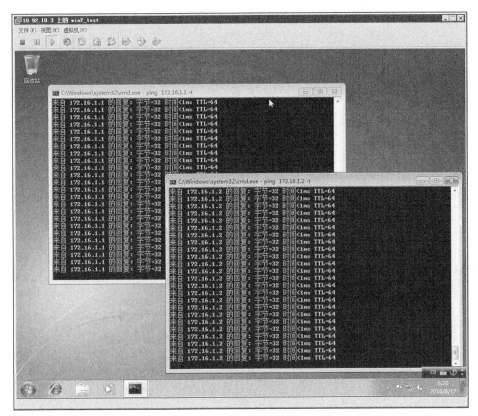

图 5-2-10 Port Channel 模式上行链路端口断开后状态

5.3 配置使用 Nexus vPC

vPC，全称为 Virtual Port Channel，简单来说就是可以跨越交换机使用的 Port Channel，是 Cisco Nexus 系列交换机特有的技术。

5.3.1 Nexus vPC 技术介绍

在配置 vPC 交换机之前，需要对 vPC 技术进行一些了解。

1. 什么是 vPC

vPC 在 2 台独立的交换机（Nexus N5K）上扩展聚合链路，基于虚拟链路技术，使 2 层网络环境更稳定和可靠，最大程度减少生成树阻塞口和最大程度利用所有现行的上联口。

图 5-3-1 显示了非 vPC 网络和使用 vPC 网络拓扑。在非 vPC 网络环境中，下游交换机连接到上游两台核心交换机使用 STP 为避免环路会阻塞端口；在 vPC 网络环境中，下游交换机连接到上游两台核心交换机可以使用 vPC 技术形成一条虚拟的聚合链路，不存在阻塞端口问题。

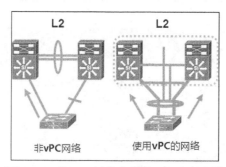

图 5-3-1 非 vPC 网络和使用 vPC 网络拓扑

2. vPC 优点

在网络环境中使用 vPC 技术，具有很多的优点。这些优点包括：提供虚拟机/服务器集群的无缝迁移；2 层的带宽扩展更容易；2 台交换机提供独立的控制层（control planes）；在有链路和设备损坏的情况下提供更快速的收敛；网络设计更简单。

3. vPC 术语

在 Nexus 交换机中使用 vPC 技术，需要了解其基本的术语（如图 5-3-2 所示）。

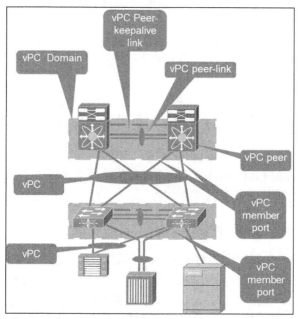

图 5-3-2 vPC 术语

vPC Domain：在 vPC 系统中的这一对 vPC 交换机。
vPC Peer：在这一对交换机中的对方。
vPC member port：vPC 聚合链路中的成员端口。
vPC：与上联或者下联设备相连的聚合链路，可以连接交换机也可以连接服务器。

vPC peer-link：在两个 vPC 交换机之间用来同步信息和状态的链路。该链路必须使用 10GE 网络端口。

vPC peer-keepalive link：2 台 vPC 交换机之间的心跳检测链路，确保 2 台 vPC 交换机在线。该链路可使用 1GE 或 10GE 网络端口。

5.3.2 配置 Nexus vPC

对 vPC 进行了解后即可进行 vPC 配置。本小节介绍基本的 vPC 配置。

1. 配置基本 vPC

第 1 步，在 N5K-01 交换机上启用 vPC 特性并创建 vPC Domain。

```
BDNETLAB-N5K-01(config)# feature vpc     //启用 vPC 特性
BDNETLAB-N5K-01(config)# vpc domain 50      //创建 vPC Domain
BDNETLAB-N5K-01(config-vpc-domain)# role priority 49    //配置优先级
Warning:
  !!:: vPCs will be flapped on current primary vPC switch while attempting role change ::!!
Note:
  --------:: Change will take effect after user has re-initd the vPC peer-link   ::--------
BDNETLAB-N5K-01(config-vpc-domain)# system-priority 2500        //配置 system 优先级
BDNETLAB-N5K-01(config-vpc-domain)# peer-keepalive destination 10.92.30.246 source 10.92.30.245 vrf management     //配置 peer-keepalive
BDNETLAB-N5K-01(config-vpc-domain)# exit
```

第 2 步，完成 N5K-01 配置后进行结果验证。

```
BDNETLAB-N5K-01(config)# show vpc
Legend:
                (*) - local vPC is down, forwarding via vPC peer-link

vPC domain id                     : 50
Peer status                       : peer link not configured
vPC keep-alive status             : Suspended (Destination IP not reachable)
Configuration consistency status  : failed
Per-vlan consistency status       : failed
Configuration inconsistency reason: vPC peer-link does not exist
Type-2 consistency status         : failed
Type-2 inconsistency reason       : vPC peer-link does not exist
vPC role                          : none established
Number of vPCs configured         : 0
Peer Gateway                      : Disabled
Dual-active excluded VLANs        : -
Graceful Consistency Check        : Disabled (due to peer configuration)
Auto-recovery status              : Disabled
```

第 3 步，在 N5K-02 交换机上启用 vPC 特性并创建 vPC Domain。注意，2 台 N5K 交换机 vPC Domain 配置必须一致。

```
BDNETLAB-N5K-02(config)# feature vpc         //启用 vPC 特性
BDNETLAB-N5K-02(config)# vpc domain 50        //创建 vPC Domain
BDNETLAB-N5K-02(config-vpc-domain)# role priority 50   //配置优先级
Warning:
  !!:: vPCs will be flapped on current primary vPC switch while attempting role change ::!!
Note:
  --------:: Change will take effect after user has re-initd the vPC peer-link   ::--------
BDNETLAB-N5K-02(config-vpc-domain)# peer-keepalive destination 10.92.30.245 source 10.92.30.246 vrf management         //配置 peer-keepalive
BDNETLAB-N5K-02(config-vpc-domain)# exit
```

第 4 步，完成 N5K-02 配置后进行结果验证。

```
BDNETLAB-N5K-02(config)# show vpc
Legend:
                (*) - local vPC is down, forwarding via vPC peer-link

vPC domain id                      : 50
Peer status                        : peer link not configured
vPC keep-alive status              : peer is alive        //当 N5K 交换机配置 keep-alive 后，状态正常
Configuration consistency status   : failed
Per-vlan consistency status        : failed
Configuration inconsistency reason : vPC peer-link does not exist
Type-2 consistency status          : failed
Type-2 inconsistency reason        : vPC peer-link does not exist
vPC role                           : none established
Number of vPCs configured          : 0
Peer Gateway                       : Disabled
Dual-active excluded VLANs         : -
Graceful Consistency Check         : Disabled (due to peer configuration)
Auto-recovery status               : Disabled
```

第 5 步，配置 2 台 N5K 交换机 vPC Peer Link 链路。

```
BDNETLAB-N5K-01(config)# feature lacp      //启用 LACP 特性
BDNETLAB-N5K-01(config)# interface e1/5-6    //使用 E1/5-6 端口作为 Peer Link 链路
BDNETLAB-N5K-01(config-if-range)# channel-group 200 mode active     //配置 Port Channel
BDNETLAB-N5K-01(config-if-range)# no shutdown
BDNETLAB-N5K-01(config)# interface port-channel 200
BDNETLAB-N5K-01(config-if)# switchport mode trunk       //配置 Port Channel 为 TRUNK 模式
BDNETLAB-N5K-01(config-if)# switchport trunk allowed vlan 10,20,130    //允许 VLAN 通过
BDNETLAB-N5K-01(config-if)# vpc peer-link    //将 Port Channel 配置为 Peer Link
Please note that spanning tree port type is changed to "network" port type on vPC peer-link.
This will enable spanning tree Bridge Assurance on vPC peer-link provided the STP Bridge Assurance
(which is enabled by default) is not disabled.
BDNETLAB-N5K-01(config-if)# no shutdown


BDNETLAB-N5K-02(config)# feature lacp
```

```
BDNETLAB-N5K-02(config)# interface e1/5-6
BDNETLAB-N5K-02(config-if-range)# channel-group 200 mode active
BDNETLAB-N5K-02(config-if-range)# no shutdown
BDNETLAB-N5K-02(config)# interface port-channel 200
BDNETLAB-N5K-02(config-if)# switchport mode trunk
BDNETLAB-N5K-02(config-if)# switchport trunk allowed vlan 10,20,130
BDNETLAB-N5K-02(config-if)# vpc peer-link
Please note that spanning tree port type is changed to "network" port type on vPC peer-link.
This will enable spanning tree Bridge Assurance on vPC peer-link provided the STP Bridge Assurance
(which is enabled by default) is not disabled.
BDNETLAB-N5K-02(config-if)# no shutdown
```

第 6 步，配置完 Peer Link 后查看状态。一定要保证 vPC 相关状态为 success，才能说明 N5K 交换机 vPC 配置正常。

```
BDNETLAB-N5K-01(config)# show vpc
Legend:
                (*) - local vPC is down, forwarding via vPC peer-link

vPC domain id                    : 50
Peer status                      : peer adjacency formed ok
vPC keep-alive status            : peer is alive
Configuration consistency status : success
Per-vlan consistency status      : success
Type-2 consistency status        : success
vPC role                         : primary
Number of vPCs configured        : 0
Peer Gateway                     : Disabled
Dual-active excluded VLANs       : -
Graceful Consistency Check       : Enabled
Auto-recovery status             : Disabled
vPC Peer-link status
---------------------------------------------------------------------
id    Port   Status Active vlans
--    ----   ------ -------------------------------------------------
1     Po200  up     -


BDNETLAB-N5K-02(config)# show vpc
Legend:
                (*) - local vPC is down, forwarding via vPC peer-link

vPC domain id                    : 50
Peer status                      : peer adjacency formed ok
vPC keep-alive status            : peer is alive
Configuration consistency status : success
Per-vlan consistency status      : success
Type-2 consistency status        : success
vPC role                         : secondary
Number of vPCs configured        : 0
```

```
Peer Gateway                              : Disabled
Dual-active excluded VLANs                : -
Graceful Consistency Check                : Enabled
Auto-recovery status                      : Disabled
vPC Peer-link status
---------------------------------------------------------------
id      Port    Status Active vlans
--      ----    ------ -------------------------------------
1       Po200   up      -
```

2. 将 N2K 交换机加入 vPC

实验环境中的 N2K 具有 4 个上行链路端口，可以分别连接至两台 N5K 交换机以实现冗余，同时可以再使用 vPC 提高链路带宽以及简化网络设计。

第 1 步，将 N2K 交换机使用的 Port Channel 加入到 vPC。

```
BDNETLAB-N5K-01(config)# interface e1/1-2
BDNETLAB-N5K-01(config-if-range)# channel-group 103
BDNETLAB-N5K-01(config-if-range)# no shutdown
BDNETLAB-N5K-01(config-if-range)# int e1/3-4
BDNETLAB-N5K-01(config-if-range)# channel-group 104
BDNETLAB-N5K-01(config-if-range)# no shutdown
BDNETLAB-N5K-01(config)# interface port-channel 103
BDNETLAB-N5K-01(config-if)# switchport mode fex-fabric
BDNETLAB-N5K-01(config-if)# fex associate 103
BDNETLAB-N5K-01(config-if)# vpc 103       //将 port channel 103 关联使用 vPC
BDNETLAB-N5K-01(config-if)# interface port-channel 104
BDNETLAB-N5K-01(config-if)# switchport mode fex-fabric
BDNETLAB-N5K-01(config-if)# fex associate 104
BDNETLAB-N5K-01(config-if)# vpc 104       //将 port channel 104 关联使用 vPC

BDNETLAB-N5K-01(config)# interface e1/1-2
BDNETLAB-N5K-01(config-if-range)# channel-group 103
BDNETLAB-N5K-01(config-if-range)# no shutdown
BDNETLAB-N5K-01(config-if-range)# int e1/3-4
BDNETLAB-N5K-01(config-if-range)# channel-group 104
BDNETLAB-N5K-01(config-if-range)# no shutdown
BDNETLAB-N5K-01(config)# interface port-channel 103
BDNETLAB-N5K-01(config-if)# switchport mode fex-fabric
BDNETLAB-N5K-01(config-if)# fex associate 103
BDNETLAB-N5K-01(config-if)# vpc 103
BDNETLAB-N5K-01(config-if)# interface port-channel 104
BDNETLAB-N5K-01(config-if)# switchport mode fex-fabric
BDNETLAB-N5K-01(config-if)# fex associate 104
BDNETLAB-N5K-01(config-if)# vpc 104
```

第 2 步，验证 N2K 交换机使用 vPC。

```
BDNETLAB-N5K-01(config)# show vpc
Legend:
                (*) - local vPC is down, forwarding via vPC peer-link
vPC domain id                     : 50
Peer status                       : peer adjacency formed ok
vPC keep-alive status             : peer is alive
Configuration consistency status  : success
Per-vlan consistency status       : success
Type-2 consistency status         : success
vPC role                          : primary
Number of vPCs configured         : 2
Peer Gateway                      : Disabled
Dual-active excluded VLANs        : -
Graceful Consistency Check        : Enabled
Auto-recovery status              : Disabled
vPC Peer-link status
---------------------------------------------------------------
id    Port    Status Active vlans
--    ----    ------ -----------------------------------------
1     Po200   up     -
vPC status
---------------------------------------------------------------
id      Port      Status Consistency Reason                Active vlans
------  --------  ------ ----------- ----------------      -----------
103     Po103     up     success     success               -         //N2K 交换机 port
channel 103 成功使用 vPC
104     Po104     up     success     success               -         //N2K 交换机 port
channel 103 成功使用 vPC
104448  Eth103/1/1 down* Not         Consistency Check Not -
                        Applicable  Performed

BDNETLAB-N5K-02(config)# show vpc
Legend:
                (*) - local vPC is down, forwarding via vPC peer-link
vPC domain id                     : 50
Peer status                       : peer adjacency formed ok
vPC keep-alive status             : peer is alive
Configuration consistency status  : success
Per-vlan consistency status       : success
Type-2 consistency status         : success
vPC role                          : secondary
Number of vPCs configured         : 2
Peer Gateway                      : Disabled
Dual-active excluded VLANs        : -
Graceful Consistency Check        : Enabled
Auto-recovery status              : Disabled
vPC Peer-link status
```

```
id    Port     Status  Active vlans
--    ----     ------  ------------
1     Po200    up      -
vPC status
------------------------------------------------------------
id    Port          Status  Consistency  Reason                    Active vlans
----  ------------  ------  -----------  ------------------------  ------------
103   Po103         up      success      success                   -
104   Po104         up      success      success                   -
104448 Eth103/1/1   down*   Not          Consistency Check Not     -
                            Applicable   Performed
```

5.4 虚拟化架构使用 N5K&N2K

在了解 Nexus 系列交换机的配置后，与传统物理交换机比较，考虑如何将其应用到虚拟化环境。本节介绍比较经典的设计以及应用。

5.4.1 N5K&N2K 连接设计

在连接服务器之前，网络核心交换机的连接设计非常重要，特别是使用 Nexus 各种特性的交换机。图 5-4-1 显示了经典的 N5K 与 N2K 连接设计，设计解释如下。

N5K-01 与 N5K-02 交换机之间使用 2 条 10GE 网络连接配置 Port Channel 并启用 vPC，任意一条链路出现故障，可自动切换到正常链路通信。

N2K-01 交换机每 2 条上行链路进行捆绑并启用 vPC，跨越两台 N5K 交换机形成 vPC，任意一条链路出现故障，可自动切换到正常链路通信，同时任意一台 N5K 交换机出现故障，可自动切换到正常的 N5K 交换机通信。

N2K-02 交换机每 2 条上行链路进行捆绑并启用 vPC，跨越两台 N5K 交换机形成 vPC，任意一条链路出现故障，可自动切换到正常链路通信，同时任意一台 N5K 交换机出现故障，可自动切换到正常的 N5K 交换机通信。

如果将 N2K 交换机更换为 ESXi 主机，配合 ESXi 主机网络 NIC Teaming 技术，能够解决在传统物理上使用 NIC Teaming 技术不能跨物理交换机配置 Port Channel 问题，更好地实现 ESXi 主机网络冗余。

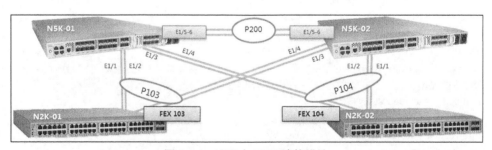

图 5-4-1　N5K 与 N2K 连接设计

5.4.2 ESXi 主机应用配置

本小节实验将在 ESXi 主机上网络配置标准交换机启用 NIC Teaming，并使用 IP 哈希作为负载均衡的方式来验证 VPC 技术在 ESXi 主机上的应用。

第 1 步，使用 Client 登录 ESXi 主机，IP 地址为 10.92.10.12 的 ESXi 主机配置有两个 10GE 的以太网接口（如图 5-4-2 所示）。

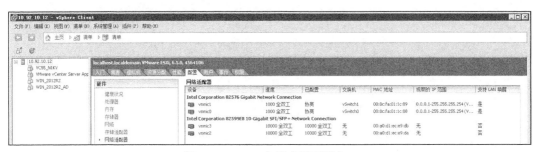

图 5-4-2　ESXi 主机应用配置之一

第 2 步，创建一个新的标准交换机使用，单击"添加网络"（如图 5-4-3 所示）。

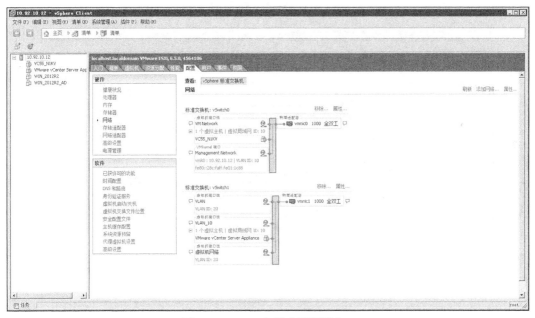

图 5-4-3　ESXi 主机应用配置之二

第 3 步，创建标准交换机，连接类型选择"虚拟机"（如图 5-4-4 所示），单击"下一步"按钮。

第 4 步，勾选 vmnic2 和 vmnic3 作为标准交换机的上行链路（如图 5-4-5 所示），单击"下一步"按钮。

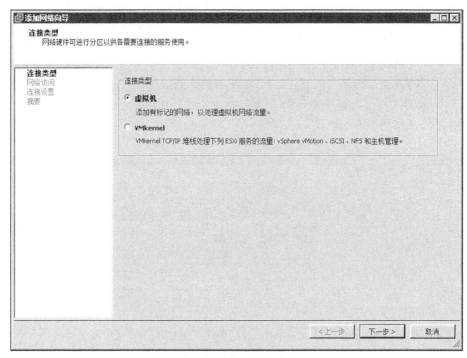

图 5-4-4　ESXi 主机应用配置之三

图 5-4-5　ESXi 主机应用配置之四

第 5 步，对标准交换机端口组进行命名，配置 VLAN ID（如图 5-4-6 所示），单击"下一步"按钮。

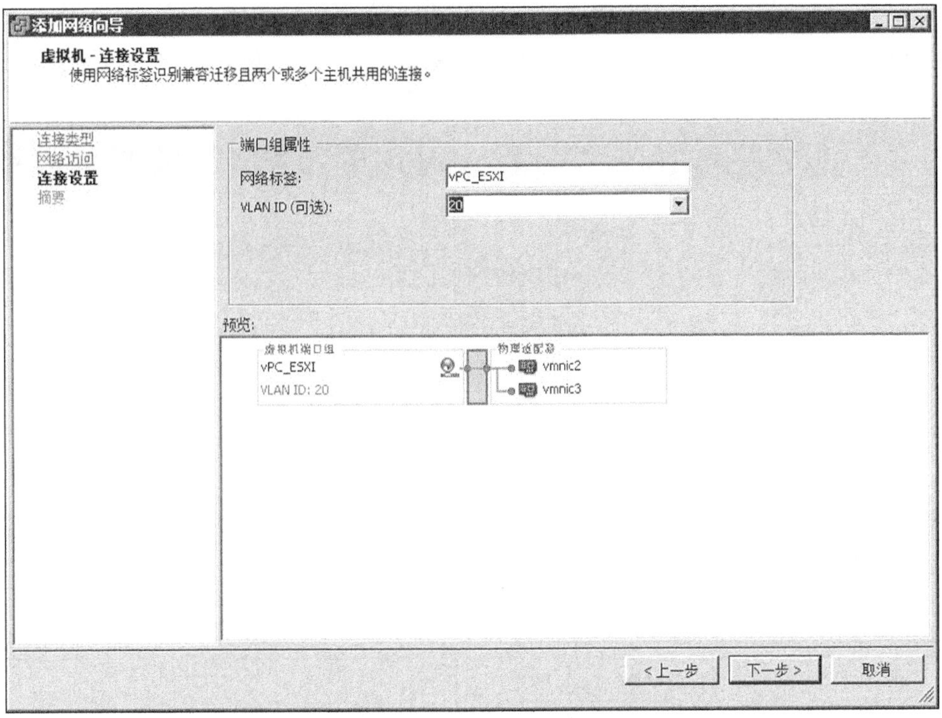

图 5-4-6　ESXi 主机应用配置之五

第 6 步，确认参数设置正确（如图 5-4-7 所示），单击"完成"按钮。

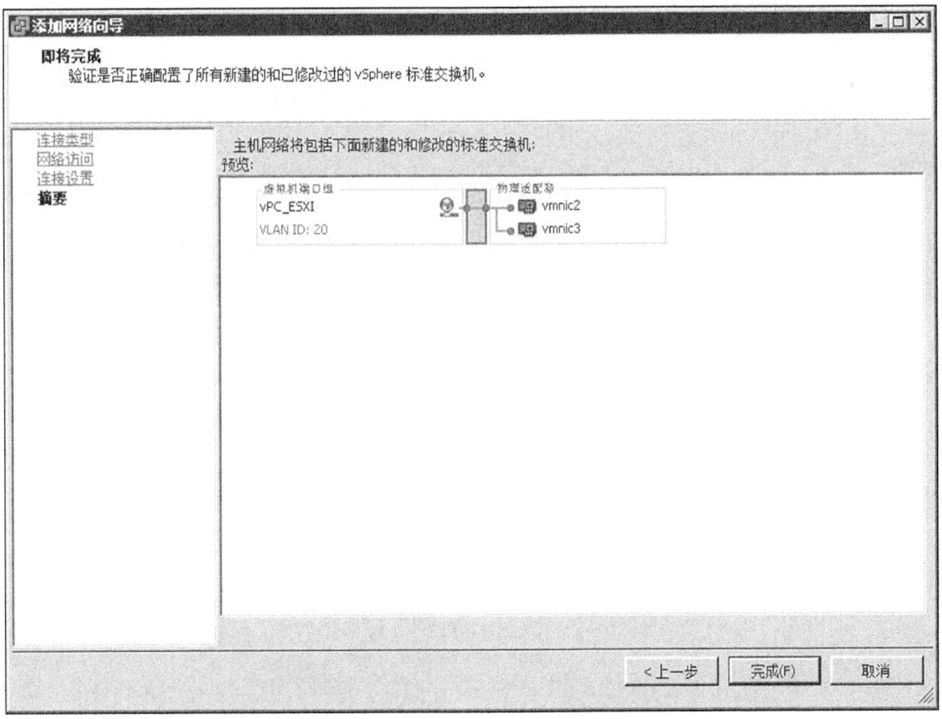

图 5-4-7　ESXi 主机应用配置之六

第 7 步，标准交换机创建完成（如图 5-4-8 所示）。

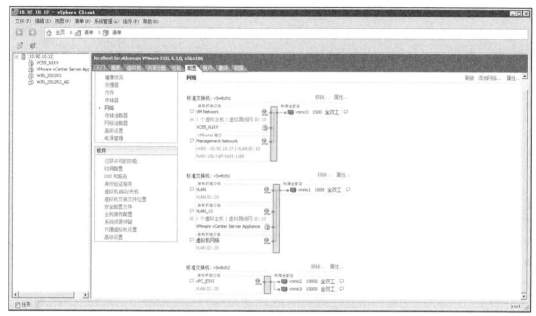

图 5-4-8　ESXi 主机应用配置之七

第 8 步，使用 Cisco 发现协议可以看到 vmnic2 连接到 N5K-01 交换机 E1/12 端口（如图 5-4-9 所示）。

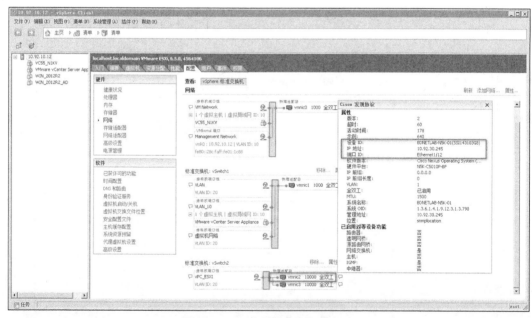

图 5-4-9　ESXi 主机应用配置之八

第 9 步，使用 Cisco 发现协议可以看到 vmnic3 连接到 N5K-02 交换机 E1/12 端口（如图 5-4-10 所示）。

5.4 虚拟化架构使用 N5K&N2K 253

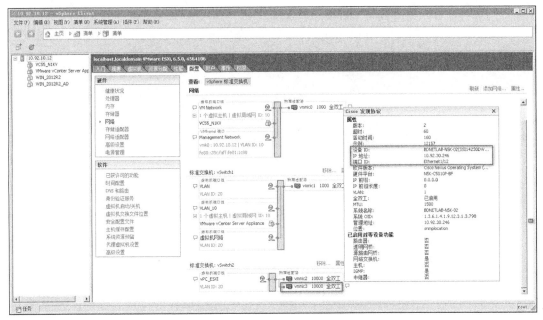

图 5-4-10 ESXi 主机应用配置之九

第 10 步，查看标准交换机端口组负载均衡方式，默认情况为端口 ID（如图 5-4-11 所示）。

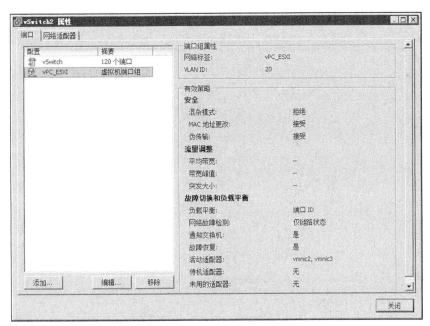

图 5-4-11 ESXi 主机应用配置之十

第 11 步，将端口组的负载均衡方式调整为"基于 IP 哈希的路由"（如图 5-4-12 所示），单击"确定"按钮。

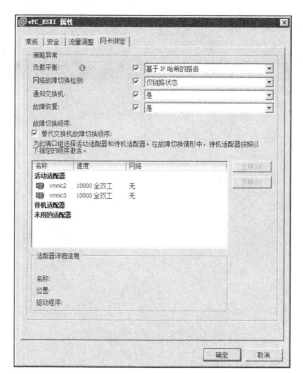

图 5-4-12　ESXi 主机应用配置之十一

第 12 步，确认端口组调整正确（如图 5-4-13 所示）。

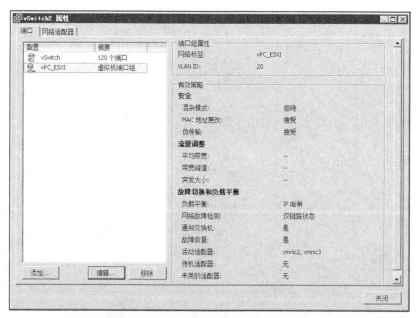

图 5-4-13　ESXi 主机应用配置之十二

第 13 步，将名为 WIN_2012R2 的虚拟机网络调整到新创建的 vPC_ESXi 端口组（如图 5-4-14 所示）。

5.4 虚拟化架构使用 N5K&N2K

图 5-4-14 ESXi 主机应用配置之十三

第 14 步，配置 N5K-01 交换机 E1/12 端口。

```
BDNETLAB-N5K-01(config)# interface e1/12
BDNETLAB-N5K-01(config-if)# channel-group 12    //端口启用 port channel
BDNETLAB-N5K-01(config-if)# no shutdown
BDNETLAB-N5K-01(config-if)# exit
BDNETLAB-N5K-01(config-if)# interface port-channel 12
BDNETLAB-N5K-01(config-if)# switchport mode trunk
BDNETLAB-N5K-01(config-if)# switchport trunk allowed vlan 10,20
BDNETLAB-N5K-01(config-if)# vpc 12    //将 port channel 12 关联使用 vPC
BDNETLAB-N5K-02(config-if)# no shutdown
BDNETLAB-N5K-01(config-if)# exit
```

第 15 步，配置 N5K-02 交换机 E1/12 端口。

```
BDNETLAB-N5K-02(config)# interface e1/12
BDNETLAB-N5K-02(config-if)# channel-group 12    //端口启用 port channel
BDNETLAB-N5K-02(config-if)# no shutdown
BDNETLAB-N5K-02(config-if)# exit
BDNETLAB-N5K-02(config)# interface port-channel 12
BDNETLAB-N5K-02(config-if)# switchport mode trunk
BDNETLAB-N5K-02(config-if)# switchport trunk allowed vlan 10,20
BDNETLAB-N5K-02(config-if)# vpc 12    //将 port channel 12 关联使用 vPC
BDNETLAB-N5K-02(config-if)# no shutdown
BDNETLAB-N5K-02(config-if)# exit
```

第 16 步，查看 N5K-01 交换机 port channel 信息以及 vPC 状态。

```
BDNETLAB-N5K-01(config)# show port-channel summary    //查看 port channel 信息
Flags:   D - Down         P - Up in port-channel (members)
         I - Individual   H - Hot-standby (LACP only)
         s - Suspended    r - Module-removed
         S - Switched     R - Routed
         U - Up (port-channel)
         M - Not in use. Min-links not met
```

```
-------------------------------------------------------------------
Group  Port-         Type     Protocol   Member Ports
       Channel
-------------------------------------------------------------------
12     Po12(SU)      Eth      NONE       Eth1/12(P)              //ESXi 主机使用的 E1/12 端口状态正常
103    Po103(SU)     Eth      NONE       Eth1/1(P)    Eth1/2(P)
104    Po104(SU)     Eth      NONE       Eth1/3(P)    Eth1/4(P)
200    Po200(SU)     Eth      LACP       Eth1/5(P)    Eth1/6(P)

BDNETLAB-N5K-01(config)# show vpc      //查看 vpc 信息
Legend:
                (*) - local vPC is down, forwarding via vPC peer-link

vPC domain id                    : 50
Peer status                      : peer adjacency formed ok
vPC keep-alive status            : peer is alive
Configuration consistency status : success
Per-vlan consistency status      : success
Type-2 consistency status        : success
vPC role                         : primary
Number of vPCs configured        : 99
Peer Gateway                     : Disabled
Dual-active excluded VLANs       : -
Graceful Consistency Check       : Enabled
Auto-recovery status             : Disabled

vPC Peer-link status
---------------------------------------------------------------
id    Port    Status  Active vlans
--    ----    ------  --------------------------------------------------
1     Po200   up      10,20,130

vPC status
----------------------------------------------------------------------------
id    Port    Status  Consistency  Reason                    Active vlans
----  ------  ------  -----------  ------------------------  -----------
12    Po12    up      success      success                   10,20       //vpc 状态正常
103   Po103   up      success      success                   -
104   Po104   up      success      success                   -
```

第 17 步，查看 N5K-02 交换机 port channel 信息以及 vPC 状态。

```
BDNETLAB-N5K-02(config)# show port-channel summary
Flags:  D - Down         P - Up in port-channel (members)
        I - Individual   H - Hot-standby (LACP only)
        s - Suspended    r - Module-removed
        S - Switched     R - Routed
```

```
              U - Up (port-channel)
              M - Not in use. Min-links not met
--------------------------------------------------------------------------
Group Port-        Type     Protocol   Member Ports
      Channel
--------------------------------------------------------------------------
12    Po12(SU)     Eth      NONE       Eth1/12(P)              //ESXi 主机使用的 E1/12 端口状态正常
20    Po20(SU)     Eth      LACP       Eth1/7(P)    Eth1/8(P)
103   Po103(SU)    Eth      NONE       Eth1/3(P)    Eth1/4(P)
104   Po104(SU)    Eth      NONE       Eth1/1(P)    Eth1/2(P)
200   Po200(SU)    Eth      LACP       Eth1/5(P)    Eth1/6(P)

BDNETLAB-N5K-02(config)# sh vpc   //查看 vpc 信息
Legend:
                (*) - local vPC is down, forwarding via vPC peer-link

vPC domain id                     : 50
Peer status                       : peer adjacency formed ok
vPC keep-alive status             : peer is alive
Configuration consistency status  : success
Per-vlan consistency status       : success
Type-2 consistency status         : success
vPC role                          : secondary
Number of vPCs configured         : 99
Peer Gateway                      : Disabled
Dual-active excluded VLANs        : -
Graceful Consistency Check        : Enabled
Auto-recovery status              : Disabled

vPC Peer-link status
---------------------------------------------------------------------
id   Port    Status Active vlans
--   ----    ------------------------------------------------
1    Po200   up     10,20,130

vPC status
---------------------------------------------------------------------
id    Port     Status Consistency Reason               Active vlans
----  -------- ------ ----------- -----------------    ------------
12    Po12     up     success     success              10,20    //vpc 状态正常
103   Po103    up     success     success              -
104   Po104    up     success     success              -
```

第 18 步，使用 VMware Remote Console 工具登录名为 WIN_2012R2 的虚拟机，查看获取的 IP 地址以及到网关的连通性（如图 5-4-15 所示）。

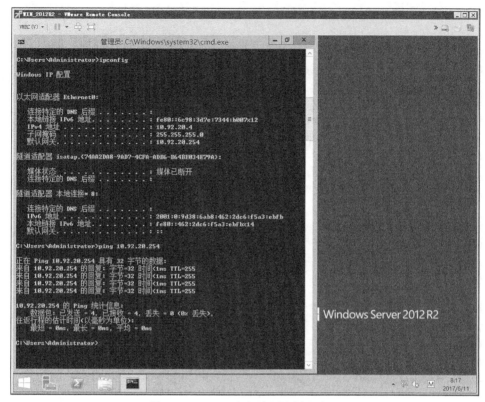

图 5-4-15　ESXi 主机应用配置之十四

第 19 步,模拟故障,将 vmnic2 对应的 N5K-01 交换机 E1/12 端口 shutdown(如图 5-4-16 所示)。

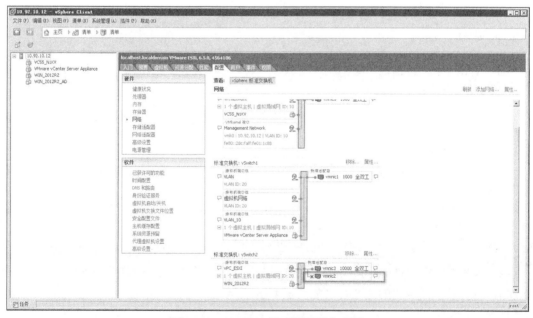

图 5-4-16　ESXi 主机应用配置之十五

第 20 步，使用 VMware Remote Console 工具登录名为 WIN_2012R2 的虚拟机，检查到网关的连通性，确认网络未发生任何变化（如图 5-4-17 所示）。通过这样的测试说明 vPC 实现负载均衡，不会因为线路端口而导致网络中断的情况。

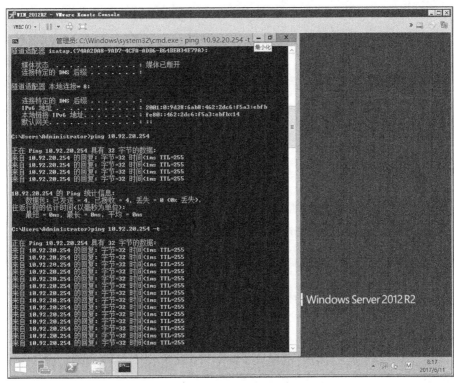

图 5-4-17　ESXi 主机应用配置之十六

第 21 步，查看 N5K-01 交换机 vPC 状态。因为将 E1/12 端口 shutdown，所以 N5K-01 交换机上的 vPC 状态为 down，无法转发数据。

```
BDNETLAB-N5K-01(config)# show vpc
Legend:
                (*) - local vPC is down, forwarding via vPC peer-link

vPC domain id                      : 50
Peer status                        : peer adjacency formed ok
vPC keep-alive status              : peer is alive
Configuration consistency status   : success
Per-vlan consistency status        : success
Type-2 consistency status          : success
vPC role                           : primary
Number of vPCs configured          : 99
Peer Gateway                       : Disabled
Dual-active excluded VLANs         : -
Graceful Consistency Check         : Enabled
Auto-recovery status               : Disabled
```

```
vPC Peer-link status
---------------------------------------------------------------------
id    Port    Status Active vlans
--    ----    ------ ------------------------------------------------
1     Po200   up     10,20,130


vPC status
---------------------------------------------------------------------
id     Port      Status Consistency Reason                    Active vlans
-----  --------  ------ ----------- ---------------------     ------------
12     Po12      down*  success     success                   -
103    Po103     up     success     success                   -
104    Po104     up     success     success                   -
```

第 22 步，查看 N5K-02 交换机 vPC 状态。因为 N5K-02 交换机上的 vPC 状态为 up，可以正常转发数据，所以虚拟机网络通信正常。

```
BDNETLAB-N5K-02(config)# show vpc
Legend:
                (*) - local vPC is down, forwarding via vPC peer-link

vPC domain id                     : 50
Peer status                       : peer adjacency formed ok
vPC keep-alive status             : peer is alive
Configuration consistency status  : success
Per-vlan consistency status       : success
Type-2 consistency status         : success
vPC role                          : secondary
Number of vPCs configured         : 99
Peer Gateway                      : Disabled
Dual-active excluded VLANs        : -
Graceful Consistency Check        : Enabled
Auto-recovery status              : Disabled

vPC Peer-link status
---------------------------------------------------------------------
id    Port    Status Active vlans
--    ----    ------ ------------------------------------------------
1     Po200   up     10,20,130


vPC status
---------------------------------------------------------------------
id     Port      Status Consistency Reason                    Active vlans
-----  --------  ------ ----------- ---------------------     ------------
```

12	Po12	up	success	success	10,20
103	Po103	up	success	success	-
104	Po104	up	success	success	-

5.5 本章小结

本章对 Cisco Nexus N5K&N2K 交换机进行了详细的介绍，包括常用的 FEX 技术、vPC 技术以及如何在虚拟化架构中使用这些技术。

对于生产环境来说，使用 Cisco Nexus N5K&N2K 系列交换机除了可以简化管理、提供一些新的特性外，还可以提供融合网络架构（后续章节会介绍），与传统交换机相比多了一些选择，其核心的 FEX 以及 vPC 技术目前在生产环境中大规模使用。

第 6 章 部署存储服务器

无论是传统数据中心还是 VMware vSphere 虚拟化数据中心，存储设备是数据中心正常运行的关键设备。作为企业虚拟化架构实施人员或者管理人员，必须考虑如何在企业生产环境构建高可用存储环境，以保证虚拟化架构的正常运行。IBM、HP、EMC 等专业级存储设备可以提供大容量、高容错、多台存储实时同步等功能，但相对来说价格昂贵，VMware 也推出了自己的软件定义存储 vSAN 解决方案。本章介绍生产环境中比较常用的 Open-E 存储服务器以及 DELL MD3620F 存储服务器的部署，完成存储服务器部署后，即可为 VMware vSphere 平台提供存储服务。

本章要点
- VMware vSphere 支持的存储介绍
- 部署使用 Open-E 存储服务器
- 部署使用 DELL MD3620F 存储服务器

6.1 VMware vSphere 支持的存储介绍

VMware vSphere 架构对于存储的支持是非常完善的，不仅支持传统存储，例如 FC SAN、iSCSI、NFS 等，而且提供最新的 VMware Virtual SAN 软件定义存储支持。

6.1.1 常见存储类型

从早期的 VMware vSphere 版本开始，VMware vSphere 支持的存储非常多，目前支持的类型如下。

1. 本地存储

传统的服务器都配置有本地硬盘，对于 ESXi 主机来说，这就是本地存储，也是基本存储之一。本地存储可以安装 ESXi，可以放置虚拟机等，但使用本地存储，虚拟化架构所有的高级特性，vMotion、HA、DRS 等功能均无法使用。

2. FC SAN 存储

FC SAN 是 VMware 官方推荐的存储之一，最大限度发挥虚拟化架构的优势，虚拟化架构所有的高级特性，vMotion、HA、DRS 等功能均可实现。同时，FC SAN 存储可以支持 ESXi 主机 FC SAN BOOT，缺点是需要 FC HBA 卡、FC 交换机、FC 存储支持，投入成本较高。

3. iSCSI 存储

相对 FC SAN 存储来说，iSCSI 是相对便宜的 IP SAN 解决方案，也被称为 VMware

vSphere 存储性价比最高的解决方案。可以使用普通服务器安装 iSCSI Target Software 来实现，同时支持 SAN BOOT 引导（取决于 iSCSI HBA 卡是否支持 BOOT）。部分观点认为，iSCSI 存储存在传输速率较慢、CPU 占用率较高等问题。如果使用 10GE 以太网络、硬件 iSCSI HBA 卡，可以在一定程度上解决此问题。

4. NFS 存储

NFS 是中小企业使用最多的网络文件系统，最大的优点是配置管理简单，虚拟化架构主要的高级特性，vMotion、HA、DRS 等功能均可实现。

6.1.2 FC SAN 存储介绍

FC 全称为 Fibre Channel，目前多数的翻译为"光纤通道"（包括 VMware vSphere 中文版本），实际上比较准确的翻译应为"网状通道"。FC 最早是作为 HP、SUN、IBM 等公司组成的 R&D 实验室中一项研究项目出现的，早期采用同轴电缆进行连接，后来发展到使用光纤连接，因此也就习惯将其称为光纤通道。

FC SAN 全称为 Fibre Channel Storage Area Network，中文翻译为"光纤/网状通道存储局域网络"，是一种将存储设备、连接设备和接口集成在一个高速网络中的技术。SAN 本身就是一个存储网络，承担了数据存储任务，SAN 网络与 LAN 业务网络相隔离，存储数据流不会占用业务网络带宽，使存储空间得到更加充分的利用以及安装和管理更加有效。

FC SAN 存储一般包括几个部分。

1. FC SAN 服务器

如果要使用 FC SAN 存储，网络中必须存在一台 FC SAN 服务器，用于提供存储服务。目前主流的存储厂商 IBM、HP、DELL 等都可以提供专业的 FC SAN 服务器，价格根据控制器型号、存储容量以及其他可以使用的高级特性来决定。存储厂商提供的 FC SAN 服务器一般来说价格比较昂贵。图 6-1-1 为 DELL 公司 DELL MD3800F 存储服务器，图 6-1-2 为华为 OceanStor 2200 存储服务器。

另外一种做法是购置普通的 PC 服务器、安装 FC SAN 存储软件以及 FC HBA 卡来提供 FC SAN 存储服务，这样的实现方式价格相对便宜。

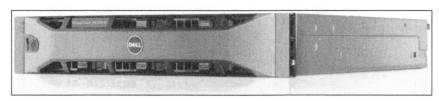

图 6-1-1　DELL MD3800F 存储服务器

图 6-1-2　华为 OceanStor 2200 存储服务器

2. FC HBA 卡

无论是 FC SAN 服务器还是需要连接 FC SAN 存储的客户端服务器，都需要配置 FC HBA 卡，用于连接 FC SAN 交换机。目前市面上常用的 FC HBA 卡分为单口（如图 6-1-3 所示）和双口（如图 6-1-4 所示）两种，也有满足特殊需求的多口 FC HBA 卡。比较主流的 FC HBA 卡速率为 4GB 或 8GB。16GB FC HBA 卡由于价格相对较高，因此使用相对较少。

图 6-1-3　单口 FC HBA 卡

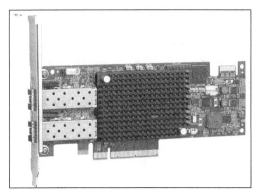

图 6-1-4　双口 FC HBA 卡

3. FC SAN 交换机

对于 FC SAN 服务器以及需要连接 FC SAN 存储的客户端服务器来说，很少会直接进行连接，大多数环境会使用 FC SAN 交换机，这样可以增加 FC SAN 的安全性并且提供冗余等特性。目前市面上常用的 FC SAN 交换机主要有博科（如图 6-1-5 所示）、Cisco（如图 6-1-6 所示）以及国产华为等品牌（如图 6-1-7 所示）。FC SAN 端口数和支持的速率需要参考 FC SAN 交换机相关文档。

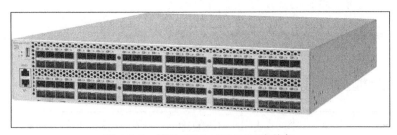

图 6-1-5　博科 6520　FC SAN 交换机

图 6-1-6　CiscoMDS 9124　FC SAN 交换机

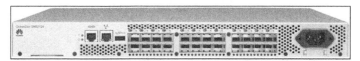

图 6-1-7　华为 OceanStor SNS2124　FC SAN 交换机

6.1.3　FCoE 介绍

FCoE，全称为 Fibre Channel over Ethernet，中文翻译为"以太网光纤通道"。FCoE 技术标准可以将光纤通道映射到以太网，可以将光纤通道信息插入以太网信息包内，从而让服务器至 SAN 存储设备的光纤通道请求和数据可以通过以太网连接来传输，而无需专门的光纤通道结构，从而可以在以太网上传输 SAN 数据。FCoE 允许在一根通信线缆上实现 LAN 和 FC SAN 通信，融合网络可以支持 LAN 和 FC SAN 数据类型，减少数据中心设备和线缆数量，同时降低供电和制冷负载，收敛成一个统一的网络后，需要支持的点也跟着减少，有助于降低管理负担。

FCoE 面向的是 10GE 以太网，其应用的优点是在维持原有服务的基础上，可以大幅减少服务器上的网络接口数量（同时减少了电缆、节省了交换机端口和管理员需要管理的控制点数量），从而降低功耗，给管理带来方便。FCoE 是通过增强的 10GE 以太网技术变成现实的，通常称为数据中心桥接（Data Center Bridging，DCB）或融合增强型以太网（Converged Enhanced Ethernet，CEE），使用隧道协议，如 FCiP 和 iFCP 传输长距离 FC 通信，但 FCoE 是一个二层封装协议，本质上使用的是以太网物理传输协议传输 FC 数据。

在生产环境使用 FCoE，一般来说需要使用比较特殊的交换机，不但能够承载 10GE 以太网流量，而且需要承载 FC 流量。图 6-1-8 为 Cisco 公司生产的 Nexus 5548P 系列交换机，可以通过扩展插槽增加 FC 模块，并直接通过命令的方式将端口修改为以太网接口或 FC 接口（后续章节将介绍）。

图 6-1-8　Cisco Nexus 5548P 交换机

6.1.4　iSCSI 存储介绍

iSCSI，全称为 Internet Small Computer System Interface，中文翻译为"小型计算机系统接口"。基于 TCP/IP 的协议，iSCSI 用来建立和管理 IP 存储设备、主机和客户机等之间的相互连接，并创建存储区域网络（SAN）。SAN 使得 SCSI 协议应用于高速数据传输网络成为可能，这种传输以数据块级别（block-level）在多个数据存储网络间进行。

iSCSI 存储的最大好处是能够在不增加专业设备的情况下，利用已有服务器以及以太网环境快速搭建。虽然其性能和带宽与 FC SAN 存储还有一些差距，但整体能为企业节省 30%～40% 的成本。相对 FC SAN 存储来说，iSCSI 存储是便宜的 IP SAN 解决方案，也被称为 Vmware vSphere 存储性价比最高的解决方案。如果企业没有预算 FC SAN 存储的费用，可以使用普通服务器安装 iSCSI Target Software 来实现 iSCSI 存储，iSCSI 存储同时支持 SAN BOOT 引导（取决于 iSCSI Target Software 以及 iSCSI HBA 卡是否支持 BOOT）。

需要注意的是，目前 85% 的 iSCSI 存储在部署过程中只采用 iSCSI Initiator 软件方式实施，对于 iSCSI 传输的数据将使用服务器 CPU 进行处理，这样会额外增加服务器 CPU 的使用率。所以，在服务器方面，使用 TCP 卸载引擎（TOE）和 iSCSI HBA 卡可以有效节省 CPU 的使用，尤其是对速度较慢但注重性能的应用程序服务器。

6.1.5　NFS 介绍

NFS，全称为 Network System，中文翻译为"网络文件系统"。是由 SUN 公司研制的 UNIX 表示层协议（pressentation layer protocol），能使使用者访问网络上别处的文件，就像在使用自己的计算机一样。NFS 是基于 UDP/IP 协议的应用，其实现主要是采用远程过程调用 RPC 机制，RPC 提供了一组与机器、操作系统以及低层传送协议无关的存取远程文件的操作。RPC 采用了 XDR 的支持。XDR 是一种与机器无关的数据描述编码的协议，以独立于任意机器体系结构的格式对网上传送的数据进行编码和解码，支持在异构系统之间数据的传送。

NFS 是 UNIX 和 Linux 系统中最流行的网络文件系统，Windows Server 也将 NFS 作为一个组件，添加配置后可以让 Windows Server 提供 NFS 存储服务。

6.2　部署使用 Open-E 存储服务器

在生产环境中，除使用专业级存储外，基于 Linux 内核的开源存储服务器也受到中小企业的喜爱，其构建方便快捷，管理成本相对低。Open-E 存储服务器在这个领域占据不小的市场份额。除 Open-E 以外，还存在其他的开源存储服务器，比较常见的有 Nexentastor、FreeNAS 等。本节介绍 Open-E 存储服务器的部署使用。

6.2.1　Open-E 存储服务器介绍

Open-E 存储服务器 V7 是一个基于 Linux 内核存储服务器软件，用于构建和管理数据的集中存储服务器，Open-E 存储服务器通过 VMware 和 Citrix Ready 认证。Open-E DSS V7 版本提供以下内置企业级功能。

- 容错透明故障转移高可用存储集群；
- 对于 NFS 和 iSCSI 主动故障转移；
- 对于 NFS 和 iSCSI 主被动切换；
- 基于域 HA 存储集群；
- 高可用集群维护模式；
- 可选使用虚拟存储设备（VSA）为超融合基础设施；
- 连续数据保护（同步卷复制）；
- 调度和基于快照的数据（文件）复制（异步复制）；
- 快照；
- 存储虚拟化；
- Hyper-V 集群的支持。

Open-E 存储服务器属于开源免费与商业收费集合产品。对于中小企业用户，可以考虑使用免费的版本，免费版本有功能以及存储容量的限制；也可考虑收费版本，收费版本提供完整的功能。需要特别说明的是，Open-E 存储服务器完整功能版本提供 FC 存储服务器功能，其他开源存储服务器无法提供 FC 存储功能，而专业 FC 存储服务器又价格不菲，相比之下，Open-E 存储服务器是一个不错的选择。关于 Open-E 存储服务器的详细介绍可以访问官方网站。

6.2.2 Open-E 存储服务器安装

Open-E 存储服务器的安装配置较为简单，访问 Open-E 下载最新的 ISO 安装文件即可，Open-E 支持物理服务器以及虚拟机安装方式。本小节的实战操作采用虚拟机安装方式进行。

第 1 步，挂载 Open-E 安装 ISO 镜像文件，运行安装程序，选择安装的版本（如图 6-2-1 所示），按"回车"键或等待自动引导。

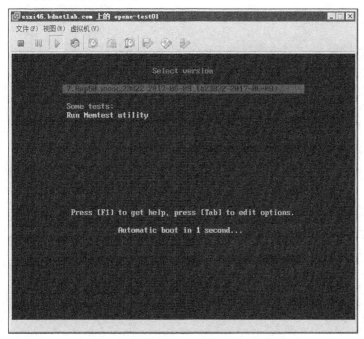

图 6-2-1　Open-E 存储服务器安装之一

第 2 步，选择引导的选项，选择"Launch software installer"开始安装 Open-E（如图 6-2-2 所示）。

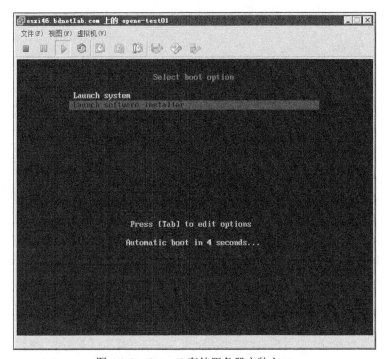

图 6-2-2　Open-E 存储服务器安装之二

第 3 步，开始加载安装程序（如图 6-2-3 所示）。

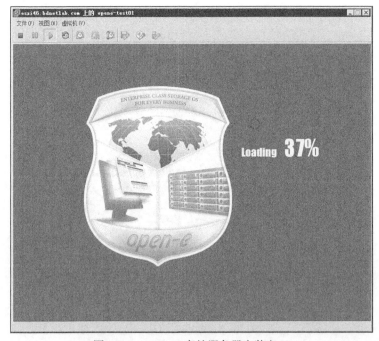

图 6-2-3　Open-E 存储服务器安装之三

第 4 步，安装程序提示 License agreement（如图 6-2-4 所示），选择"Agree"。

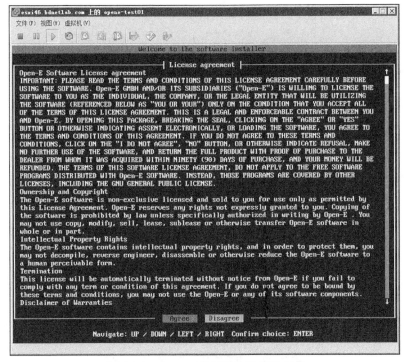

图 6-2-4　Open-E 存储服务器安装之四

第 5 步，选择 Open-E 安装的硬盘（如图 6-2-5 所示），选择"Select drive"。

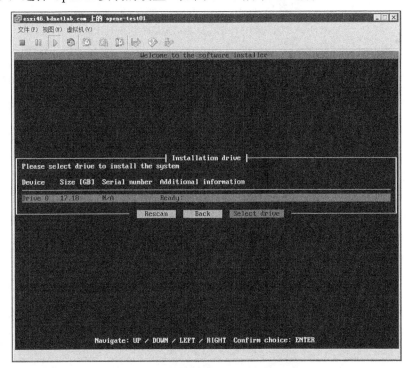

图 6-2-5　Open-E 存储服务器安装之五

第 6 步，系统对源文件进行检验后开始安装（如图 6-2-6 所示）。

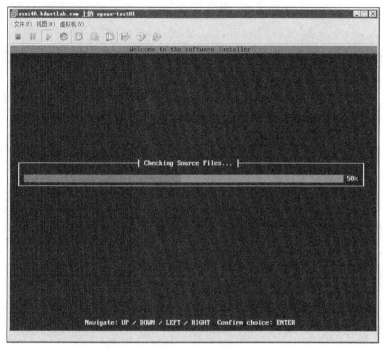

图 6-2-6　Open-E 存储服务器安装之六

第 7 步，整个安装过程时间很短，一般情况下几分钟即可完成 Open-E 安装。如图 6-2-7 所示，选择"Reboot"，重启服务器。

图 6-2-7　Open-E 存储服务器安装之七

第 8 步，Open-E 重新启动（如图 6-2-8 所示）。

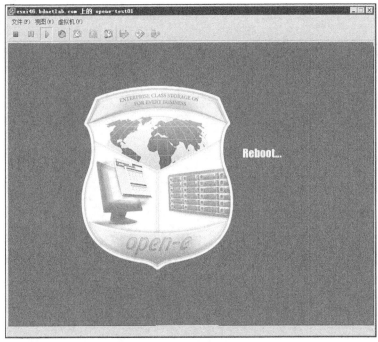

图 6-2-8　Open-E 存储服务器安装之八

第 9 步，Open-E 启动完成（如图 6-2-9 所示）。系统安装完成后 IP 地址为 192.168.0.220/24，如果需要修改 IP 地址，使用 Ctrl+Alt+N 键进行修改。

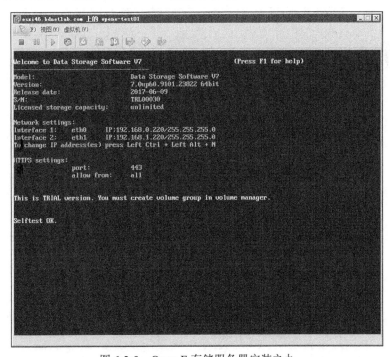

图 6-2-9　Open-E 存储服务器安装之九

第 10 步，选择"eth0"修改 IP 地址（如图 6-2-10 所示），选择"OK"。

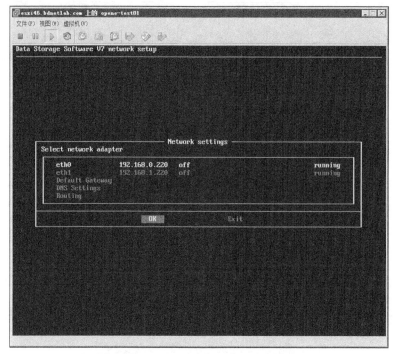

图 6-2-10　Open-E 存储服务器安装之十

第 11 步，进入 eth0 适配器配置选项，将光标移动到 IP 处（如图 6-2-11 所示），选择"Edit"。

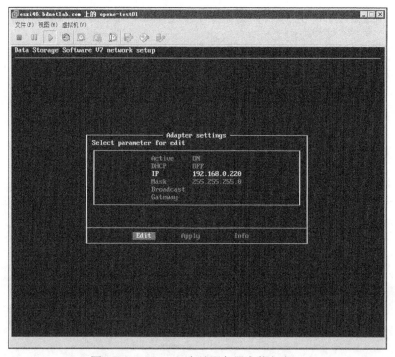

图 6-2-11　Open-E 存储服务器安装之十一

第 12 步，将 IP 修改为需要使用的 IP 地址（如图 6-2-12 所示），选择"OK"。

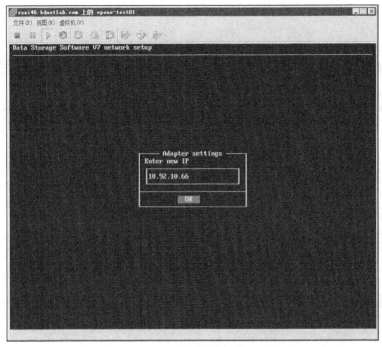

图 6-2-12　Open-E 存储服务器安装之十二

第 13 步，修改完成后回到 eth0 适配器配置选项，以同样的方式修改 Gateway 地址（如图 6-2-13 所示），修改完成后选择"Apply"。

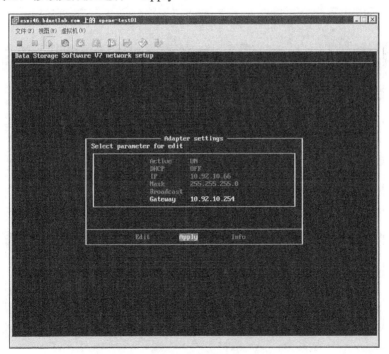

图 6-2-13　Open-E 存储服务器安装之十三

第 14 步，确认 IP 地址的修改（如图 6-2-14 所示），选择"OK"。

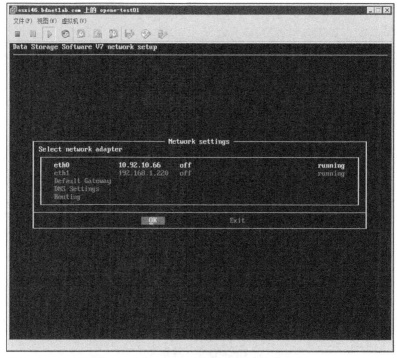

图 6-2-14　Open-E 存储服务器安装之十四

第 15 步，完成 IP 地址的修改（如图 6-2-15 所示）。eth1 适配器的 IP 地址暂不修改。

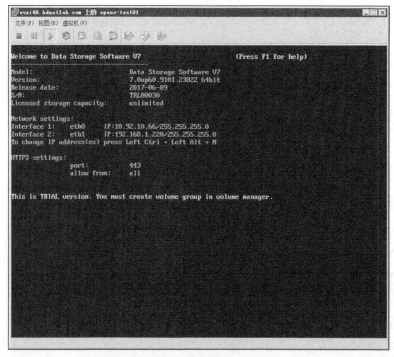

图 6-2-15　Open-E 存储服务器安装之十五

6.2 部署使用 Open-E 存储服务器　　275

第 16 步，使用浏览器访问 Open-E 存储服务器，系统提示需要 product key（如图 6-2-16 所示），可以根据需求申请不同的 product key。关于限制可以访问 Open-E 官方网站查看详细的说明。

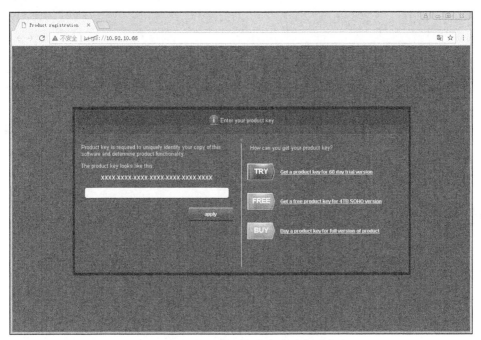

图 6-2-16　Open-E 存储服务器安装之十六

第 17 步，输入接收 product key 的邮箱（如图 6-2-17 所示）。

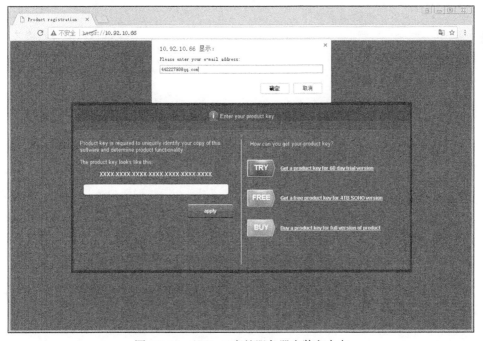

图 6-2-17　Open-E 存储服务器安装之十七

第 18 步，系统跳转到 Open-E 官方网站并提示 product key 已发送到邮箱（如图 6-2-18 所示）。

图 6-2-18　Open-E 存储服务器安装之十八

第 19 步，登录邮箱查看 product key（如图 6-2-19 所示）。

图 6-2-19　Open-E 存储服务器安装之十九

第 20 步，再次访问 Open-E 存储，输入获取的 product key（如图 6-2-20 所示），单击"Apply"。

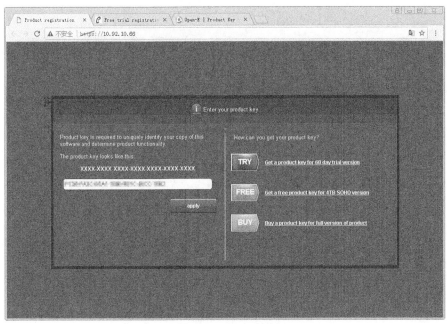

图 6-2-20　Open-E 存储服务器安装之二十

第 21 步，系统出现 Open-E Software License agreement（如图 6-2-21 所示），单击"Agree"接受。

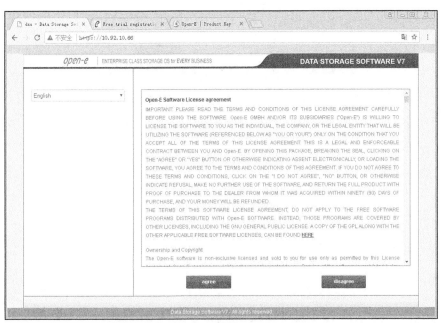

图 6-2-21　Open-E 存储服务器安装之二十一

第 22 步，进入 Open-E 登录界面，输入密码即可登录，初始完全访问密码为 admin（如

图 6-2-22 所示),单击"LOGIN"按钮。

图 6-2-22　Open-E 存储服务器安装之二十二

第 23 步,初始配置 Open-E 进入向导模式,可以根据熟悉程度确定是否使用,勾选"Do not show me the wizard at logon"(登录时不显示向导,如图 6-2-23 所示),单击"Cancel"按钮。

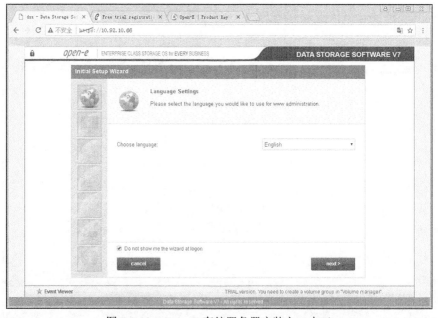

图 6-2-23　Open-E 存储服务器安装之二十三

第 24 步,登录 Open-E 存储服务器(如图 6-2-24 所示),可以看到安装的版本为 V7,使用的是 60 天评估无限制版本。

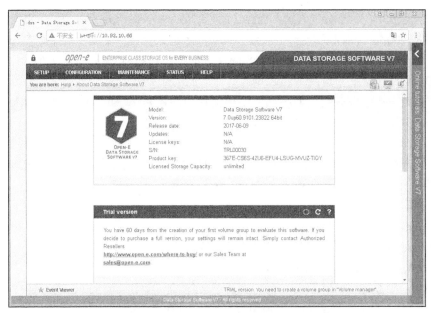

图 6-2-24　Open-E 存储服务器安装之二十四

6.2.3　生产环境部署 Open-E 存储服务器建议

由于开源的特性以及对常用 PC 服务器良好的兼容性，Open-E 存储服务器得到了不少企业的支持。对于在生产环境中部署使用 Open-E 存储服务器，作者根据多年的项目实施经验，给出一些建议，以便在设计过程中有一些参考。由于不同的项目存在不同的需求，因此也请读者不要原封不动地照搬。

1. Open-E 存储服务器硬件的选择

对于生产环境部署 Open-E 存储服务器，在一定程度是基于成本的考虑，但由于存储服务器在生产环境处于核心的地位，推荐使用大厂品牌服务器。大厂品牌服务器的稳定性能够得保证，如果需要使用大容量，推荐选择多硬盘位的服务器。当然，如果预算有限，也可以考虑 DIY 服务器，但在主板、电源等核心部件的选择上推荐选择大厂品牌。

其次，Open-E 存储服务器所使用的 CPU 以及内存要求并不高，入门级 INTEL E3 系列的 CPU 能够满足多数企业的存储服务器需求，内存推荐使用 8GB 或以上容量即可。

2. Open-E 存储服务器阵列的选择

无论是品牌服务器还是 DIY 服务器，基本上会使用阵列卡，Open-E 支持硬件 RAID 以及软件 RAID 两种模式。如果使用硬件 RAID，需要考虑 Open-E 是否包括其驱动；如果使用软件 RAID，则不考虑此问题。

确定使用硬件或软件 RAID 后，需要考虑 RAID 的模式。目前企业比较常用的是 RAID 5 以及 RAID 10 两种模式。对于写入较小，而对 I/O 读取比较多的环境，推荐使用 RAID 5 阵列，RAID 5 阵列对硬盘容量的利用率也相对较高；对于需要高性能以及高可靠性的环境来说，推荐使用 RAID 10 阵列，其同时使用了条带以及镜像，但可以使用的硬盘容量仅为阵列硬盘容量总和的一半。

确定 RAID 模式后，需要考虑硬盘类型，主要考虑的是硬盘转速以及容量。SAS 硬盘可以提供 1 万转/分或 1.5 万转/分的转速，优点是读写的效率高，缺点是容量小且价格贵；SATA 硬盘可以提供 7200 转/分或 1 万转/分的转速，优点是容量大且价格便宜，缺点是读写效率一般；SSD 硬盘的优点是读写效率远远超过 SAS 硬盘以及 SATA 硬盘，缺点是容量小且价格昂贵。

3. Open-E 存储服务器网络的选择

对于多数中小企业来说，其核心网络架构使用的还是传统的 1GE 网络。Open-E 存储服务器提供对 1GE 以及 10GE 网络的支持，同时支持基于 Linux 系统的端口绑定功能，以便实现负载均衡以及冗余。

对于生产环境使用 Open-E 作为 iSCSI 存储服务器来说，在预算成本允许的情况推荐使用 10GE 网络进行传输。经过测试，10GE iSCSI 效率远超 1GE iSCSI。如果预算成本不允许，建议使用 1GE 网络进行传输，同时使用端口绑定。

4. Open-E 存储服务器授权

目前 Open-E 存储服务器有 Open-E Jovian DSS、Open-E DSS V7、Open-E DSS V7 SOHO 三个版本。其中，Open-E Jovian DSS 用于大中型企业、云服务供应商等环境，基于 ZFS 存储系统；Open-E DSS V7 用于中小企业，基于 XFS 存储系统；Open-E DSS V7 SOHO 用于小微企业，基于 XFS 存储系统，目前可以申请 4TB 免费容量。

除 Open-E DSS V7 SOHO 外，其余两个版本均提供完整的企业级功能，但是需要购买授权，否则只能使用 60 天评估版本；Open-E DSS V7 SOHO 可以使用 4TB 免费容量，仅能使用 NFS、iSCSI 等基本存储功能，数据保护、VAAI、缓存、Fibre Channel 等企业级功能受限。

对于生产环境来说，需要考虑免费版本是否能够满足已有应用。如果不能满足，需要购买什么样的授权等。

6.3 部署使用 DELL MD 3620 存储服务器

Dell PowerVault MD 3200/3600 系列存储服务器可以优化数据存储体系结构、确保数据可用性，进而简化 IT 管理。这样，就可以将宝贵的资源释放出来，有助于降低成本，并使创新成为日常行为。DELL MD 3200/3600/1200 系列存储阵列支持 iSCSI、光纤通道或 SAS 技术。

6.3.1 DELL MD 3620F 存储服务器介绍

为了还原生产环境实际配置，实验环境使用一台 DELL PowerVault MD 3620F 存储服务器进行操作。在开始部署前，简要了解 DELL PowerVault MD 3 系列存储服务器的特点。

1. 支持 VMware VAAI，为虚拟化优化

VMware VAAI 将特定的存储操作从服务器转移到存储阵列。这提升了扩展能力和虚拟化的效能，因为操作在存储内直接进行，而不需要再通过主机，根据生产环境的测试，性能提升最多可达 46%。

2. 经验证的 MD 系列阵列，可提供持续的灵活性、可扩展性和价值

DELL PowerVault MD 3600F 和 MD 3620F 阵列将光纤通道引入了 MD 系列。凭借每控制器四个 8 Gb/s 端口，可提高数据密集型应用程序的吞吐量。同时，通过远程复制功能，

可扩展业务连续性。D 3600F 和 MD 3620F 秉承了经验证的 MD 系列阵列一贯的向后兼容性。通过 MD 1200 或 MD 1220 扩展盘柜，可以对 MD 3600 系列阵列进行扩展，最多可支持 64 台主机和 96 个驱动器。此外，多协议、多代式 MD Storage Manager 软件提供简单直观的阵列管理界面，可实现卓越的灾难恢复与业务连续性平台。

3. 超凡的整合与虚拟化

基于 iSCSI 的 PowerVault MD 3200i/MD 3220i 和 MD 3600i/MD 3620i 系列存储区域网络(SAN) 阵列可提供强大的系统功能、性能和灵活性，适用于存储整合与虚拟化部署。MD 3600i 系列采用紧凑的 2U 外形规格，支持高带宽、10GE 直接连接或 SAN 连接以及多达 64 台主机；MD 3200i 系列支持 1GE 直接连接或 SAN 连接以及多达 32 台物理服务器。

4. 高可用性存储

DELL PowerVault MD 3200 系列存储阵列采用 6 Gb/s SAS 技术，可在 Dell PowerEdge 服务器上提供卓越的灵活性、可扩展性和性能。双控制器机型可连接多达 4 台高可用性服务器或 8 台非冗余服务器，从而在混合虚拟化环境中实现均衡的性能。

5. 利用灵活的扩展模块，提高性能或容量

通过连接 PowerVault MD 1200 和 MD 1220 直连式 6 Gb/s SAS 扩展盘柜，可以轻松灵活地扩展 MD 3200/3600 系列阵列。PowerVault MD 1220 提供的速度、灵活性和可靠性可满足处理活动或动态数据存储的性能密集型应用程序的需求。这款高性能 2U 阵列可容纳多达 24 个 2.5 英寸 SAS 硬盘或 SAS 固态硬盘，提供高达 12 TB 的存储容量。PowerVault MD 1200 扩展盘柜非常适合移动大量数据并需要扩展存储容量的应用程序。这款 2U 阵列可容纳多达 12 个 3.5 英寸 SAS 硬盘或固态硬盘，提供高达 24 TB 的存储容量。可以在单个 PowerEdge RAID 控制器中结合使用这两种尺寸的驱动器。借助 PowerVault MD 系列存储阵列的模块化特性，用户可以轻松升级已有系统并改进当前的存储基础架构，以满足不断变化的业务需求。例如，要从标准单控制器迁移到高度可用的双控制器 PowerVault MD 3200/3600 系列阵列，只需加入第二个控制器模块即可，不需要进行数据迁移，也不会导致长时间停机。而且，这种模块化特性还支持对 Dell PowerEdge 服务器和 PowerVault MD 存储阵列进行简单而经济的扩展。

6. 带强大管理功能的多功能存储

DELL PowerVault MD 3200/3600/1200 系列阵列可支持多种功能强大、简单易用的智能存储管理软件。对于 MD 3200 和 MD 3600 系列阵列，多协议、多代式 Modular Disk StorageManager 集直观的界面、有向导指引的工具和基于任务的管理结构于一身，极大地降低了安装、配置、管理和诊断任务的复杂性。

此外，Dell OpenManage Storage Management 为所有 PowerVault MD 3200/3600/1200 系列阵列提供了基于标准的一整套主动信息管理工具和控制功能，在优化部署、运行状态监控、故障恢复和变更管理方面，这些工具和功能都是必不可少的。

DELL PowerVault MD 3200/3600/1200 系列阵列旨在提供通用性，让存储驱动器与不同层次的应用程序数据性能和空间要求相匹配，从而以最低的每 GB 成本最大限度地提高容量。为了满足大型机和服务器应用程序几乎呈线性增长的速度、性能和可靠性要求，可以在所有 MD 系列阵列中使用固态硬盘、10K 和 15K RPM SAS 以及近线 SAS 驱动器。如果需要优先考虑安全性，则可以在 PowerVault MD 3200 和 MD 3600 系列阵列中配备自我加密驱动器（SED）。

6.3.2 DELL MD 3620F 存储服务器基本操作

DELL 公司为简化存储服务器配置，专门开发有管理存储软件（可以访问 DELL 官方网站进行下载）。本小节介绍 DELL MD 3620F 存储服务器的基本操作。

第 1 步，下载 DELL PowerVault MD 存储软件并安装（如图 6-3-1 所示）。

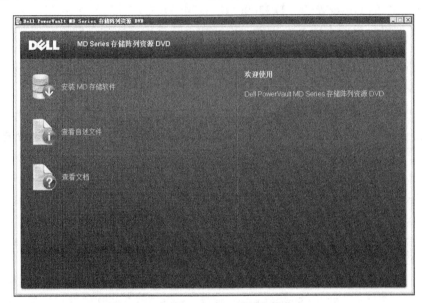

图 6-3-1 DELL MD 3620F 基本操作之一

第 2 步，在运行 DELL PowerVault MD 存储软件正式配置前，确认 DELL MD 3620F 存储电源已经打开同时控制器接入网络，由于存储软件刚安装完成，系统会提示需要进行初始化查找存储阵列（如图 6-3-2 所示），单击"否"按钮。

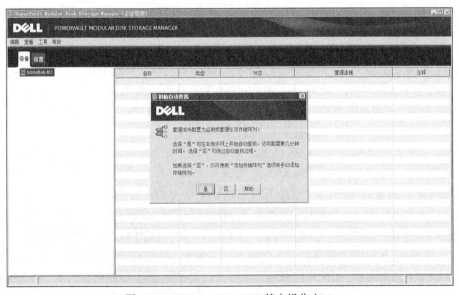

图 6-3-2 DELL MD 3620F 基本操作之二

6.3 部署使用 DELL MD 3620 存储服务器

第 3 步，点击"设置"选项，可以进入初始设置任务界面，一些基本的操作可以通过该界面完成（如图 6-3-3 所示），单击"添加存储阵列"。

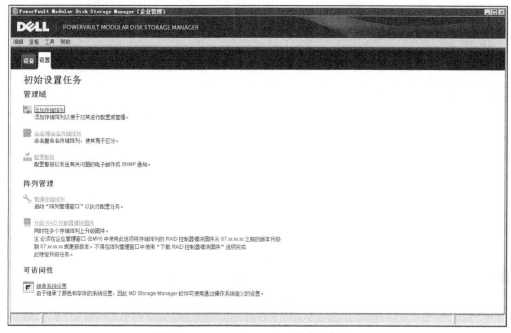

图 6-3-3　DELL MD 3620F 基本操作之三

第 4 步，选择添加存储阵列的方式，可以使用自动或手动两种进行。需要注意，客户端与存储阵列在本地子网，本小节实验选择"自动"（如图 6-3-4 所示），单击"确定"按钮。

第 5 步，系统开始自动查找存储阵列，需要一定的时间（如图 6-3-5 所示），单击"确定"按钮。

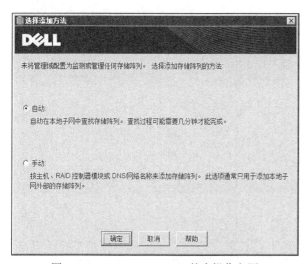

图 6-3-4　DELL MD 3620F 基本操作之四

图 6-3-5　DELL MD 3620F 基本操作之五

第 6 步，系统开始进行查找，右下角会显示进度条（如图 6-3-6 所示）。

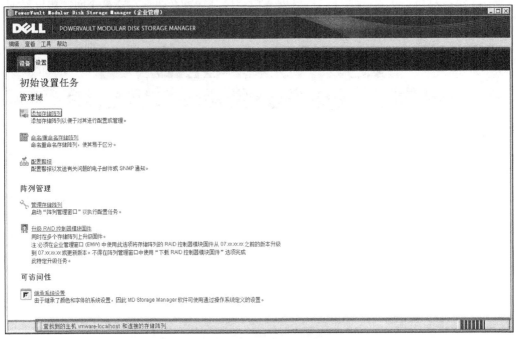

图 6-3-6　DELL MD 3620F 基本操作之六

第 7 步，系统已经查找到本地子网中的 DELL MD 3620F 存储阵列（如图 6-3-7 所示）。

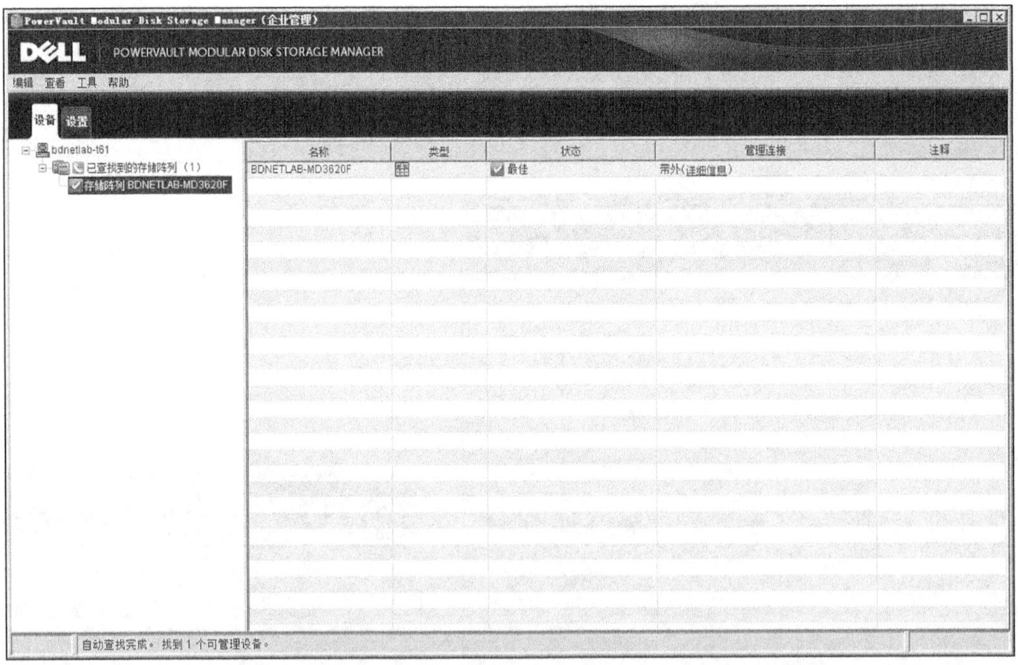

图 6-3-7　DELL MD 3620F 基本操作之七

第 8 步，通过管理连接窗口，可以看到存储阵列有两个控制器模块，分别为 0，1，每个控制器分配有一个 IP 地址（如图 6-3-8 所示）。

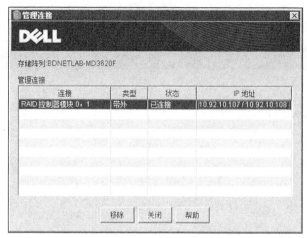

图 6-3-8 DELL MD 3620F 基本操作之八

第 9 步，在查找到的存储阵列上单击右键，选择"管理存储阵列"（如图 6-3-9 所示）。

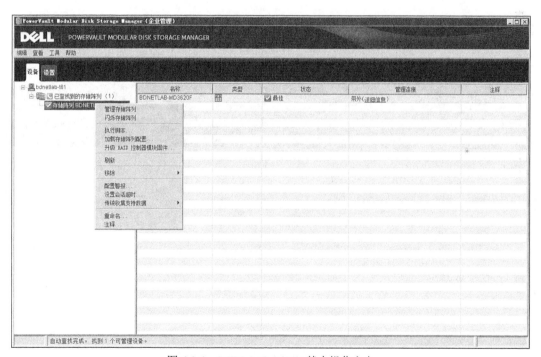

图 6-3-9 DELL MD 3620F 基本操作之九

第 10 步，进入存储阵列的管理界面，这也是日常使用的管理界面，通过"摘要"选项可以看到存储阵列的基本信息，包括管理软件版本、控制器版本、存储容量、物理磁盘等（如图 6-3-10 所示）。

第 11 步，单击"性能"选项，可以看到存储阵列的性能情况（如图 6-3-11 所示）。

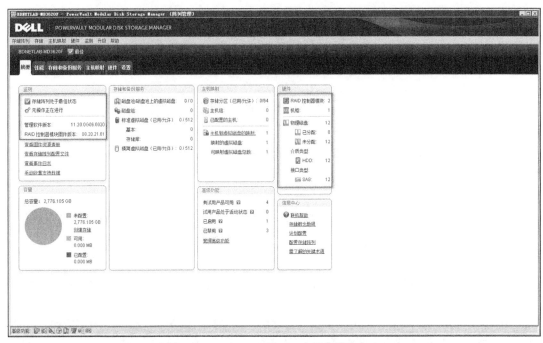

图 6-3-10　DELL MD 3620F 基本操作之十

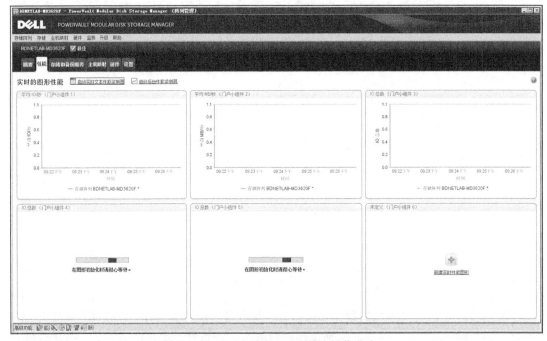

图 6-3-11　DELL MD 3620F 基本操作之十一

第 12 步，单击"存储和备份服务"选项，可以对存储阵列磁盘进行操作，后续的实验将介绍详细操作（如图 6-3-12 所示）。

6.3 部署使用 DELL MD 3620 存储服务器　　287

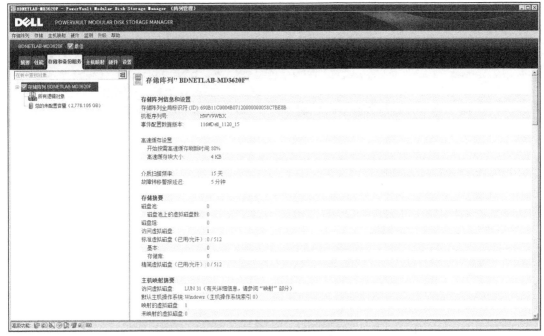

图 6-3-12　DELL MD 3620F 基本操作之十二

第 13 步，单击"主机映射"选项，可以对使用存储阵列的主机进行配置，后续的实验将介绍详细操作（如图 6-3-13 所示）。

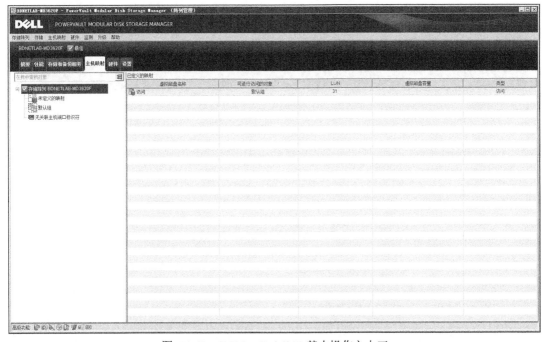

图 6-3-13　DELL MD 3620F 基本操作之十三

第 14 步，单击"硬件"选项，可以看到存储阵列物理磁盘以及控制器信息，实验所使用的存储配置有 4 块 15K SAS 硬盘、8 块 10K SAS 硬盘以及两个 RAID 控制器模块（如图 6-3-14 所示）。

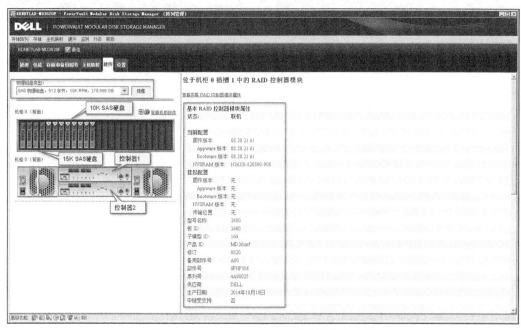

图 6-3-14　DELL MD 3620F 基本操作之十四

第 15 步，单击"设置"选项，可以对存储阵列基本参数以及高级功能等进行配置（如图 6-3-15 所示）。

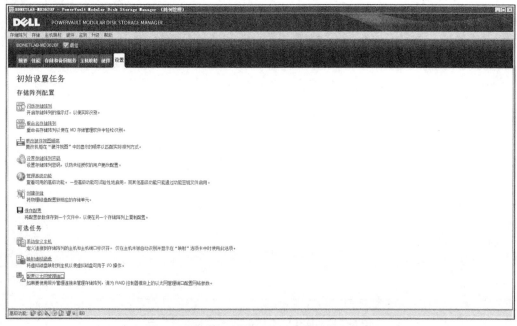

图 6-3-15　DELL MD 3620F 基本操作之十五

6.3.3 生产环境部署 DELL MD 存储服务器建议

对于生产环境使用 DELL MD 系列存储来说，其本身就是为企业生产环境设计的，建议主要在配件以及配置上。

1. 选择接口类型

DELL MD 系列存储提供 iSCSI 以及 FC 接口类型，直连存储不在本书的讨论范围中，其接口速率也决定其价格。10GE iSCSI 以及 16GB FC 接口价格相对较贵，应根据预算成本进行考虑以及选择。

2. 选择控制器

DELL MD 系列存储提供双控制器，双控制器的主要目的是避免单点故障。DELL MD 系列存储也支持单控制器运行。对于生产环境来说，推荐使用双控制器，当其中一个控制器故障后，备用的控制器会接替工作；如果使用单控制器，控制器一旦出现故障将导致存储不可用。

3. 选择存储硬盘

对于 DELL MD 系列存储来说，其内置阵列卡不存在阵列卡的选择，但需要考虑阵列的模式。目前企业比较常用的是 RAID 5 以及 RAID 10 两种模式。对于写入较小而对 I/O 读取比较多的环境，推荐使用 RAID 5 阵列，RAID 5 阵列对硬盘容量的利用率也相对较高；对于需要高性能以及高可靠的环境，推荐使用 RAID 10 阵列，其同时使用了条带以及镜像，但可以使用的硬盘容量仅为阵列硬盘容量总和的一半。

确定阵列模式后，需要考虑硬盘类型，主要考虑的是硬盘转速以及容量。SAS 硬盘可以提供 1 万转/分或 1.5 万转/分的转速，优点是读写的效率高，缺点是容量小且价格贵，专业级存储推荐使用 SAS 1.5 万转/分硬盘。需要特别注意的是，专业级存储存在"硬盘微码"的概念，意思是非存储厂商的硬盘插入到存储硬盘仓位后可能无法识别，这一点需要重视。

6.4 本章小结

本章对 VMware vSphere 支持的存储类型进行了介绍，同时介绍了比较常用的开源 Open-E 存储服务器的安装配置。为了还原生产环境的真实存储服务器环境，特别介绍了 DELL MD 3620F 存储以及基本操作，在后续的实验操作环节，我们将使用 Open-E 以及 DELL MD 3620F 进行配置，快速提升读者的动手能力。

第 7 章　部署使用 FC SAN 存储

Fibre Channel 这个概念于 1984 年提出，最初的构想是用来进行数据传输服务，但由于设计以及实现的复杂性，以太网得到快速发展，Fibre Channel 则被用于服务器与存储设备之间通信。经过多年的发展，Fibre Channel 已经成为企业和运营商数据中心存储网络最流行的协议。本章介绍 FC SAN 的基本概念以及如何在虚拟化平台上使用 FC SAN 存储。

本章要点
- FC SAN 存储介绍
- FCoE 存储介绍
- 配置 DELL MD 系列企业级存储
- 配置 Cisco MDS 系列企业级存储交换机
- 配置 ESXi 主机使用 FC 存储
- 配置 ESXi 主机使用 FCoE 存储

7.1　FC SAN 存储介绍

Fibre Channel 是由国际标准化组织 T11 制定、美国国家标准学会（ANSI）批准的标准，用于提供高速的数据传输服务。

7.1.1　FC SAN 基本概念

FC 全称为 Fibre Channel，目前多数的翻译为"光纤通道"（包括 VMware vSphere 中文版本），实际上比较准确的翻译应为"网状通道"。FC 最早是作为 HP、SUN、IBM 等公司组成的 R&D 实验室中一项研究项目出现的，早期采用同轴电缆进行连接，后来发展到使用光纤连接，因此也习惯将其称为光纤通道。

FC SAN 全称为 Fibre Channel Storage Area Network，中文翻译为"光纤/网状通道存储局域网络"，是一种将存储设备、连接设备和接口集成在一个高速网络中的技术。SAN 本身就是一个存储网络，承担了数据存储任务，SAN 网络与 LAN 业务网络相隔离，存储数据流不会占用业务网络带宽，使存储空间得到更加充分的利用以及安装和管理更加有效。

7.1.2　FC SAN 的组成

1. FC SAN 服务器

生产环境中使用 FC SAN 存储，必须存在一台 FC SAN 服务器，用于提供存储服务。

目前国外厂商如 IBM、HP、DELL 等，国内厂商如华为、浪潮等都可以提供专业的 FC SAN 服务器，价格根据控制器型号、存储容量以及其他可以使用的高级特性来决定。存储厂商提供的 FC SAN 服务器一般来说价格比较昂贵。

另外一种做法是购置普通的 PC 服务器、安装 FC SAN 存储软件以及 FC HBA 卡来提供 FC SAN 存储服务，这样的实现方式价格相对便宜。

2. FC HBA 卡

无论是 FC SAN 服务器本身还是需要连接 FC SAN 存储的服务器，都需要配置 FC HBA 卡，用于连接 FC SAN 交换机。目前市面上常用的 FC HBA 卡分为单口和双口两种，也有满足特殊需求的多口 FC HBA 卡。比较主流的 FC HBA 卡速率为 4GB 或 8GB，速率为 16GB 的 FC HBA 卡由于价格相对较高，因此使用相对较少。

3. FC SAN 交换机

对于 FC SAN 服务器以及需要连接 FC SAN 存储的服务器来说，很少直接进行连接，大多数环境会使用 FC SAN 交换机，这样可以增加 FC SAN 的安全性以及提供冗余等特性。目前市面上常用的 FC SAN 交换机主要有博科和 Cisco 两个品牌，FC SAN 端口数和支持的速率需要参考 FC SAN 交换机相关文档。

7.1.3　FC 协议介绍

FC 协议在 1988 年由博科等公司共同提出，最初的用途是扩展硬盘传输带宽。与 ISO 七层模型 TCP/IP 协议栈类似，但 FC 具有自己独立的协议集，分层如下。

1. FC-0（Physical Interface）

类似于以太网的物理层，主要定义传输介质和接口规范、电气规范等。物理接口定义 FC 使用的传输介质，常用的是光纤，也可以使用同轴电线等。不同的传输介质速度也不一样，常用的光纤传输可以达到 2GB、4GB 以及 8GB，目前市面有支持 16GB 的 FC 光纤设备。

2. FC-1（Encode/Decode）

与以太网一样，FC 协议在链路层也是成帧的，因为需要使用编码规则，FC 协议不使用其他协议的编码规则，自定义一个 8B 或 10B 的编码规则，FC-1 层通信所使用的编码规则。

3. FC-2（Framing Protocol）

类似于以太网的数据链路层，主要进行成帧和链路控制等功能。FC 协议规定了 24 字节的帧头，帧头包括寻址和传输保障功能，网络和传输都使用这 24 字节。

4. FC-3（Common Services）

通用服务层，T11 组织定义该层为保留使用，主要是为各个厂家提供扩展的功能。

5. FC-4（Transport Layer）

上层协议层，也可以称为映射层，定义了上层协议至 FC 协议的映射，上层协议包括 IP 协议、SCSI 协议、HIPPI 协议、ATM 协议等。

从 FC-0 到 FC-2 的三个协议层称为 FC-PH，也就是所谓的 FC 协议的"物理层"。其中，FC-2 是 FC 核心协议层，包括了 FC 传输最主要的协议和结构定义。

7.1.4　FC 拓扑介绍

FC 拓扑分为三种，在生产环境中主要使用的为两种。

1. 点到点

FC 存储与服务器之间互相直连，称为点到点拓扑结构。该拓扑一般情况下用于一些没有 FC 交换机的环境或特殊环境，如图 7-1-1 所示。

2. 交换式（类似于以太网中的星型网）

FC 存储以及服务器均连接至 FC 交换机，通过 FC 交换机相互进行访问。这是生产环境中常用的拓扑，如图 7-1-2 所示。

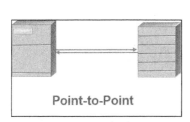

图 7-1-1　点对点拓扑

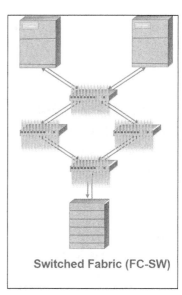

图 7-1-2　交换机拓扑

3. 仲裁环

在 FC 仲裁环拓扑中，FC 存储以及主机连接到一个共享的环，每个设备都与其他设备争用信道以进行 I/O 操作，在环上的设备必须被仲裁才能获得环的控制权。在某个给定的时间点，只有一个设备可以在环上进行 I/O 操作，如图 7-1-3 所示。目前该拓扑基本不使用。

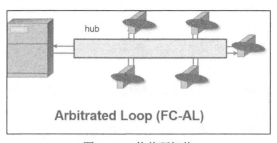

图 7-1-3　仲裁环拓扑

7.1.5　FC 端口介绍

由于 FC SAN 的特殊性，FC 存储与 FC 交换机以及服务器之间的通信通过各种类型的端口来实现，FC 网络根据不同的角色定义很多的接口（如图 7-1-4 所示）。

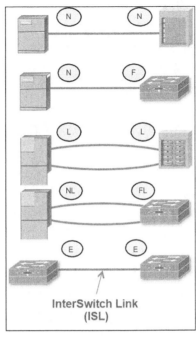

图 7-1-4　标准 FC 接口

1. N port（Node Port）

节点端口，所有存储服务器自身的端口都是 N 口。N port 可以点到点连接其他主机或盘阵，或者连接交换机。

2. F port（Fabric Port）

矩阵端口，所有 FC 交换机连接主机或者盘阵的端口都是 F 口。

3. L port（Loop Port）

环回端口，主要在仲裁环使用。

4. E port（Expansion Port）

扩展端口，主要用于连接其他 FC 交换机的链路（这个链路称为 ISL，inter switch link）

5. G port（General Port）

通用端口，兼容于 E、F、L 口，是一种自动协商的端口。

6. FL port（Fabric + Loop）

兼容于 F 和 L 口。

7. TE port（TRUNKing Expansion）

支持 VSAN trunking 的接口（支持 vsan 标记），支持传输 QOS 的属性，支持 fctrace 特性等。这样的链路又称为 EISL（Enhance Inter Switch Link）。

8. SD port（SPAN Destination Port）

在 FC 网络上进行抓包，将抓到的数据发送给分析器。这个分析器所在的端口称为 SD Port。

9. ST Port（SPAN Tunnel Port）

在 FC 网络上抓包，且做了 RPSAN 的情况下，抓到的数据包需要发送到另外一个交换机上,这个连接另外的交换机的端口叫作 ST Port。

10. Fx Port

兼容 F Port 和 FL Port。MDS 默认的接口的类型就是 Fx Port。

11. Auto Port

可以兼容 F 口、FL 口、E 口、TE 口等。

12. NP Port（Node Proxy Port）

在 NPV/NPIV 环境下进行使用。NP 口也可以支持 TRUNK。

7.1.6 WWN/FCID 介绍

FC 网络中每个设备都具有一个 WWN（World Wide Name），可以理解为以太网络的 MAC 地址。WWN 又分为 WWNN（World Wide Node Name）或 WWPN（World Wide Port Name）两种类型，WWNN 也可称为 NWWN，WWPN 也可称为 PWWN。

一个不可拆分的独立的设备有 WWNN，一个端口有 WWPN。例如 FC 交换机，有一个 WWNN，FC 交换机由若干端口，每个端口有一个 WWPN；一块多口 FC HBA，卡本身有一个 WWNN，每个端口有一个 WWPN，单口的 HBA 也是，不过只有一个 WWNN 和一个 WWPN。但主机没有 WWNN，因为卡和主机是可以分离的，单纯一个主机本身并不一定是 FC SAN 环境中的设备。图 7-1-5 获取的是 FC 交换机上 WWNN 以及 WWPN 的信息。

```
BDNETLAB-MDS9124-1# show flogi database
--------------------------------------------------------------------------------
INTERFACE     VSAN    FCID       PORT NAME               NODE NAME
--------------------------------------------------------------------------------
fc1/1         1       0x240200   21:01:00:1b:32:a2:01:9b 20:01:00:1b:32:a2:01:9b
fc1/3         1       0x240000   21:00:00:1b:32:11:0a:5a 20:00:00:1b:32:11:0a:5a
fc1/4         1       0x240100   21:00:00:1b:32:11:ae:5d 20:00:00:1b:32:11:ae:5d

Total number of flogi = 3.
```

图 7-1-5　FC 交换机 WWPN 以及 WWNN 信息

FC 协议使用 24 位地址进行数据转发，这个存储网络地址称为 FCID，可以理解为以太网络的 IP 地址，FCID 由 Domain ID（用于区分 FC 网络中每个 FC 交换机本身）、Area ID（用于区分同一台交换机上不同的端口组）、Port ID（用于区分一个区域中不同的端口）三部组成，每部分 8 位，总编址位数为 2^{24}。一台 FC 交换机以及该交换机连接的所有 N 端口都用一个相同的 Domain 表示；一台 FC 交换机上的 N 端口可以划分为多个 Area，用 Area ID 进行标识；一个 N 端口则通过 Port ID 来标识；FCID 是 FC 交换机分配给 N 端口的，E 端口以及 F 端口都没有 FCID。FCID 的申请和获取是通过 FLOGIN 过程过完成的，后面的章节将进行介绍。图 7-1-6 显示了 FC 交换机分配的 FCID 信息。

```
BDNELTAB-MDS9124-2# show flogi database
--------------------------------------------------------------------------------
INTERFACE     VSAN    FCID       PORT NAME               NODE NAME
--------------------------------------------------------------------------------
fc1/5         101     0xca0100   21:00:00:1b:32:82:5a:99 20:00:00:1b:32:82:5a:99
fc1/6         101     0xca0200   21:00:00:1b:32:95:54:23 20:00:00:1b:32:95:54:23
fc1/7         101     0xca0300   21:00:00:1b:32:8f:b7:2a 20:00:00:1b:32:8f:b7:2a
```

图 7-1-6　FC 交换机分配的 FCID 信息

7.1.7 FC 数据通信介绍

FC 网络与以太网很重要的区别在于 FLOGIN。以下是 FC 网络数据通信过程。

1. 节点注册

FC 存储服务器以及安装有 FC HBA 卡的其他服务器（例如 ESXi 主机）要通过 FLOGIN 过程向 FC 交换机进行注册，当分配到 FC 地址后，该节点设备才能与其他节点设备进行通信。

2. 分配 FC 地址

FC 网络中，设备标识都是通过 FC 地址来实现的。在访问节点设备之前，FC 交换机需要给节点设备分配 FC 地址。

3. 名称服务

FC 网络中，还有一个非常重要的 FCNS 服务器（也称为名称服务器），类似于网络中的 DNS 服务器。FC 上层协议是通过 WWN 来识别访问节点设备的，而在 FC 网络中则是使用 FC 地址，因此需要一个服务将 WWN 转换为 FC 地址，名称服务器提供的就是这个功能。名称服务器的原理为：节点设备 FC 交换机发送名称服务注册请求，并携带 WWN 和 FC 地址等名称服务信息，这些信息由 FC 交换机保存并维护，当某节点需要访问另一节点时，通过名称服务来查询 FC 交换机，就可以知道其对应的 WWN 和 FC 地址。每个 FC 交换机都是 FCNS 服务器，默认情况下，FCNS 同步到整个 FC 网络。

4. 数据交换

FC 网络数据交换是以帧为单位进行的，帧头中包含一个 FC 地址字段，FC 交换机中保存有一个转发表，当收到的数据帧需要进行数据转发时，FC 交换机以 FC 地址为选择路径依据，在转发表查找数据帧转发的下一跳，下一跳 FC 交换机收到数据帧后也进行相应的转发。这样，FC 数据帧将不断在 FC 网络中进行转发，直到目的节点。

7.1.8 VSAN 介绍

随着 SAN 技术的普及，SAN 已经不再是一个小网络或者仅仅依赖一个 SAN 网络连接众多的设备，这样可能造成稳定性以及安全性等多种问题。能否在 SAN 网络使用类似于以太网的 VLAN 技术呢？答案是可以的。为了将以太网中的 VLAN 技术延伸到 SAN 网络，Cisco 等厂商提出了 Virtual Storage Area Network（虚拟存储区域网络，也就是 VSAN），可以理解为以太网交换机上的 VLAN，其主要的功能为分割物理架构，将一台物理交换机分割成功多个 VSAN，不同 VSAN 成员不能通信，VSAN 已被 T11 委员会定义为行业标准。

图 7-1-7 简单表示了 VSAN 是如何实现分区的。在这个网络中，存储网络被分为 2 个 VSAN。如果不划分 VSAN，所有设备之间是互通的，而利用 VSAN 技术后，不同 VSAN 内的设备将无法通信。对于每一个 VSAN 来说，其本身就相当于是一个 SAN 网络，VSAN 之间是相互独立的设备，一个 VSAN 内的设备无法获得该 VSAN 之外其他 VSAN 和设备的信息，可以基于每一个 VSAN 配置 Domain ID 并运行主交换机选举。在网络层，每个 VSAN 也独立运行路由协议并独立维护路由转发表。

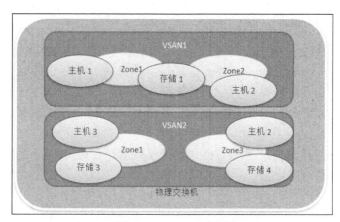

图 7-1-7　VSAN 介绍

通过 VSAN 可以把 FC 设备划分成相同或不同的 VSAN 成员，从而达到资源共享以及安全控制的目的。VSAN 的优点如下。

1. 提高安全性

每个 VSAN 之间相互隔离，在一定程度上提供了安全性。

2. 提高适合性

每个 VSAN 都可以独立运行并独立提供各种服务，不同 VSAN 可以使用重复的地址空间，在一定程度上增强了组网功能。

3. 灵活组网

通过配置 FC 交换机可以将接口加入到不同的 VSAN 并随时修改配置，不需要改变 SAN 网络的物理连接方式。

7.1.9　ZONE 介绍

ZONE 是为了保证其存储安全性而引入的一个重要功能。分割 VSAN，隔离同一 VSAN 里面不同成员；支持设备共享，允许多个 ZONE 包含同一个设备；提高安全性，同一个 VSAN 不同 ZONE 成员无法通信。

在 FC SAN 存储环境中，ZONE 和 LUN 映射是两个较为重要的概念。存储上，ZONE、LUN 映射需要与 FC 交换机功能配合起来使用，目的是使用不同的主机只能访问到不同的 LUN，从而更方便地进行存储资源的管理与调配。

VSAN 负责物理端口分配使用，在物理交换机上创建逻辑交换机，不同逻辑交换机是独立的，逻辑交换机之间完全隔离，并负责将物理端口分配到不同逻辑交换机里面去。不同 VSAN 之间设备不能直接通信，必须借助 IVR（Inter Vsan Routing）才能通信。ZONE 负责端口之间通信，只有同一个 ZONE 里面的设备才能互相通信，一个设备可以被多个 ZONE 共享。VSAN 和 ZONE 都可以增强 FC SAN 的安全性能，VSAN 还可以有效提供资源利用率。VSAN 和 ZONE 是互补关系，最佳操作是推荐同时使用。

7.1.10　NPV/NPIV 介绍

在 FC 交换环境中，有 NPV（N_Port ID Virtualization）以及 NPIV（N_Port Virtualization）

两个比较特殊的技术。

一般情况下，FC 交换机如果没有启用 NPV 或 NPIV，可以称为标准的 FC 交换机。那么为什么会使用 NPV 以及 NPIV 呢？主要有以下几个原因。

（1）FC 交换机最多只能有 239 个 DOMAIN ID，每一个 DOMAIN ID 表示一个标准模式的交换机。对于中小型环境来说，239 个 DOMAIN ID 是够用的；对于大型环境且 VSAN 较多的环境来说，DOMAIN ID 数量不够。

（2）数据中心虚拟机使用，虚拟接入要解决的问题是要把虚拟机的网络流量纳入传统网络交换设备的管理之中，让交换机可以识别来自同一链路不同虚拟机的流量，从而使 QoS 保证和安全隔离成为可能。NPIV 技术于 2006 年在 IBM 的 System z9 服务器上出现，允许管理员为每个 Linux 分区分配独立的虚拟端口名称（VirtualWWPN）。NPIV 允许单个 FCP 的端口（Port）在域名服务器注册多个 WWPN。每个虚拟的 WWPN 在登录后获得各自的端口编号（N-PortID）。每一对 WWPN/NPID 都可以用来作为 SAN 的分区（zoning）和 LUN 的屏蔽（masking）。在网络上的其他节点看来，虚拟的和物理的 WWPN 不存在任何区别。

简单来说，标准 FC 交换机使用 NPV 技术，能够有效控制 DOMAIN ID 和 FC 的管理点的数量，但是标准交换机在启用 NPV 技术后，其多种配置将被屏蔽，仅具有基本的 FC 交换功能。当 NPV 技术启用后，系统认为激活 NPV 的交换机为一个 F-port，上联交换机认为激活 NPV 的交换机也为一个 F-port。NPV 交换机上联交换机一般配置为 NPIV 模式，如图 7-1-8 所示。

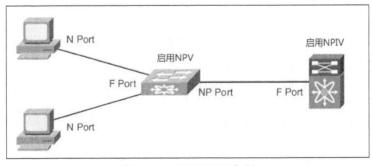

图 7-1-8　NPV/NPIV 介绍

7.2　FCoE 存储介绍

FCoE，全称为 Fibre Channel over Ethernet，中文翻译为"以太网光纤通道"。FCoE 技术标准可以将光纤通道映射到以太网，将光纤通道信息插入以太网信息包内，让服务器至 SAN 存储设备的光纤通道请求和数据通过以太网连接来传输，无需专门的光纤通道结构，从而可以在以太网上传输 SAN 数据。FCoE 允许在一根通信线缆上传输 LAN 和 FC SAN 通信，融合网络可以支持 LAN 和 FC SAN 数据类型，减少数据中心设备和线缆数量，同时降低供电和制冷负载，收敛成一个统一的网络后，需要支持的点也随着减少，有助于降低管理负担。

FCoE 面向的是 10GE 以太网，其应用的优点是在维持原有服务的基础上，可以大幅减少服务器上的网络接口数量（同时减少电缆、节省交换机端口和管理员需要管理的控制点数量），从而降低功耗，给管理带来方便。FCoE 是通过增强的 10GE 以太网技术变成现实的，通常称为数据中心桥接(Data Center Bridging，DCB)或融合增强型以太网(Converged Enhanced Ethernet，CEE)，使用隧道协议，如 FCiP 和 iFCP 传输长距离 FC 通信，但 FCoE 是一个二层封装协议，本质上使用的是以太网物理传输协议传输 FC 数据。

7.2.1 FCoE 存储组成

1. FC SAN 服务器

如果要使用 FC SAN 存储，网络必须存在一台 FC SAN 服务器，用于提供存储服务。目前主流的存储厂商 IBM、HP、DELL 等公司都可以提供专业的 FC SAN 服务器，价格根据控制器型号、存储容量以及其他可以使用的高级特性来决定。存储厂商提供的 FC SAN 服务器一般来说价格比较昂贵。

另外一种做法是购置普通的 PC 服务器、安装 FC SAN 存储软件以及 FC HBA 卡来提供 FC SAN 存储服务，这样的实现方式价格相对便宜。

2. CNA 适配器

FCoE 的适配器称为 CNA 卡，一般为 10GE（可以是光可以是电），也有双口 40G 的适配器。CNA 适配器可以虚拟出多个上行的链路，最多支持 512 个上行链路，用来解决服务器多个网络接口的需求。

3. FCoE 交换机

FCoE 交换机是一款特殊交换机，也可以称为融合网络交换机。这种交换机既支持以太网又支持 FC 网络，交换机配置有专用的芯片用于协议之间的转换。需要注意的是，FCoE 不支持 1GE 以太网，需要使用 10GE 以太网。第 6 章介绍过的 Cisco Nexus 5548UP 交换机就是一款融合网络交换机，支持 1GE\10GE 以太网，也支持 FC，同时还支持 FCoE，可以通过命令的方式将端口修改为以太网或 FC 接口类型。

7.2.2 FCoE 协议介绍

要实现 FC 与以太网之间的传输，需要使用 FCoE 协议。FCoE 包含两个协议。

1. 控制层协议

FIP 协议（FCoE 初始化协议）主要用来在以太网上处理 FC，例如 flogi 等。虽然 FCoE 运行在以太上，但是主要的控制层功能还是 FC 的控制层功能，以太网仅作为承载。

2. 数据层协议

FCoE 协议主要用于承载 FCoE 的流量，是一个传输协议。以太网络主要是通过以太类型（Ethernet Type）字段来区分不同的二层协议。FIP 协议的 ET 值为 0x8914，FCoE 协议的 ET 值为 0x8906。

FCoE 协议由 IETF 和 T11 组织共同颁布，主要体现在 FC-BB-5 标准中，在 IETF 中也有相应的标准号码。IETF 的标准和 T11 的标准负责不同的内容，T11 主要负责 FCoE 的信令通信（主要是 FC 协议的部分），IETF 主要负责底层以太网的承载。

在 FCoE 的协议定义（承载流量的数据层协议）中，802.1Q 字段是强制必须有的。换句话来说，它的以太网的底层链路必须是 TRUNK 模式而不能是 ACCESS 模式。

启用 FCoE 后，因为 FCoE 的帧较大，设备会自动调整接口 FCoE 队列的 MTU。如果启用了 FCoE 后，设备会将流量自动分为两个队列，具体是怎么分为两个队列的，后面将介绍。一个是普通以太网队列，一个是 FCoE 的队列，默认两个队列各占 50%带宽，可以手工调整。

FCoE 协议实际上就是将原来 FC 协议栈的 FC0、FC1 两个层次去除，换为以太网的物理层和数据链路层。

7.3 配置 DELL MD 系列企业级存储

第 6 章简要介绍了 DELL MD 系列企业级存储的操作界面，本节介绍 DELL MD 系列企业级存储的磁盘、映射以及其他常用的配置。

7.3.1 DELL MD 存储磁盘配置

对于任何存储来说，磁盘的配置都非常重要。本小节介绍 DELL MD 存储磁盘配置。

第 1 步，运行 DELL PowerVault MD 存储软件，选择"存储和备份服务"，系统会计算出目前未配置容量，显示的容量会根据创建的阵列变化，在"总的未配置容量"上单击右键，选择创建"创建磁盘组"（如图 7-3-1 所示）。

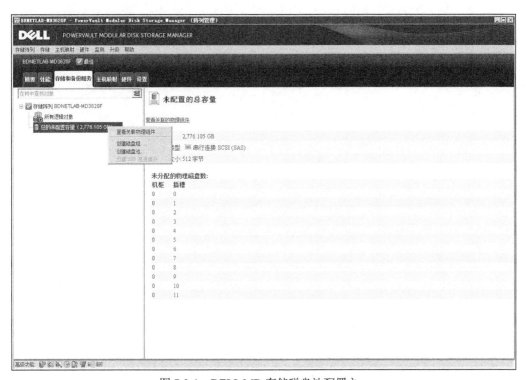

图 7-3-1　DELL MD 存储磁盘池配置之一

第 2 步，进入磁盘组创建向导，系统提示磁盘组是逻辑上组合在一起的物理磁盘以及未配置的容量（如图 7-3-2 所示），单击"下一步"按钮。

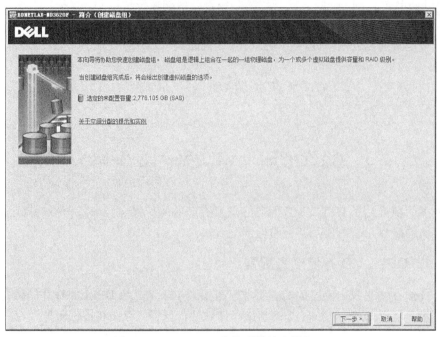

图 7-3-2　DELL MD 存储磁盘池配置之二

第 3 步，输入创建磁盘组的名称。DELL MD 存储物理磁盘有"自动"以及"手动"两种模式进行选择（如图 7-3-3 所示），先选择"自动（建议）"模式，单击"下一步"按钮。

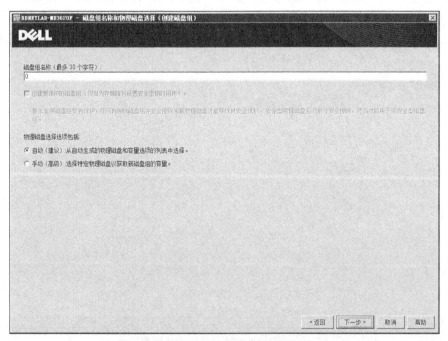

图 7-3-3　DELL MD 存储磁盘池配置之三

第 4 步，对于创建的磁盘组选择 RAID 级别，实验环境的 DELL MD 3620F 存储支持 RAID 0、RAID 1/10、RAID 5、RAID 6 几种 RAID 级别（如图 7-3-4 所示），可根据生产环境的需求进行选择。

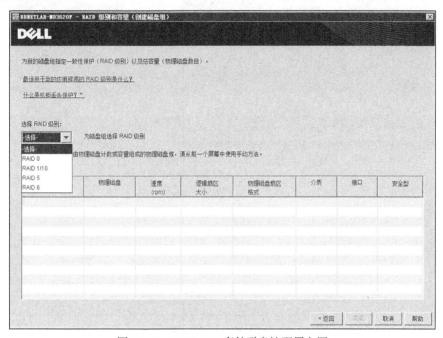

图 7-3-4　DELL MD 存储磁盘池配置之四

第 5 步，选择 RAID 级别为 RAID 1/10，DELL MD 存储会根据目前物理磁盘进行最佳组合（如图 7-3-5 所示）。

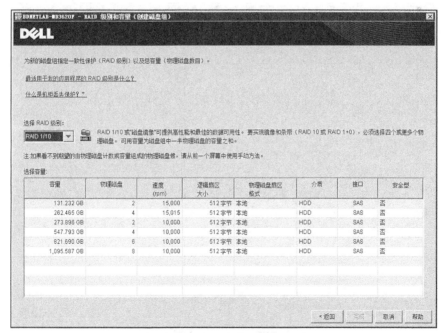

图 7-3-5　DELL MD 存储磁盘池配置之五

第 6 步，选择 RAID 级别为 RAID 5，DELL MD 存储会根据目前物理磁盘进行最佳组合（如图 7-3-6 所示）。

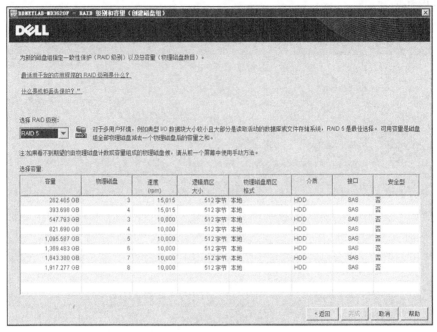

图 7-3-6　DELL MD 存储磁盘池配置之六

第 7 步，了解自动配置模式，再了解手动配置模式，选择"手动（高级）"模式（如图 7-3-7 所示），单击"下一步"按钮。

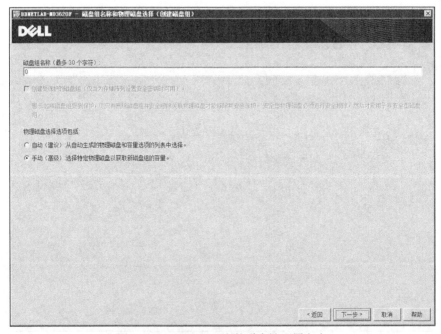

图 7-3-7　DELL MD 存储磁盘池配置之七

第 8 步，与自动模式一样，需要选择 RAID 级别，区别在于自动模式根据 RAID 级别自动添加磁盘，而手动模式需要手动选择磁盘参与的 RAID 级别（如图 7-3-8 所示），选择 0-3 号插槽的磁盘参与 RAID 1/10，单击"完成"按钮。

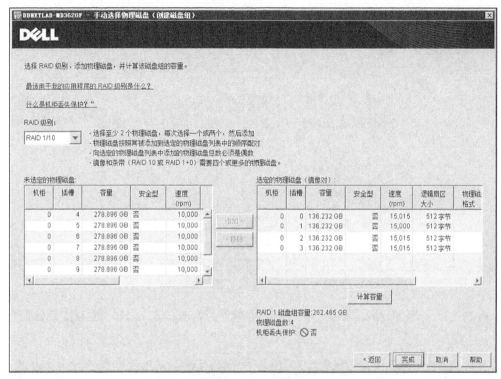

图 7-3-8　DELL MD 存储磁盘池配置之八

第 9 步，完成磁盘组的创建，系统提示必须创建一个虚拟磁盘，才能使用新磁盘组的容量（如图 7-3-9 所示），单击"否"按钮。

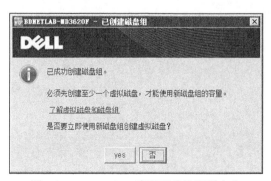

图 7-3-9　DELL MD 存储磁盘池配置之九

第 10 步，通过图 7-3-10 可以看到新创建的 RAID 10 磁盘组，容量为 262.465GB，右栏可以看到该磁盘组的详细信息。

图 7-3-10　DELL MD 存储磁盘池配置之十

第 11 步，开始创建虚拟磁盘，在"空闲容量"上单击右键，选择"创建虚拟磁盘"（如图 7-3-11 所示）。

图 7-3-11　DELL MD 存储磁盘池配置之十一

第 12 步，进入虚拟磁盘参数配置，新建一个 10GB 虚拟磁盘用于 ESXi12 主机（如图 7-3-12 所示），单击"完成"按钮。

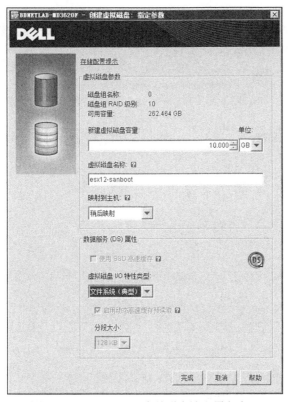

图 7-3-12　DELL MD 存储磁盘池配置之十二

第 13 步，完成虚拟磁盘创建（如图 7-3-13 所示），暂时不创建其他的虚拟磁盘，单击"否"按钮。

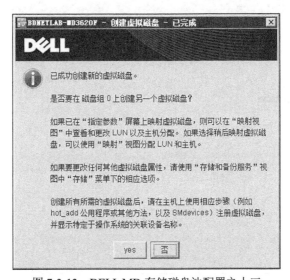

图 7-3-13　DELL MD 存储磁盘池配置之十三

第 14 步，单击名为"esx12-sanboot"的虚拟磁盘，通过图 7-3-14 可以看到，该虚拟磁盘正在初始化。

图 7-3-14　DELL MD 存储磁盘池配置之十四

第 15 步，按照上述方法再创建两个虚拟磁盘以便后续使用，虚拟磁盘分别名为"esx13-sanboot"、"esx-share"（如图 7-3-15 所示）。

图 7-3-15　DELL MD 存储磁盘池配置之十五

7.3.2 DELL MD 存储映射配置

完成虚拟磁盘的创建后，需要将虚拟磁盘映射到主机才能使用。本小节介绍如何将虚拟磁盘映射到 ESXi 主机。

第 1 步，在"主机映射"可以看到 7.3.1 小节创建的 3 块虚拟磁盘均处于未定义映射状态（如图 7-3-16 所示）。

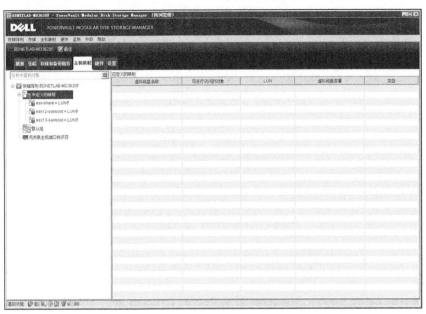

图 7-3-16　DELL MD 存储映射配置之一

第 2 步，在"默认组"上单击右键，选择"定义"→"主机"（如图 7-3-17 所示）。

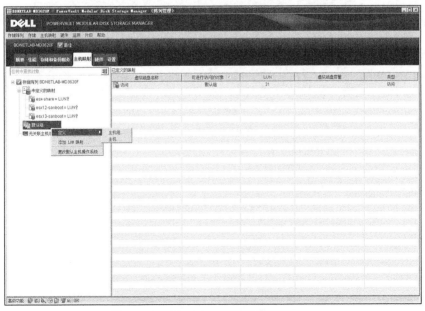

图 7-3-17　DELL MD 存储映射配置之二

第3步，指定映射虚拟磁盘的主机名，推荐使用便于识别的名称（如图7-3-18所示），单击"下一步"按钮。

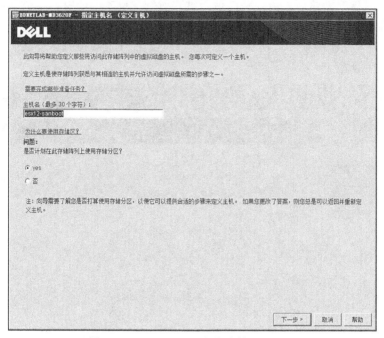

图7-3-18　DELL MD 存储映射配置之三

第4步，输入新建主机端口标识符，此处添加的是 ESXi 主机 FC HBA 卡 WWPN 地址，同时输入别名（如图7-3-19所示），单击"添加"按钮。

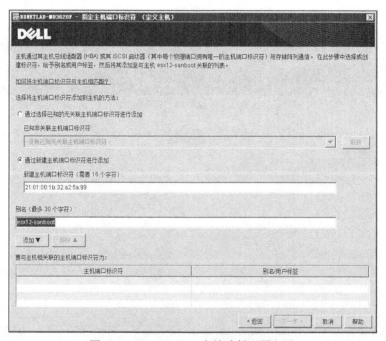

图7-3-19　DELL MD 存储映射配置之四

第 5 步，完成 ESXi 主机 FC HBA 卡 WWPN 地址添加（如图 7-3-20 所示），单击"下一步"按钮。

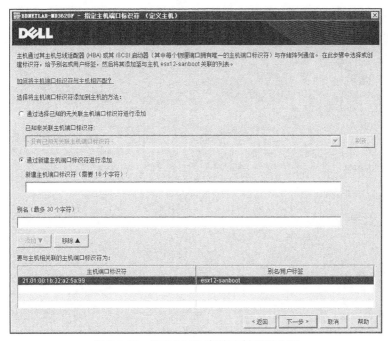

图 7-3-20　DELL MD 存储映射配置之五

第 6 步，选择主机的类型，DELL MD 系列存储支持主流的操作系统 Linux、VMware 以及 Windows（如图 7-3-21 所示），选择"VMware"，单击"下一步"按钮。

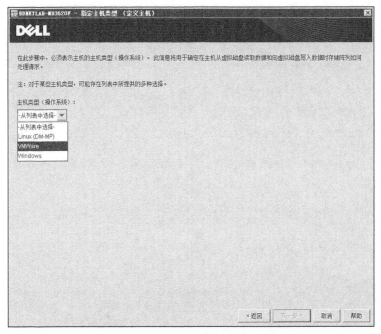

图 7-3-21　DELL MD 存储映射配置之六

第 7 步，确定主机能否与其他主机共享对相同虚拟磁盘的访问（如图 7-3-22 所示），选择"否"，单击"下一步"按钮。

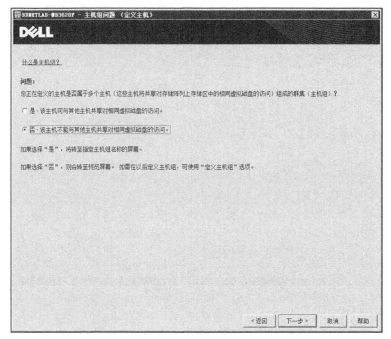

图 7-3-22　DELL MD 存储映射配置之七

第 8 步，确认虚拟磁盘到主机的映射参数设置正确（如图 7-3-23 所示），单击"完成"按钮。

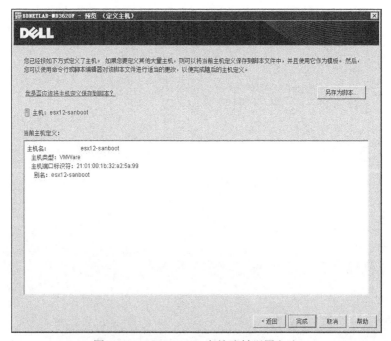

图 7-3-23　DELL MD 存储映射配置之八

第 9 步，完成虚拟磁盘到主机映射，系统提示是否定义另一个主机（如图 7-3-24 所示），单击"是"按钮。

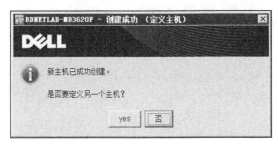

图 7-3-24　DELL MD 存储映射配置之九

第 10 步，按照上述方法定义 ESXi13 主机的映射（如图 7-3-25 所示）。

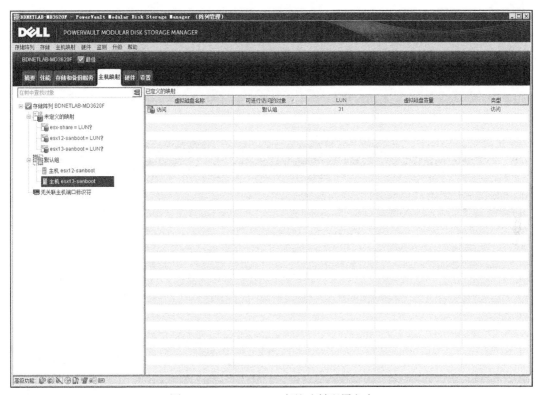

图 7-3-25　DELL MD 存储映射配置之十

第 11 步，为便于日常管理，推荐将定义的相同类型的主机添加到主机组，输入主机组名称（如图 7-3-26 所示）。

第 12 步，将 esx12-sanboot 以及 esx13-sanboot 主机添加到主机组（如图 7-3-27 所示），单击"确定"按钮。

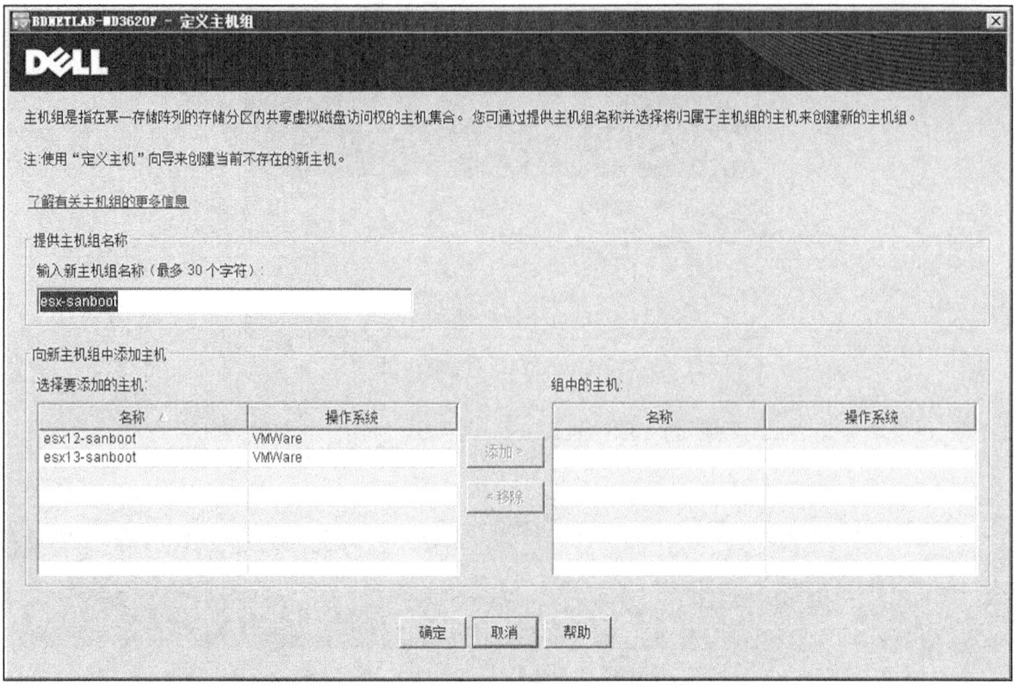

图 7-3-26 DELL MD 存储映射配置之十一

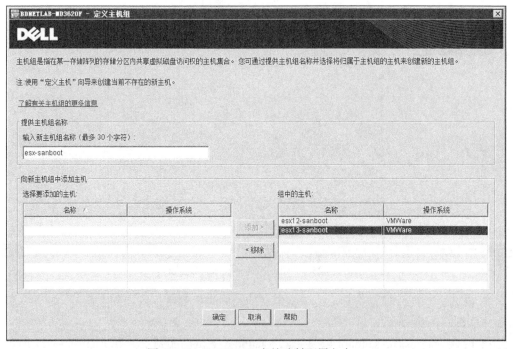

图 7-3-27 DELL MD 存储映射配置之十二

第 13 步，成功将主机添加到主机组（如图 7-3-28 所示）。

7.3 配置 DELL MD 系列企业级存储　313

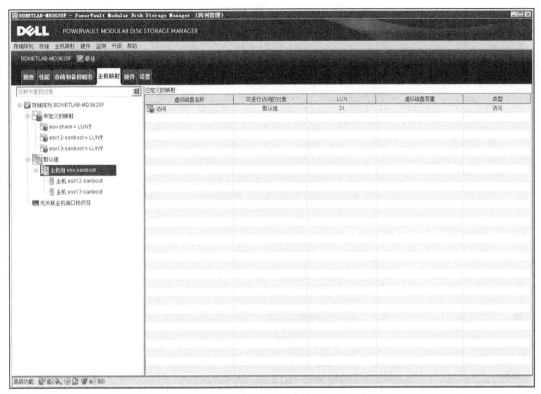

图 7-3-28　DELL MD 存储映射配置之十三

第 14 步，在创建的虚拟磁盘上单击右键，选择"添加 LUN 映射"（如图 7-3-29 所示）。

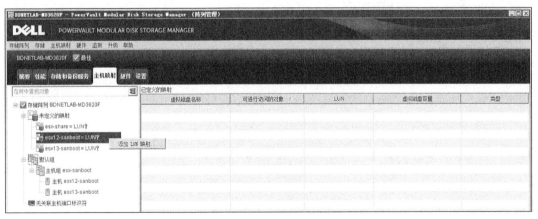

图 7-3-29　DELL MD 存储映射配置之十四

第 15 步，选择定义的主机以及虚拟磁盘（如图 7-3-30 所示），单击"添加"按钮。
第 16 步，完成主机 ESXi12-sanboot 虚拟磁盘 LUN 映射（如图 7-3-31 所示）。

314　第 7 章　部署使用 FC SAN 存储

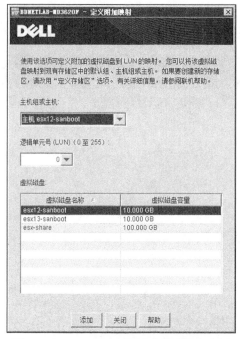

图 7-3-30　DELL MD 存储映射配置之十五

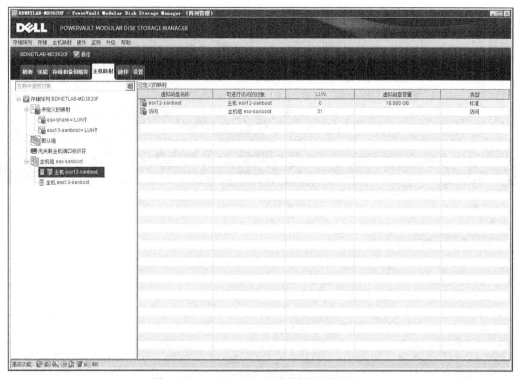

图 7-3-31　DELL MD 存储映射配置之十六

第 17 步，按照上述方法完成主机 ESXi13-sanboot 虚拟磁盘 LUN 映射（如图 7-3-32 所示）。

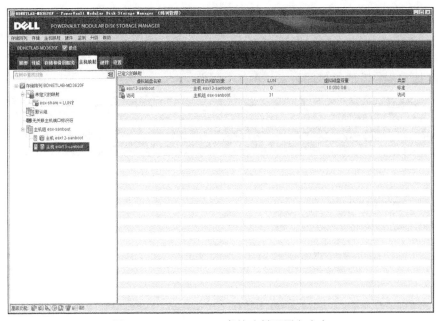

图 7-3-32　DELL MD 存储映射配置之十七

7.3.3　DELL MD 存储其他配置

对于 DELL MD 存储来说，除了基本的磁盘配置以及主机映射外，还有一些常用配置。本小节介绍一些常用配置。

1. 调整磁盘组 RAID 级别

对于 DELL MDS 存储来说，与传统存储不一样，可以在线调整 RAID 级别。在创建的磁盘组上单击右键，选择"更改"→"RAID 级别"，选择需要修改的 RAID 级别即可（如图 7-3-33 所示）。

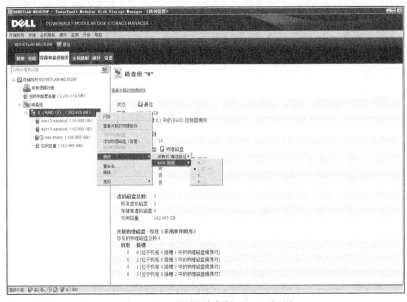

图 7-3-33　调整磁盘组 RAID 级别

2. 调整 RAID 控制器首选路径

生产环境一般使用双 RAID 控制器。一般情况下，插槽 0 的 RAID 控制器为首选，但也可能由于多种原因需要进行调整。在磁盘组上单击右键，选择"更改"→"所有权/首选路径"，选择需要修改的插槽即可（如图 7-3-34 所示）。

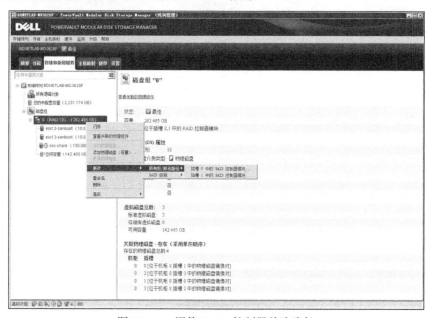

图 7-3-34　调整 RAID 控制器首选路径

3. 增加磁盘组容量

第 1 步，生产环境创建的磁盘组在使用一段时间后可能出现容量不足的情况，对于 DELL MD 系列存储来说，在磁盘组上单击右键，选择"添加物理磁盘（容量）"（如图 7-3-35 所示）。

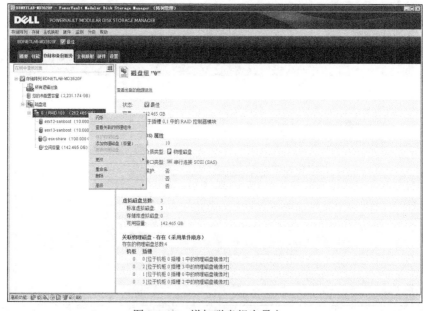

图 7-3-35　增加磁盘组容量之一

第 2 步，选择需要添加的磁盘。特别注意，不同的 RAID 级别，可添加的物理磁盘数量是不同的（如图 7-3-36 所示）。

图 7-3-36　增加磁盘组容量之二

4. 增加热备用物理磁盘

在生产环境中，对于创建的磁盘组使用热备用物理磁盘非常重要。热备用物理磁盘在 RAID 级别 1、RAID 级别 10、RAID 级别 5 或 RAID 级别 6 磁盘组上发生物理磁盘故障时提供额外的数据保护。如果在物理磁盘发生故障时有热备用物理磁盘，RAID 控制器模块会使用一致性数据将故障物理磁盘上的数据重新构建到热备用物理磁盘。物理上更换发生故障的物理磁盘之后，将发生一次回写操作，以从热备用物理磁盘将数据复制到更换后的物理磁盘。如果已指定热备用物理磁盘作为磁盘组的永久成员，则无需回写操作。对于磁盘组，机柜丢失保护和盘位丢失保护功能的可用性取决于构成该磁盘组的物理磁盘的位置。机柜丢失保护和盘位丢失保护功能可能会由于某个发生故障的物理磁盘和热备用物理磁盘的位置而不可用。要确保不影响机柜丢失保护和盘位丢失保护功能，必须更换故障物理磁盘以启动回写进程。如果使用自动配置功能，DELL MD 存储管理软件会为每 30 个特定介质类型和接口类型的物理磁盘创建一个热备用物理磁盘。如果手动配置存储阵列，则可创建在存储阵列的磁盘组中使用的热备用物理磁盘。热备用物理磁盘的容量必须等于或大于物理磁盘集中的最大物理磁盘的容量。建议为存储阵列中的每个物理磁盘集创建两个热备用物理磁盘。

第 1 步，DELL MD 存储增加热备用物理磁盘操作比较简单，选择"硬件"菜单中的"热备用容量"（如图 7-3-37 所示）。

318　第 7 章　部署使用 FC SAN 存储

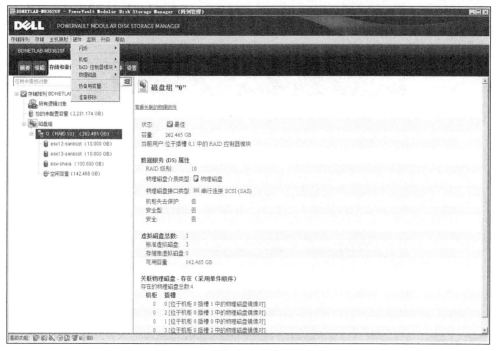

图 7-3-37　增加热备用物理磁盘之一

第 2 步，进入热备用物理磁盘选项，可以查看或自动分配物理磁盘，选择"查看/更改当前热备用容量"（如图 7-3-38 所示），单击"确定"按钮。

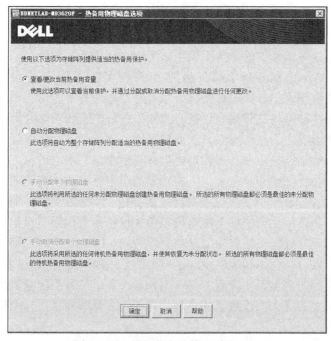

图 7-3-38　增加热备用物理磁盘之二

第 3 步，目前 DELL MD 存储创建有磁盘组且关联正在使用的虚拟磁盘，但没有分配热备用物理磁盘（如图 7-3-39 所示），单击"分配"按钮。

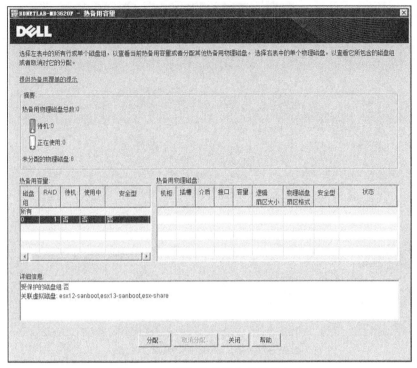

图 7-3-39　增加热备用物理磁盘之三

第 4 步，选择未使用的物理磁盘作为热备用磁盘（如图 7-3-40 所示），单击"确定"按钮。

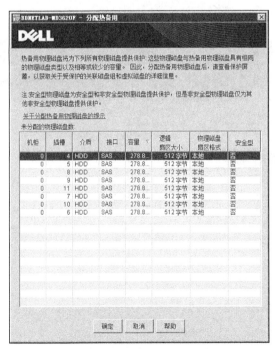

图 7-3-40　增加热备用物理磁盘之四

第 5 步，为磁盘组分配热备用磁盘完成（如图 7-3-41 所示）。

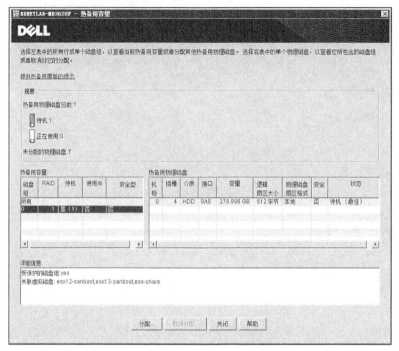

图 7-3-41　增加热备用物理磁盘之五

5. 查看存储硬件状态信息

（1）在生产环境日常维护中，需要关注存储硬件状态，如果某个硬件出现报错需要及时进行更换。选择"硬件"，可以看到存储常用的信息以及状态（如图 7-3-42 所示）；如果查看其他信息，单击"查看机柜组件"。

图 7-3-42　查看存储硬件状态之一

（2）打开机柜组件窗口，此窗口可以查看多种硬件信息（如图 7-3-43 所示）。

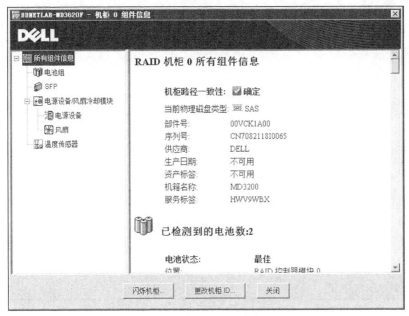

图 7-3-43　查看存储硬件状态之二

（3）查看存储电池组信息（如图 7-3-44 所示）。

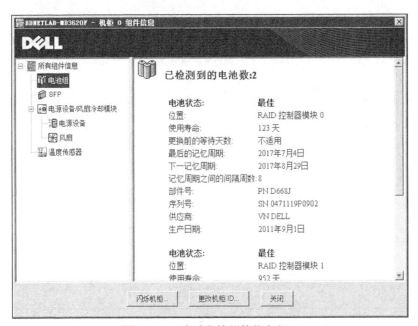

图 7-3-44　查看存储硬件状态之三

（4）查看存储 SFP 模块信息（如图 7-3-45 所示）。

图 7-3-45　DELL MD 查看存储硬件状态之四

（5）查看存储电源设备等信息（如图 7-3-46 所示）。

图 7-3-46　查看存储硬件状态之五

（6）查看存储电源设备信息（如图 7-3-47 所示）。

7.3 配置 DELL MD 系列企业级存储　　323

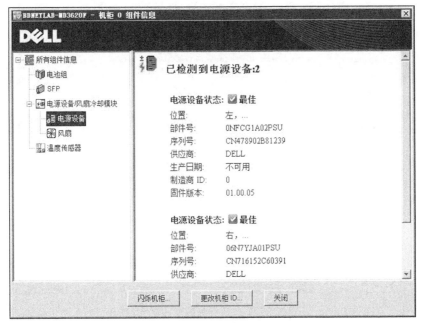

图 7-3-47　查看存储硬件状态之六

（7）查看存储风扇信息（如图 7-3-48 所示）。

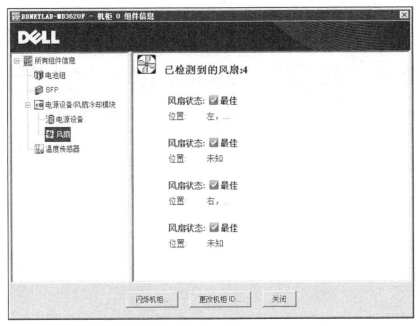

图 7-3-48　查看存储硬件状态之七

（8）查看存储温度传感器信息（如图 7-3-49 所示）。

324 第 7 章 部署使用 FC SAN 存储

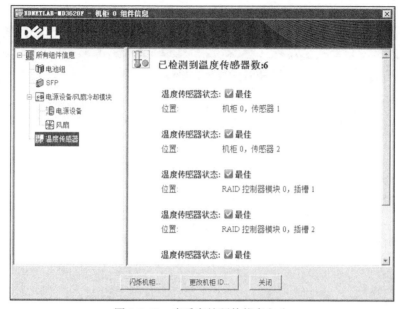

图 7-3-49 查看存储硬件状态之八

6. 存储告警处理

第 1 步，DELL MD 存储在生产环境中使用，状态要处于"最佳"，如果出现告警，需要进行问题排查（如图 7-3-50 所示）。存储出现了"需要注意"的告警提示，单击打开提示。

图 7-3-50 存储告警处理之一

第 2 步，告警提示为"电源设备-无电源输入"（如图 7-3-51 所示），说明该存储其中一个电源出现故障。

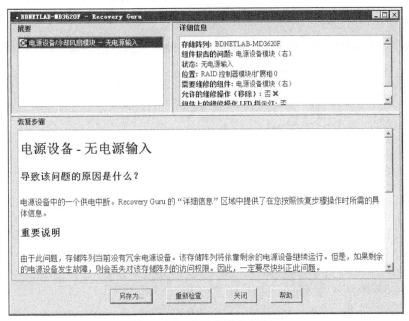

图 7-3-51　存储告警处理之二

第 3 步，往下看提示，可以看到官方给出的供参考的恢复步骤（如图 7-3-52 所示）。

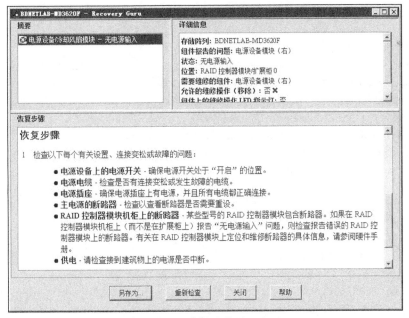

图 7-3-52　存储告警处理之三

本小节仅列举了几个简单的操作配置，读者可根据生产环境的需求进行配置。对于 DELL MD 系列存储配置,可以访问 DELL 官方网站查看其详细配置或致电 DELL 客户服务热线咨询。

7.4 配置 Cisco MDS 系列企业级存储交换机

在企业生产环境中，除了一些特殊应用 FC 存储与主机直接连接外，多数情况下 FC 存储会通过 FC 交换机进行连接，其优势是可以支持多个主机以及更加灵活地对 FC 存储进行访问控制等配置。本节介绍如何配置使用 Cisco MDS 系列 FC 交换机。

7.4.1 基本命令行介绍

1. 查看 MDS 交换机软/硬件信息

```
DC1-MDS-01# show version
Cisco Nexus Operating System (NX-OS) Software    //运行 NXOS 软件
TAC support: http://www.cisco.com/tac
Documents: http://www.cisco.com/en/US/products/ps9372/tsd_products_support_series
_home.html
Copyright (c) 2002-2011, Cisco Systems, Inc. All rights reserved.
The copyrights to certain works contained herein are owned by
other third parties and are used and distributed under license.
Some parts of this software are covered under the GNU Public
License. A copy of the license is available at
http://www.gnu.org/licenses/gpl.html.

Software
  BIOS:      version 1.0.19
  loader:    version N/A
  kickstart: version 5.2(2)      //kickstart 包括 linux 内核、基本驱动以及初始化文件系统
  system:    version 5.2(2)      //system 包括系统软件、基础以及四到七层功能
  BIOS compile time:       02/01/10
  kickstart image file is: bootflash:/m9100-s2ek9-kickstart-mz.5.2.2.bin
  kickstart compile time:  12/25/2020 12:00:00 [01/17/2012 22:12:01]
  system image file is:    bootflash:/m9100-s2ek9-mz.5.2.2.bin
  system compile time:     12/30/2011 14:00:00 [01/17/2012 23:06:53]

Hardware
  cisco MDS 9124 (1 Slot) Chassis ("1/2/4 Gbps FC/Supervisor-2")   //MDS 交换机型号
  Motorola, e500   with 516128 kB of memory.    //MDS 交换机 CPU、内存信息
  Processor Board ID JAF1211APCL

  Device name: DC1-MDS-01
  bootflash:      254464 kB
Kernel uptime is 0 day(s), 0 hour(s), 28 minute(s), 56 second(s)

Last reset
  Reason: Unknown
```

System version: 5.2(2)
Service:

2. 查看 MDS 交换机 NXOS 软件

```
DC1-MDS-01# show boot
Current Boot Variables:

kickstart variable = bootflash:/m9100-s2ek9-kickstart-mz.5.2.2.bin
system variable = bootflash:/m9100-s2ek9-mz.5.2.2.bin
No module boot variable set

Boot Variables on next reload:

kickstart variable = bootflash:/m9100-s2ek9-kickstart-mz.5.2.2.bin
system variable = bootflash:/m9100-s2ek9-mz.5.2.2.bin
No module boot variable set
```

3. 查看 MDS 交换机模块信息

```
DC1-MDS-01# show module
Mod   Ports   Module-Type                           Model              Status
---   -----   -----------------------------------   ----------------   ----------
1     24      1/2/4 Gbps FC/Supervisor-2            DS-C9124-K9-SUP    active *

Mod   Sw             Hw      World-Wide-Name(s) (WWN)
---   -------------  ------  -----------------------------------------------
1     5.2(2)         2.0     20:01:00:0d:ec:90:e0:00 to 20:18:00:0d:ec:90:e0:00

Mod   MAC-Address(es)                         Serial-Num
---   -------------------------------------   ----------
1     00-1f-ca-63-50-cc to 00-1f-ca-63-50-d0  JAF1211APCL

* this terminal session
```

4. 查看 MDS 交换机 feature 信息

```
DC1-MDS-01# show feature
Feature Name          Instance    State
-------------------   --------    --------
cimserver             1           disabled
dmm                   1           disabled
dpvm                  1           disabled
fabric-binding        1           disabled
fcsp                  1           disabled
ficon                 1           disabled
fport-channel-trunk   1           disabled
http-server           1           enabled
```

ioa	1	disabled
isapi	1	disabled
ldap	1	disabled
npiv	1	disabled
npv	1	disabled
port-security	1	disabled
port_track	1	disabled
privilege	1	disabled
qos-manager	1	disabled
santap	1	disabled
scheduler	1	disabled
scpServer	1	disabled
sdv	1	disabled
sfm	1	disabled
sftpServer	1	disabled
sshServer	1	enabled
tacacs	1	disabled
telnetServer	1	enabled
tpc	1	disabled

5. 查看 MDS 交换机 host-id 信息

```
DC1-MDS-01# show license host-id
License hostid: VDH=FOX11330HQU
```

6. 查看 MDS 交换机许可文件

```
DC1-MDS-01# show license file MDS201604190045398110.lic
MDS201604190045398110.lic:
SERVER this_host ANY
VENDOR cisco
INCREMENT PORT_ACTIVATION_PKG cisco 1.0 permanent 8 \
        VENDOR_STRING=<LIC_SOURCE>MDS_SWIFT</LIC_SOURCE><SKU>M9124PL8-4G=</SKU> \
        HOSTID=VDH=FOX11330HQU \
        NOTICE="<LicFileID>20160419004539811</LicFileID><LicLineID>1</LicLineID> \
        <PAK></PAK>" SIGN=994EDDE4FC36
```

7. 查看 MDS 交换机安装的许可授权

```
DC1-MDS-01# show license usage
Feature                      Ins  Lic    Status  Expiry Date  Comments
                                  Count
--------------------------------------------------------------------------------
FM_SERVER_PKG                No    -     Unused               -
ENTERPRISE_PKG               No    -     Unused               -
PORT_ACTIVATION_PKG          Yes   16    In use  never        -
10G_PORT_ACTIVATION_PKG      No    0     Unused               -
--------------------------------------------------------------------------------
```

7.4.2 配置 VSAN

VSAN 是 Cisco MDS 系列交换机中的一个重要概念,熟练掌握 VSAN 的配置非常重要。第 1 步,登录 MDS 交换机,进入特权配置模式。

```
DC1-MDS-01# configure terminal    //进入特权配置模式
Enter configuration commands, one per line.   End with CNTL/Z.
DC1-MDS-01(config)#
```

第 2 步,使用命令"show flogi database"查看 FC HBA 卡注册信息。其中,fc1/5 和 fc1/6 分别连接 ESXi12 主机和 ESXi13,fc1/15 和 fc1/16 连接 DELL MD 3620F 存储主控制 1 和主控制 2。

```
DC1-MDS-01# show flogi database
--------------------------------------------------------------------------------
INTERFACE        VSAN     FCID        PORT NAME                 NODE NAME
--------------------------------------------------------------------------------
fc1/5             1       0x320700    21:01:00:1b:32:a2:5a:99 20:01:00:1b:32:a2:5a:99
fc1/6             1       0x320800    21:01:00:1b:32:b5:54:23 20:01:00:1b:32:b5:54:23
fc1/15            1       0x320200    20:14:b0:83:fe:da:d4:dc 20:04:b0:83:fe:da:d4:dc
fc1/16            1       0x320900    20:15:b0:83:fe:da:d4:dc 20:04:b0:83:fe:da:d4:dc
Total number of flogi = 4.
```

第 3 步,使用命令"vsan database"进入 VSAN 数据库配置模式,创建 VSAN,VSAN 的名称为可选项,推荐在生产环境中配置以便于日常管理。

```
DC1-MDS-01(config)# vsan database    //进入 VSAN 数据库配置
DC1-MDS-01(config-vsan-db)# vsan 10 name sanboot     //创建 VSAN 10 并命名为 sanboot
DC1-MDS-01(config-vsan-db)# vsan 20 name sanshare    //创建 VSAN 20 并命名为 sanshare
```

第 4 步,使用命令"show vsan"查看创建的 VSAN 信息。

```
DC1-MDS-01(config)# show vsan
vsan 1 information
         name:VSAN0001    state:active
         interoperability mode:default
         loadbalancing:src-id/dst-id/oxid
         operational state:down

vsan 10 information
         name:sanboot    state:active
         interoperability mode:default
         loadbalancing:src-id/dst-id/oxid
         operational state:up

vsan 20 information
         name:sanshare    state:active
```

```
                    interoperability mode:default
                    loadbalancing:src-id/dst-id/oxid
                    operational state:down

vsan 4079:evfp_isolated_vsan

vsan 4094:isolated_vsan
```

第 5 步，使用命令将 FC 接口加入 VSAN。

```
DC1-MDS-01(config-vsan-db)# vsan 10 interface fc1/5
Traffic on fc1/5 may be impacted. Do you want to continue? (y/n) [n] y
DC1-MDS-01(config-vsan-db)# vsan 10 interface fc1/6
Traffic on fc1/6 may be impacted. Do you want to continue? (y/n) [n] y
DC1-MDS-01(config-vsan-db)# vsan 10 interface fc1/15
Traffic on fc1/15 may be impacted. Do you want to continue? (y/n) [n] y
DC1-MDS-01(config-vsan-db)# vsan 10 interface fc1/16
Traffic on fc1/16 may be impacted. Do you want to continue? (y/n) [n] y
```

第 6 步，使用命令 "show vsan membership" 查看 FC 接口是否加入对应的 VSAN。

```
DC1-MDS-01(config)# show vsan membership
vsan 1 interfaces:
      fc1/1          fc1/2          fc1/3          fc1/4
      fc1/7          fc1/8          fc1/9          fc1/10
      fc1/11         fc1/12         fc1/13         fc1/14
      fc1/17         fc1/18         fc1/19         fc1/20
      fc1/21         fc1/22         fc1/23         fc1/24

vsan 10 interfaces:
      fc1/5          fc1/6          fc1/15         fc1/16

vsan 20 interfaces:

vsan 4079(evfp_isolated_vsan) interfaces:

vsan 4094(isolated_vsan) interfaces:
```

第 7 步，使用命令 "show fcdomain domain-list" 查看 FCDOMAIN 信息，因为目前只开启一台 MDS 交换机，所以 DC1-MDS-01 为所有 VSAN 的主控交换机。

```
DC1-MDS-01# show fcdomain domain-list

VSAN 1
Number of domains: 1
Domain ID              WWN
---------         ----------------------
```

```
    0x32(50)        20:01:00:0d:ec:90:e0:01 [Local] [Principal]

VSAN 10
Number of domains: 1
Domain ID              WWN
---------              ----------------------
    0x05(5)         20:0a:00:0d:ec:90:e0:01 [Local] [Principal]

VSAN 20
Number of domains: 1
Domain ID              WWN
---------              ----------------------
    0xe5(229)       20:14:00:0d:ec:90:e0:01 [Local] [Principal]
```

第 8 步，使用命令"show fcdomain fcid persistent"查看 FCID 分配情况。

```
DC1-MDS-01# show fcdomain fcid persistent
Total entries 16.

Persistent FCIDs table contents:
VSAN WWN                        FCID        Mask   Used Assignment Interface
---- ----------------------- -------- ------ ---- ---------- ---------
   1 21:00:00:1b:32:82:01:9b 0x320000 SINGLE  NO   DYNAMIC    --
   1 21:00:00:e0:8b:9a:67:13 0x320100 SINGLE  NO   DYNAMIC    --
   1 20:14:b0:83:fe:da:d4:dc 0x320200 SINGLE  NO   DYNAMIC    --
   1 20:0b:00:05:9b:76:93:80 0x320300 SINGLE  NO   DYNAMIC    --
   1 21:01:00:e0:8b:ba:67:13 0x320301 SINGLE  NO   DYNAMIC    --
   1 21:01:00:1b:32:a2:01:9b 0x320401 SINGLE  NO   DYNAMIC    --
   1 21:00:00:1b:32:04:63:b8 0x320600 SINGLE  NO   DYNAMIC    --
   1 20:0c:00:05:9b:76:93:80 0x320400 SINGLE  NO   DYNAMIC    --
   1 21:01:00:1b:32:a2:5a:99 0x320700 SINGLE  NO   DYNAMIC    --
   1 24:0b:00:05:9b:76:93:80 0x320500 SINGLE  NO   DYNAMIC    --
   1 21:01:00:1b:32:b5:54:23 0x320800 SINGLE  NO   DYNAMIC    --
   1 20:15:b0:83:fe:da:d4:dc 0x320900 SINGLE  NO   DYNAMIC    --
  10 21:01:00:1b:32:a2:5a:99 0x050000 SINGLE  YES  DYNAMIC    --
  10 21:01:00:1b:32:b5:54:23 0x050100 SINGLE  YES  DYNAMIC    --
  10 20:14:b0:83:fe:da:d4:dc 0x050200 SINGLE  YES  DYNAMIC    --
  10 20:15:b0:83:fe:da:d4:dc 0x050300 SINGLE  YES  DYNAMIC    --
```

第 9 步，使用命令"show fcns database"查看 FCNS 名称服务器信息。一般来说，sisi-fcp:init 对应主机的 FC HBA 卡，scsi-fcp:both 对应 FC 存储。

```
DC1-MDS-01# show fcns database
VSAN 10:
--------------------------------------------------------------------
FCID        TYPE  PWWN                    (VENDOR)        FC4-TYPE:FEATURE
--------------------------------------------------------------------
```

```
0x050000    N    21:01:00:1b:32:a2:5a:99 (Qlogic)    scsi-fcp:init
0x050100    N    21:01:00:1b:32:b5:54:23 (Qlogic)    scsi-fcp:init
0x050200    N    20:14:b0:83:fe:da:d4:dc             scsi-fcp:both
0x050300    N    20:15:b0:83:fe:da:d4:dc             scsi-fcp:both
Total number of entries = 4
```

7.4.3 配置 ZONE

完成 VSAN 的基本配置以及将 FC 接口加入 VSAN 后就可以进行 ZONE 配置，ZONE 配置完成后 FC 存储就可以在 ESXi 主机使用。

第 1 步，创建用于 ESXi 主机 sanboot 启动的 ZONE，ZONE 关联的信息可以为 PWWN，也可以关联设备连接至 FC 交换机的具体接口，可以根据生产环境的情况自行决定。

```
DC1-MDS-01(config)# zone name sanboot vsan 10         //创建名为 sanboot 的 ZONE 并关联到 VSAN 10
DC1-MDS-01(config-zone)# member pwwn 21:01:00:1b:32:a2:5a:99    //将 ESXi12 主机加入到 ZONE
DC1-MDS-01(config-zone)# member pwwn 21:01:00:1b:32:b5:54:23    //将 ESXi13 主机加入到 ZONE
DC1-MDS-01(config-zone)# member interface fc1/15      //将 DELL MD 存储主控 01 加入到 ZONE
DC1-MDS-01(config-zone)# member interface fc1/16      //将 DELL MD 存储主控 02 加入到 ZONE
```

第 2 步，使用命令"show zone"查看 ZONE 关联的信息。

```
DC1-MDS-01(config)# show zone
zone name sanboot vsan 10
    pwwn 21:01:00:1b:32:a2:5a:99
    pwwn 21:01:00:1b:32:b5:54:23
      interface fc1/15 swwn 20:00:00:0d:ec:90:e0:00
      interface fc1/16 swwn 20:00:00:0d:ec:90:e0:00
```

第 3 步，配置 ZONESET 并关联 ZONE。

```
DC1-MDS-01(config)# zoneset name zs_sanboot vsan 10
DC1-MDS-01(config-zoneset)# member sanboot      //调用关联 ZONE
```

第 4 步，使用命令"show zoneset"查看 ZONESET 配置以及关联 ZONE 信息。

```
DC1-MDS-01(config)# show zoneset
zoneset name zs_sanboot vsan 10
  zone name sanboot vsan 10
    pwwn 21:01:00:1b:32:a2:5a:99
    pwwn 21:01:00:1b:32:b5:54:23
      interface fc1/15 swwn 20:00:00:0d:ec:90:e0:00
      interface fc1/16 swwn 20:00:00:0d:ec:90:e0:00
```

第 5 步，使用命令"zoneset activate name"激活 ZONESET。

```
DC1-MDS-01(config)# zoneset activate name zs_sanboot vsan 10
Zoneset activation initiated. check zone status
```

第 6 步，使用命令"show zoneset active"查看激活状态。fcid 前有"*"表示 ESXi 主机与存储之间建立好访问，如果没有则应检查配置。

```
DC1-MDS-01# show zoneset active
zoneset name zs_sanboot vsan 10
  zone name sanboot vsan 10
  * fcid 0x050000 [pwwn 21:01:00:1b:32:a2:5a:99]
  * fcid 0x050100 [pwwn 21:01:00:1b:32:b5:54:23]
  * fcid 0x050200 [interface fc1/15 swwn 20:00:00:0d:ec:90:e0:00]
  * fcid 0x050300 [interface fc1/16 swwn 20:00:00:0d:ec:90:e0:00]
```

7.4.4 配置多台 FC 交换机级联

由于各种需求，在生产环境中 FC 交换机之间的级联比较常见，如何配置多台 FC 交换机 TRUNK 是需要掌握的技能。本小节介绍多台 FC 交换机级联配置。实验环境中 DC1-MDS-01 与 DC1-MDS-02 两台交换机使用 FC1/11、FC1/12 级联。

第 1 步，登录 DC1-MDS-02 交换机。由于 VSAN 不能同步，所以需要手动创建相同 VSAN。

```
DC1-MDS-02(config)# vsan database
DC1-MDS-02(config-vsan-db)# vsan 10
DC1-MDS-02(config-vsan-db)# vsan 20
```

第 2 步，使用命令"show interface"查看级联接口信息，在不配置的情况下，两台 Cisco MDS 交换机级联后，接口自动协商为 TRUNK 模式。

```
DC1-MDS-01# show interface fc1/11
  fc1/11 is trunking (Not all VSANs UP on the trunk)    //自动协商为 TRUNK 模式，Not all VSANs UP on the trunk 提示后续介绍如何处理
    Hardware is Fibre Channel, SFP is short wave laser w/o OFC (SN)
    Port WWN is 20:0b:00:0d:ec:90:e0:00
    Peer port WWN is 20:0b:00:05:9b:76:93:80
    Admin port mode is auto, trunk mode is on
    snmp link state traps are enabled
    Port mode is TE
    Port vsan is 1
    Speed is 4 Gbps
    Rate mode is dedicated
    Transmit B2B Credit is 16
    Receive B2B Credit is 16
    B2B State Change Number is 14
    Receive data field Size is 2112
    Beacon is turned off
    Trunk vsans (admin allowed and active) (1,10,20)
    Trunk vsans (up)                        (10,20)      //VSAN10，20 状态为 UP
    Trunk vsans (isolated)                  (1)
    Trunk vsans (initializing)              ()
```

```
        5 minutes input rate 656 bits/sec, 82 bytes/sec, 0 frames/sec
        5 minutes output rate 704 bits/sec, 88 bytes/sec, 1 frames/sec
            286 frames input, 22696 bytes
                0 discards, 0 errors
                0 CRC,   0 unknown class
                0 too long, 0 too short
            301 frames output, 24228 bytes
                0 discards, 0 errors
            0 input OLS, 0 LRR, 0 NOS, 2 loop inits
            1 output OLS, 2 LRR, 0 NOS, 1 loop inits
            16 receive B2B credit remaining
            16 transmit B2B credit remaining
            14 low priority transmit B2B credit remaining
Interface last changed at Sat Jul 22 13:38:35 2017


DC1-MDS-01# show interface fc1/12
fc1/12 is trunking (Not all VSANs UP on the trunk)    //自动协商为 TRUNK 模式
        Hardware is Fibre Channel, SFP is short wave laser w/o OFC (SN)
        Port WWN is 20:0c:00:0d:ec:90:e0:00
        Peer port WWN is 20:0c:00:05:9b:76:93:80
        Admin port mode is auto, trunk mode is on
        snmp link state traps are enabled
        Port mode is TE
        Port vsan is 1
        Speed is 4 Gbps
        Rate mode is dedicated
        Transmit B2B Credit is 16
        Receive B2B Credit is 16
        B2B State Change Number is 14
        Receive data field Size is 2112
        Beacon is turned off
        Trunk vsans (admin allowed and active) (1,10,20)
        Trunk vsans (up)                       (10,20)
        Trunk vsans (isolated)                 (1)
        Trunk vsans (initializing)             ()
        5 minutes input rate 624 bits/sec, 78 bytes/sec, 0 frames/sec
        5 minutes output rate 648 bits/sec, 81 bytes/sec, 0 frames/sec
            271 frames input, 21376 bytes
                0 discards, 0 errors
                0 CRC,   0 unknown class
                0 too long, 0 too short
            260 frames output, 22548 bytes
                0 discards, 0 errors
            0 input OLS, 0 LRR, 0 NOS, 3 loop inits
            1 output OLS, 2 LRR, 0 NOS, 1 loop inits
            16 receive B2B credit remaining
            16 transmit B2B credit remaining
```

        14 low priority transmit B2B credit remaining
Interface last changed at Sat Jul 22 13:38:35 2017

DC1-MDS-02# show interface fc1/11
fc1/11 is trunking (Not all VSANs UP on the trunk)   //自动协商为 TRUNK 模式
    Hardware is Fibre Channel, SFP is short wave laser w/o OFC (SN)
    Port WWN is 20:0b:00:05:9b:76:93:80
    Peer port WWN is 20:0b:00:0d:ec:90:e0:00
    Admin port mode is auto, trunk mode is on
    snmp link state traps are enabled
    Port mode is TE
    Port vsan is 1
    Speed is 4 Gbps
    Rate mode is dedicated
    Transmit B2B Credit is 16
    Receive B2B Credit is 16
    B2B State Change Number is 14
    Receive data field Size is 2112
    Beacon is turned off
    Trunk vsans (admin allowed and active) (1,10,20)
    Trunk vsans (up)   (10,20)
    Trunk vsans (isolated)   (1)
    Trunk vsans (initializing)   ()
    5 minutes input rate 720 bits/sec, 90 bytes/sec, 1 frames/sec
    5 minutes output rate 672 bits/sec, 84 bytes/sec, 0 frames/sec
        315 frames input, 25044 bytes
            0 discards, 0 errors
            0 CRC,   0 unknown class
            0 too long, 0 too short
        297 frames output, 23280 bytes
            0 discards, 0 errors
        1 input OLS, 2 LRR, 0 NOS, 2 loop inits
        1 output OLS, 0 LRR, 0 NOS, 1 loop inits
        16 receive B2B credit remaining
        16 transmit B2B credit remaining
        15 low priority transmit B2B credit remaining
Interface last changed at Thu Aug 30 14:23:17 2001

DC1-MDS-02# show interface fc1/12
fc1/12 is trunking (Not all VSANs UP on the trunk)   //自动协商为 TRUNK 模式
    Hardware is Fibre Channel, SFP is short wave laser w/o OFC (SN)
    Port WWN is 20:0c:00:05:9b:76:93:80
    Peer port WWN is 20:0c:00:0d:ec:90:e0:00
    Admin port mode is auto, trunk mode is on
    snmp link state traps are enabled
    Port mode is TE
    Port vsan is 1

```
        Speed is 4 Gbps
        Rate mode is dedicated
        Transmit B2B Credit is 16
        Receive B2B Credit is 16
        B2B State Change Number is 14
        Receive data field Size is 2112
        Beacon is turned off
        Trunk vsans (admin allowed and active) (1,10,20)
        Trunk vsans (up)                        (10,20)
        Trunk vsans (isolated)                  (1)
        Trunk vsans (initializing)              ()
        5 minutes input rate 656 bits/sec, 82 bytes/sec, 0 frames/sec
        5 minutes output rate 632 bits/sec, 79 bytes/sec, 0 frames/sec
            263 frames input, 22816 bytes
              0 discards, 0 errors
              0 CRC,   0 unknown class
              0 too long, 0 too short
            279 frames output, 21884 bytes
              0 discards, 0 errors
            1 input OLS, 2 LRR, 0 NOS, 1 loop inits
            1 output OLS, 0 LRR, 0 NOS, 1 loop inits
            16 receive B2B credit remaining
            16 transmit B2B credit remaining
            15 low priority transmit B2B credit remaining
        Interface last changed at Thu Aug 30 14:23:17 2001
```

第 3 步，在 DC1-MDS-02 交换机上使用命令"show fcns database"查看 FCNS 名称服务器同步信息。可以看到，DC1-MDS-01 上的信息已经同步到 DC1-MDS-02，说明 FCNS 名称服务器是全网同步的。

```
DC1-MDS-02# show fcns database
VSAN 10:
--------------------------------------------------------------
FCID         TYPE   PWWN                      (VENDOR)        FC4-TYPE:FEATURE
--------------------------------------------------------------
0x050000     N      21:01:00:1b:32:a2:5a:99 (Qlogic)          scsi-fcp:init
0x050100     N      21:01:00:1b:32:b5:54:23 (Qlogic)          scsi-fcp:init
0x050200     N      20:14:b0:83:fe:da:d4:dc                   scsi-fcp:both
0x050300     N      20:15:b0:83:fe:da:d4:dc                   scsi-fcp:both
Total number of entries = 4
```

第 4 步，在 DC1-MDS-02 交换机上使用命令"show fcdomain domain-list"查看 FCDOMAIN 信息。由于 VSAN 1 没有配置，所以 DC1-MDS-02 成为了 VSAN 10 以及 VSAN 20 的主控交换机。

```
DC1-MDS-02# show fcdomain domain-list

VSAN 1
```

```
 Number of domains: 1
 Domain ID              WWN
 ---------         ----------------------
 0x88(136)         20:01:00:05:9b:76:93:81 [Local] [Principal]


VSAN 10
 Number of domains: 2
 Domain ID              WWN
 ---------         ----------------------
 0x52(82)          20:0a:00:05:9b:76:93:81 [Local] [Principal]
 0x05(5)           20:0a:00:0d:ec:90:e0:01


VSAN 20
 Number of domains: 2
 Domain ID              WWN
 ---------         ----------------------
 0x5d(93)          20:14:00:05:9b:76:93:81 [Local] [Principal]
 0xe5(229)         20:14:00:0d:ec:90:e0:01
```

第 5 步，在 DC1-MDS-02 交换机上使用命令"show zoneset active"查看 ZONESET 激活情况，DC1-MDS-01 上激活名为 zs_sanboot 同步传递到 DC1-MDS-02。

```
DC1-MDS-02# show zoneset active
zoneset name zs_sanboot vsan 10
  zone name sanboot vsan 10
  * fcid 0x050000 [pwwn 21:01:00:1b:32:a2:5a:99]
  * fcid 0x050100 [pwwn 21:01:00:1b:32:b5:54:23]
  * fcid 0x050200 [interface fc1/15 swwn 20:00:00:0d:ec:90:e0:00]
  * fcid 0x050300 [interface fc1/16 swwn 20:00:00:0d:ec:90:e0:00]
```

第 6 步，综上所述，如果不配置使用自动协商，两台 Cisco MDS 交换机可以正常工作，但在生产环境中，推荐手动进行配置。

```
DC1-MDS-02(config)# interface fc1/11-12      //指定级联使用的 FC 接口
DC1-MDS-02(config-if)# shut                  //重新配置前建议关闭 FC 接口
DC1-MDS-02(config-if)# channel-group 10 force     //启用 FC 接口的 port-channel
fc1/11 fc1/12 added to port-channel 10 and disabled
please do the same operation on the switch at the other end of the port-channel,
then do "no shutdown" at both ends to bring it up
DC1-MDS-02(config-if)# no shutdown
DC1-MDS-02(config-if)# exit
DC1-MDS-02(config)# interface port-channel 10    //进入 port-channel 接口
DC1-MDS-02(config-if)# switchport mode e         //配置 port-channel 模式为 TRUNK
DC1-MDS-02(config-if)# no shutdown
DC1-MDS-02(config-if)# exit
DC1-MDS-02(config)# vsan database        //进入 VSAN 配置数据库
DC1-MDS-02(config-vsan-db)# vsan 10 interface port-channel 10    //将 port-channel 接口加入 VSAN 10
```

第 7 步，按照上述操作配置 DC1-MDS-02 交换机。

```
DC1-MDS-01(config)# interface fc1/11-12    //指定级联使用的 FC 接口
DC1-MDS-01(config-if)# shut                //重新配置前建议关闭 FC 接口
DC1-MDS-01(config-if)# channel-group 10 force    //启用 FC 接口的 port-channel
fc1/11 fc1/12 added to port-channel 10 and disabled
please do the same operation on the switch at the other end of the port-channel,
then do "no shutdown" at both ends to bring it up
DC1-MDS-01(config-if)# no shutdown
DC1-MDS-01(config-if)# exit
DC1-MDS-01(config)# interface port-channel 10    //进入 port-channel 接口
DC1-MDS-01(config-if)# switchport mode e    //配置 port-channel 模式为 TRUNK
DC1-MDS-01(config-if)# no shutdown
DC1-MDS-01(config)# vsan database    //进入 VSAN 配置数据库
DC1-MDS-01(config-vsan-db)# vsan 10 interface port-channel 10    //将 port-channel 接口加入 VSAN 10
```

第 8 步，使用命令"show interface port-channel"查看 port-channel 状态。

```
DC1-MDS-01# show interface port-channel 10
port-channel 10 is trunking (Not all VSANs UP on the trunk)    //port-channel 为 TRUNK 模式
    Hardware is Fibre Channel
    Port WWN is 24:0a:00:0d:ec:90:e0:00
    Admin port mode is E, trunk mode is on
    snmp link state traps are enabled
    Port mode is TE
    Port vsan is 10
    Speed is 8 Gbps
    Trunk vsans (admin allowed and active) (1,10,20)
    Trunk vsans (up)                       (10,20)
    Trunk vsans (isolated)                 (1)
    Trunk vsans (initializing)             ()
    5 minutes input rate 1680 bits/sec, 210 bytes/sec, 2 frames/sec
    5 minutes output rate 1680 bits/sec, 210 bytes/sec, 2 frames/sec
        1420 frames input, 112908 bytes
            0 discards, 0 errors
            0 CRC,    0 unknown class
            0 too long, 0 too short
        1424 frames output, 113380 bytes
            0 discards, 0 errors
        4 input OLS, 4 LRR, 2 NOS, 10 loop inits
        10 output OLS, 14 LRR, 0 NOS, 4 loop inits
    Member[1] : fc1/11
    Member[2] : fc1/12
Interface last changed at Sat Jul 22 13:47:39 2017


DC1-MDS-02(config)# show interface port-channel 10
port-channel 10 is trunking (Not all VSANs UP on the trunk)    //port-channel 为 TRUNK 模式
```

```
Hardware is Fibre Channel
Port WWN is 24:0a:00:05:9b:76:93:80
Admin port mode is E, trunk mode is on
snmp link state traps are enabled
Port mode is TE
Port vsan is 10
Speed is 8 Gbps
Trunk vsans (admin allowed and active) (1,10,20)
Trunk vsans (up)                        (10,20)
Trunk vsans (isolated)                  (1)
Trunk vsans (initializing)              ()
5 minutes input rate 1120 bits/sec, 140 bytes/sec, 1 frames/sec
5 minutes output rate 1120 bits/sec, 140 bytes/sec, 1 frames/sec
    1151 frames input, 92316 bytes
      0 discards, 0 errors
      0 CRC,   0 unknown class
      0 too long, 0 too short
    1161 frames output, 91596 bytes
      0 discards, 0 errors
    12 input OLS, 12 LRR, 2 NOS, 6 loop inits
    6 output OLS, 4 LRR, 6 NOS, 4 loop inits
Member[1] : fc1/11
Member[2] : fc1/12
Interface last changed at Thu Aug 30 14:32:20 2001
```

第 9 步，出现 "Not all VSANs UP on the trunk" 提示的原因，是因为 port-channel 没有配置 allowed 选项，进入 DC1-MDS-01 交换机配置。

```
DC1-MDS-02(config)# interface port-channel 10
DC1-MDS-02(config-if)# switchport trunk allowed vsan 10      //允许 VSAN 10 通过 port-channel 10
DC1-MDS-02(config-if)# switchport trunk allowed vsan add 20  //增加 VSAN 20 通过 port-channel 10
```

第 10 步，再次使用命令 "show interface port-channel" 查看 port-channel 状态。

```
DC1-MDS-01# show interface port-channel 10
port-channel 10 is trunking    // 不存在 Not all VSANs UP on the trunk 提示
    Hardware is Fibre Channel
    Port WWN is 24:0a:00:0d:ec:90:e0:00
    Admin port mode is E, trunk mode is on
    snmp link state traps are enabled
    Port mode is TE
    Port vsan is 10
    Speed is 8 Gbps
    Trunk vsans (admin allowed and active) (1,10,20)
    Trunk vsans (up)                        (10,20)
    Trunk vsans (isolated)                  (1)
    Trunk vsans (initializing)              ()
```

```
        5 minutes input rate 1656 bits/sec, 207 bytes/sec, 2 frames/sec
        5 minutes output rate 1664 bits/sec, 208 bytes/sec, 2 frames/sec
            1428 frames input, 113232 bytes
                0 discards, 0 errors
                0 CRC,   0 unknown class
                0 too long, 0 too short
            1432 frames output, 113716 bytes
                0 discards, 0 errors
            4 input OLS, 4 LRR, 2 NOS, 10 loop inits
            10 output OLS, 14 LRR, 0 NOS, 4 loop inits
        Member[1] : fc1/11
        Member[2] : fc1/12
    Interface last changed at Sat Jul 22 13:47:39 2017

DC1-MDS-02(config)# show interface port-channel 10
port-channel 10 is trunking
    Hardware is Fibre Channel
    Port WWN is 24:0a:00:05:9b:76:93:80
    Admin port mode is E, trunk mode is on
    snmp link state traps are enabled
    Port mode is TE
    Port vsan is 10
    Speed is 8 Gbps
    Trunk vsans (admin allowed and active) (10,20)
    Trunk vsans (up)                       (10,20)
    Trunk vsans (isolated)                 ()
    Trunk vsans (initializing)             ()
    5 minutes input rate 1688 bits/sec, 211 bytes/sec, 2 frames/sec
    5 minutes output rate 1688 bits/sec, 211 bytes/sec, 2 frames/sec
        1432 frames input, 113952 bytes
            0 discards, 0 errors
            0 CRC,   0 unknown class
            0 too long, 0 too short
        1442 frames output, 114040 bytes
            0 discards, 0 errors
        12 input OLS, 12 LRR, 2 NOS, 6 loop inits
        6 output OLS, 4 LRR, 6 NOS, 4 loop inits
    Member[1] : fc1/11
    Member[2] : fc1/12
    Interface last changed at Thu Aug 30 14:32:20 2001
```

第 11 步，使用命令"show interface"查看 FC 接口状态。

```
DC1-MDS-01# show interface fc1/11
fc1/11 is trunking
    Hardware is Fibre Channel, SFP is short wave laser w/o OFC (SN)
    Port WWN is 20:0b:00:0d:ec:90:e0:00
```

        Peer port WWN is 20:0b:00:05:9b:76:93:80
        Admin port mode is E, trunk mode is on
        snmp link state traps are enabled
        Port mode is TE
        Port vsan is 10
        Speed is 4 Gbps
        Rate mode is dedicated
        Transmit B2B Credit is 16
        Receive B2B Credit is 16
        B2B State Change Number is 14
        Receive data field Size is 2112
        Beacon is turned off
        Belongs to port-channel 10
        Trunk vsans (admin allowed and active) (1,10,20)
        Trunk vsans (up)   (10,20)
        Trunk vsans (isolated)   (1)
        Trunk vsans (initializing)   ()
        5 minutes input rate 256 bits/sec, 32 bytes/sec, 0 frames/sec
        5 minutes output rate 192 bits/sec, 24 bytes/sec, 0 frames/sec
            32466 frames input, 2415052 bytes
              0 discards, 0 errors
              0 CRC,   0 unknown class
              0 too long, 0 too short
            32535 frames output, 1788256 bytes
              0 discards, 0 errors
            2 input OLS, 2 LRR, 1 NOS, 5 loop inits
            5 output OLS, 7 LRR, 0 NOS, 2 loop inits
            16 receive B2B credit remaining
            16 transmit B2B credit remaining
            14 low priority transmit B2B credit remaining
Interface last changed at Sat Jul 22 13:44:25 2017

DC1-MDS-01# show interface fc1/12
fc1/12 is trunking
        Hardware is Fibre Channel, SFP is short wave laser w/o OFC (SN)
        Port WWN is 20:0c:00:0d:ec:90:e0:00
        Peer port WWN is 20:0c:00:05:9b:76:93:80
        Admin port mode is E, trunk mode is on
        snmp link state traps are enabled
        Port mode is TE
        Port vsan is 10
        Speed is 4 Gbps
        Rate mode is dedicated
        Transmit B2B Credit is 16
        Receive B2B Credit is 16
        B2B State Change Number is 14
        Receive data field Size is 2112

```
        Beacon is turned off
        Belongs to port-channel 10
        Trunk vsans (admin allowed and active) (1,10,20)
        Trunk vsans (up)                        (10,20)
        Trunk vsans (isolated)                  (1)
        Trunk vsans (initializing)              ()
        5 minutes input rate 0 bits/sec, 0 bytes/sec, 0 frames/sec
        5 minutes output rate 0 bits/sec, 0 bytes/sec, 0 frames/sec
          410 frames input, 33292 bytes
            0 discards, 0 errors
            0 CRC,   0 unknown class
            0 too long, 0 too short
          345 frames output, 28124 bytes
            0 discards, 0 errors
          2 input OLS, 2 LRR, 1 NOS, 5 loop inits
          5 output OLS, 7 LRR, 0 NOS, 2 loop inits
          16 receive B2B credit remaining
          16 transmit B2B credit remaining
          14 low priority transmit B2B credit remaining
        Interface last changed at Sat Jul 22 13:44:23 2017

DC1-MDS-02# show interface fc1/11
fc1/11 is trunking
        Hardware is Fibre Channel, SFP is short wave laser w/o OFC (SN)
        Port WWN is 20:0b:00:05:9b:76:93:80
        Peer port WWN is 20:0b:00:0d:ec:90:e0:00
        Admin port mode is E, trunk mode is on
        snmp link state traps are enabled
        Port mode is TE
        Port vsan is 10
        Speed is 4 Gbps
        Rate mode is dedicated
        Transmit B2B Credit is 16
        Receive B2B Credit is 16
        B2B State Change Number is 14
        Receive data field Size is 2112
        Beacon is turned off
        Belongs to port-channel 10
        Trunk vsans (admin allowed and active) (10,20)
        Trunk vsans (up)                        (10,20)
        Trunk vsans (isolated)                  ()
        Trunk vsans (initializing)              ()
        5 minutes input rate 192 bits/sec, 24 bytes/sec, 0 frames/sec
        5 minutes output rate 256 bits/sec, 32 bytes/sec, 0 frames/sec
          32533 frames input, 1788184 bytes
            0 discards, 0 errors
            0 CRC,   0 unknown class
```

             0 too long, 0 too short
          32470 frames output, 2415016 bytes
             0 discards, 0 errors
          6 input OLS, 6 LRR, 1 NOS, 3 loop inits
          3 output OLS, 2 LRR, 3 NOS, 2 loop inits
          16 receive B2B credit remaining
          16 transmit B2B credit remaining
          15 low priority transmit B2B credit remaining
       Interface last changed at Thu Aug 30 14:29:07 2001

DC1-MDS-02# show interface fc1/12
fc1/12 is trunking
       Hardware is Fibre Channel, SFP is short wave laser w/o OFC (SN)
       Port WWN is 20:0c:00:05:9b:76:93:80
       Peer port WWN is 20:0c:00:0d:ec:90:e0:00
       Admin port mode is E, trunk mode is on
       snmp link state traps are enabled
       Port mode is TE
       Port vsan is 10
       Speed is 4 Gbps
       Rate mode is dedicated
       Transmit B2B Credit is 16
       Receive B2B Credit is 16
       B2B State Change Number is 14
       Receive data field Size is 2112
       Beacon is turned off
       Belongs to port-channel 10
       Trunk vsans (admin allowed and active) (10,20)
       Trunk vsans (up)   (10,20)
       Trunk vsans (isolated)   ()
       Trunk vsans (initializing)   ()
       5 minutes input rate 0 bits/sec, 0 bytes/sec, 0 frames/sec
       5 minutes output rate 0 bits/sec, 0 bytes/sec, 0 frames/sec
          350 frames input, 28476 bytes
             0 discards, 0 errors
             0 CRC,   0 unknown class
             0 too long, 0 too short
          423 frames output, 33964 bytes
             0 discards, 0 errors
          6 input OLS, 6 LRR, 1 NOS, 3 loop inits
          3 output OLS, 2 LRR, 3 NOS, 2 loop inits
          16 receive B2B credit remaining
          16 transmit B2B credit remaining
          15 low priority transmit B2B credit remaining
       Interface last changed at Thu Aug 30 14:29:04 2001

7.4.5 配置 NPV/NPIV

在大中型环境中 NPV 以及 NPIV 的使用比较多。本小节介绍如何配置 NPV 以及 NPIV，将 DC1-MDS-01 交换机配置为 NPIV，DC1-MDS-02 配置为 NPV。

第 1 步，登录 DC1-MDS-02 交换机，使用命令 "show feature" 查看高级特性的启用情况，默认情况下，NPV/NPIV 均处于 disabled 状态，同时，标准 FC 交换机可以使用的特性较多。使用命令 "feature npv" 启用 NPV 技术，需要注意的是，启用 NPV 技术后，交换机会清空配置并重新启动。

```
DC1-MDS-02# show feature
Feature Name          Instance    State
--------------------  --------    --------
assoc_mgr             1           disabled
cimserver             1           disabled
dmm                   1           disabled
dpvm                  1           disabled
fabric-binding        1           disabled
fcoe_mgr              1           disabled
fcsp                  1           disabled
ficon                 1           disabled
fport-channel-trunk   1           disabled
http-server           1           enabled
ioa                   1           disabled
isapi                 1           disabled
ldap                  1           disabled
npiv                  1           disabled
npv                   1           disabled
port-security         1           disabled
port_track            1           disabled
privilege             1           disabled
qos-manager           1           disabled
santap                1           disabled
scheduler             1           disabled
sdv                   1           disabled
sfm                   1           disabled
sshServer             1           enabled
tacacs                1           disabled
telnetServer          1           enabled
tpc                   1           disabled
DC1-MDS-02# conf t
Enter configuration commands, one per line.  End with CNTL/Z.
DC1-MDS-02(config)# feature npv    //启用 NPV
Verify that boot variables are set and the changes are saved. Changing to npv mode erases the current configuration and reboots the switch in npv mode. Do you want to continue? (y/n):y
```

第 2 步，启用 NPV 后重启成功，再使用命令 "show feature" 查看高级特性的启用情况，可以看到 FC 交换机仅具有基本的功能。

```
DC1-MDS-02# show feature
Feature Name          Instance    State
--------------------  ---------   --------
http-server               1        enabled
npiv                      1        disabled
npv                       1        enabled
port_track                1        disabled
scheduler                 1        disabled
sshServer                 1        enabled
tacacs                    1        disabled
telnetServer              1        enabled
```

第 3 步，使用命令"show npv status"查看 NPV 状态，NPIV 状态处于 disabled 状态，由于目前还未配置 NPV 接口，因此 External Interfaces 为 0。

```
DC1-MDS-02# show npv status
npiv is disabled

disruptive load balancing is disabled
External Interfaces:
=====================
    Number of External Interfaces: 0
Server Interfaces:
===================
    Number of Server Interfaces: 0
```

第 4 步，在 DC1-MDS-02 上创建 VSAN 并将 FC 接口划入 VSAN。

```
DC1-MDS-02(config)# vsan database
DC1-MDS-02(config-vsan-db)# vsan 10
DC1-MDS-02(config-vsan-db)# vsan 20
DC1-MDS-02(config-vsan-db)# vsan 20 interface fc1/11    //将连接 DC1-MDS-01 接口划入 VSAN 20
DC1-MDS-02(config-vsan-db)# vsan 20 interface fc1/12    //将连接 DC1-MDS-01 接口划入 VSAN 20
```

第 5 步，在 DC1-MDS-02 上使用命令"show vsan membership"查看 FC 接口划入 VSAN 情况。

```
DC1-MDS-02(config)# show vsan membership
vsan 1 interfaces:
    fc1/1           fc1/2           fc1/3           fc1/4
    fc1/5           fc1/6           fc1/7           fc1/8
    fc1/9           fc1/10          fc1/13          fc1/14
    fc1/15          fc1/16          fc1/17          fc1/18
    fc1/19          fc1/20          fc1/21          fc1/22
    fc1/23          fc1/24
```

```
vsan 10 interfaces:

vsan 20 interfaces:
    fc1/11           fc1/12
vsan 4079(evfp_isolated_vsan) interfaces:
vsan 4094(isolated_vsan) interfaces:
```

第 6 步，在 DC1-MDS-02 上将连接 DC1-MDS-01 交换机的 FC 接口配置为 NP 模式。

```
DC1-MDS-02(config)# interface fc1/11-12
DC1-MDS-02(config-if)# switchport mode ?     //FC 接口可以选择多种配置模式
    E      F       FL      Fx      NP      SD      ST      auto
DC1-MDS-02(config-if)# switchport mode np
DC1-MDS-02(config-if)# no shutdown
```

第 7 步，在 DC1-MDS-01 上启用 NPIV 模式并将 FC 接口划入 VSAN。

```
DC1-MDS-01(config)# feature npiv       //启用 NPIV
DC1-MDS-01(config)# vsan database
DC1-MDS-01(config-vsan-db)# vsan 20 interface fc1/11
Traffic on fc1/11 may be impacted. Do you want to continue? (y/n) [n] y
DC1-MDS-01(config-vsan-db)# vsan 20 interface fc1/12
Traffic on fc1/12 may be impacted. Do you want to continue? (y/n) [n] y
```

第 8 步，DC1-MDS-01 上将连接 DC1-MDS-02 交换机的 FC 接口配置为 F 模式。

```
DC1-MDS-01(config)# interface fc1/11-12
DC1-MDS-01(config-if)# switchport mode f
DC1-MDS-01(config-if)# no shutdown
```

第 9 步，在 DC1-MDS-02 上再次使用命令 "show npv status" 查看 NPV 状态，External Interfaces 为 2 并关联到 fc1/11 和 fc1/12 接口。

```
DC1-MDS-02(config)# show npv status
npiv is disabled
disruptive load balancing is disabled

External Interfaces:
====================
    Interface: fc1/11, VSAN:    20, FCID: 0xe50000, State: Up
    Interface: fc1/12, VSAN:    20, FCID: 0xe50100, State: Up
    Number of External Interfaces: 2
Server Interfaces:
==================
    Number of Server Interfaces: 0
```

第 10 步，在 DC1-MDS-02 交换机上使用命令 "show interface" 查看 FC 接口信息。

```
DC1-MDS-02(config)# show interface fc1/11
fc1/11 is up
```

        Hardware is Fibre Channel, SFP is short wave laser w/o OFC (SN)
        Port WWN is 20:0b:00:05:9b:76:93:80
        Admin port mode is NP, trunk mode is off
        snmp link state traps are enabled
        Port mode is NP
        Port vsan is 20
        Speed is 4 Gbps
        Rate mode is dedicated
        Transmit B2B Credit is 16
        Receive B2B Credit is 16
        Receive data field Size is 2112
        Beacon is turned off
        5 minutes input rate 192 bits/sec, 24 bytes/sec, 0 frames/sec
        5 minutes output rate 168 bits/sec, 21 bytes/sec, 0 frames/sec
            84 frames input, 6992 bytes
                0 discards, 0 errors
                0 CRC,   0 unknown class
                0 too long, 0 too short
            68 frames output, 6276 bytes
                0 discards, 0 errors
            8 input OLS, 5 LRR, 3 NOS, 0 loop inits
            4 output OLS, 5 LRR, 3 NOS, 0 loop inits
            16 receive B2B credit remaining
            16 transmit B2B credit remaining
            16 low priority transmit B2B credit remaining
        Interface last changed at Fri Aug 31 12:57:27 2001

DC1-MDS-02(config)# show interface fc1/12
fc1/12 is up
        Hardware is Fibre Channel, SFP is short wave laser w/o OFC (SN)
        Port WWN is 20:0c:00:05:9b:76:93:80
        Admin port mode is NP, trunk mode is off
        snmp link state traps are enabled
        Port mode is NP
        Port vsan is 20
        Speed is 4 Gbps
        Rate mode is dedicated
        Transmit B2B Credit is 16
        Receive B2B Credit is 16
        Receive data field Size is 2112
        Beacon is turned off
        5 minutes input rate 72 bits/sec, 9 bytes/sec, 0 frames/sec
        5 minutes output rate 48 bits/sec, 6 bytes/sec, 0 frames/sec
            28 frames input, 2820 bytes
                0 discards, 0 errors
                0 CRC,   0 unknown class
                0 too long, 0 too short

```
        20 frames output, 1872 bytes
          0 discards, 0 errors
        7 input OLS, 4 LRR, 2 NOS, 0 loop inits
        2 output OLS, 2 LRR, 3 NOS, 0 loop inits
        16 receive B2B credit remaining
        16 transmit B2B credit remaining
        16 low priority transmit B2B credit remaining
    Interface last changed at Fri Aug 31 12:57:29 2001
```

第 11 步，将 ESXi 主机以及 DELL MD 存储接口划入 VSAN。

```
DC1-MDS-02(config)# vsan database
DC1-MDS-02(config-vsan-db)# vsan 20 interface fc1/5
DC1-MDS-02(config-vsan-db)# vsan 20 interface fc1/6
DC1-MDS-02(config-vsan-db)# vsan 20 interface fc1/15
DC1-MDS-02(config-vsan-db)# vsan 20 interface fc1/16
```

第 12 步，使用命令"show vsan membership"查看 FC 接口划入信息。

```
DC1-MDS-02(config)# show vsan membership
vsan 1 interfaces:
    fc1/1           fc1/2           fc1/3           fc1/4
    fc1/7           fc1/8           fc1/9           fc1/10
    fc1/13          fc1/14          fc1/17          fc1/18
    fc1/19          fc1/20          fc1/21          fc1/22
    fc1/23          fc1/24
vsan 10 interfaces:
vsan 20 interfaces:
    fc1/5           fc1/6           fc1/11          fc1/12
    fc1/15          fc1/16
vsan 4079(evfp_isolated_vsan) interfaces:
vsan 4094(isolated_vsan) interfaces:
```

第 13 步，使用命令"show npv flogi-table"查看 ESXi 主机 FC HBA 卡注册信息，注意在 NPV 模式下命令的变化。

```
DC1-MDS-02(config)# show npv flogi-table
--------------------------------------------------------------------------------
SERVER                                                                   EXTERNAL
INTERFACE VSAN FCID       PORT NAME                 NODE NAME             INTERFACE
--------------------------------------------------------------------------------
fc1/5     20   0xe50102   21:00:00:1b:32:82:5a:99   20:00:00:1b:32:82:5a:99 fc1/12
fc1/6     20   0xe50001   21:00:00:1b:32:95:54:23   20:00:00:1b:32:95:54:23 fc1/11
fc1/15    20   0xe50101   20:24:b0:83:fe:da:d4:dc   20:04:b0:83:fe:da:d4:dc fc1/12
fc1/16    20   0xe50002   20:25:b0:83:fe:da:d4:dc   20:04:b0:83:fe:da:d4:dc fc1/11
Total number of flogi = 4.
```

第 14 步，使用命令"show npv status"查看 NPV 状态信息，配置完成后 Server Interfaces

数量为 4。

```
DC1-MDS-02(config)# show npv status
npiv is disabled
disruptive load balancing is disabled
External Interfaces:
====================
    Interface: fc1/11, VSAN:    20, FCID: 0xe50000, State: Up
    Interface: fc1/12, VSAN:    20, FCID: 0xe50100, State: Up
    Number of External Interfaces: 2
Server Interfaces:
==================
    Interface:   fc1/5, VSAN:     20, State: Up
    Interface:   fc1/6, VSAN:     20, State: Up
    Interface:   fc1/15, VSAN:    20, State: Up
    Interface:   fc1/16, VSAN:    20, State: Up
    Number of Server Interfaces: 4
```

第 15 步，使用命令 "show npv external-interface-usage" 查看 ESXi 主机关联上行接口信息。

```
DC1-MDS-02(config)# show npv external-interface-usage
NPV Traffic Usage Information:
--------------------------------------
Server-If        External-If
--------------------------------------
fc1/5            fc1/12
fc1/6            fc1/11
fc1/15           fc1/12
fc1/16           fc1/11
--------------------------------------
```

7.5　配置 ESXi 主机使用 FC 存储

生产环境对于 FC 存储的使用是多样化，比较常见的就是远程引导以及共享使用。本节介绍服务器如何使用 sanboot 引导并安装 ESXi 6.5 系统以及共享 FC 存储的使用。

7.5.1　配置 ESXi 主机 SANBOOT 启动

第 1 步，确认 ESXi 主机已经安装好 FC HBA 卡，打开服务器电源，当检测到 FC HBA 卡时，进入相关的配置，实战环境使用的是 QLE 246 的 FC HBA 卡（如图 7-5-1 所示），按 "CTRL+Q" 键进入配置。

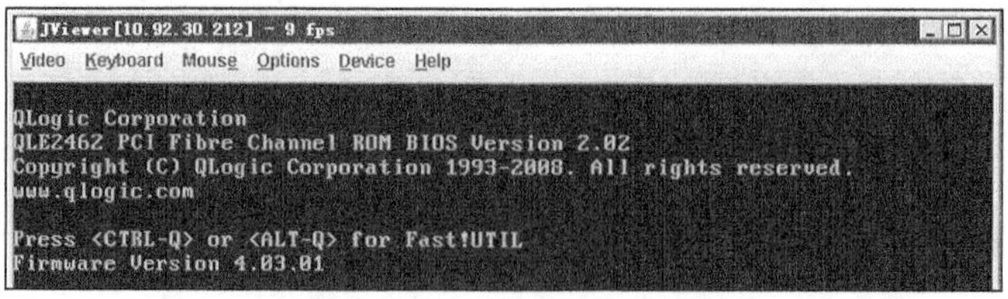

图 7-5-1 配置 ESXi 主机 SANBOOT 启动之一

第 2 步，进入 QLE 2462 FC HBA 卡配置界面（如图 7-5-2 所示），选择"Configuration Settings"。

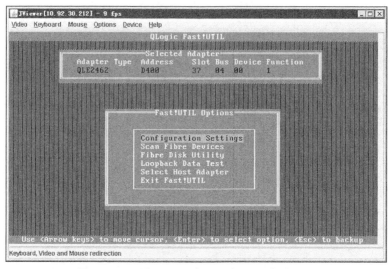

图 7-5-2 配置 ESXi 主机 SANBOOT 启动之二

第 3 步，选择"Adapter Settings"对 FC HBA 卡进行配置（如图 7-5-3 所示），按"回车"键。

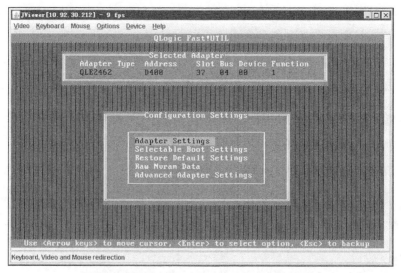

图 7-5-3 配置 ESXi 主机 SANBOOT 启动之三

第 4 步，确认"Host Adapter BIOS"状态为"Enabled"（如图 7-5-4 所示）。

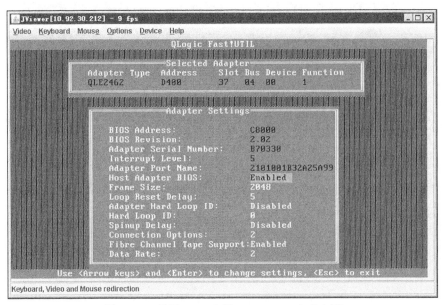

图 7-5-4　配置 ESXi 主机 SANBOOT 启动之四

第 5 步，开始配置 SAN BOOT，选择"Selectable Boot Settings"（如图 7-5-5 所示），按"回车"键。

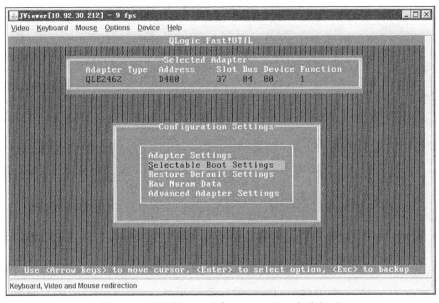

图 7-5-5　配置 ESXi 主机 SANBOOT 启动之五

第 6 步，将"Selectable Boot"状态修改为"Enabled"（如图 7-5-6 所示）。

第 7 步，默认情况下用于 BOOT 引导的 LUN 处于未配置状态，选择第一项（如图 7-5-7 所示），按"回车"键。

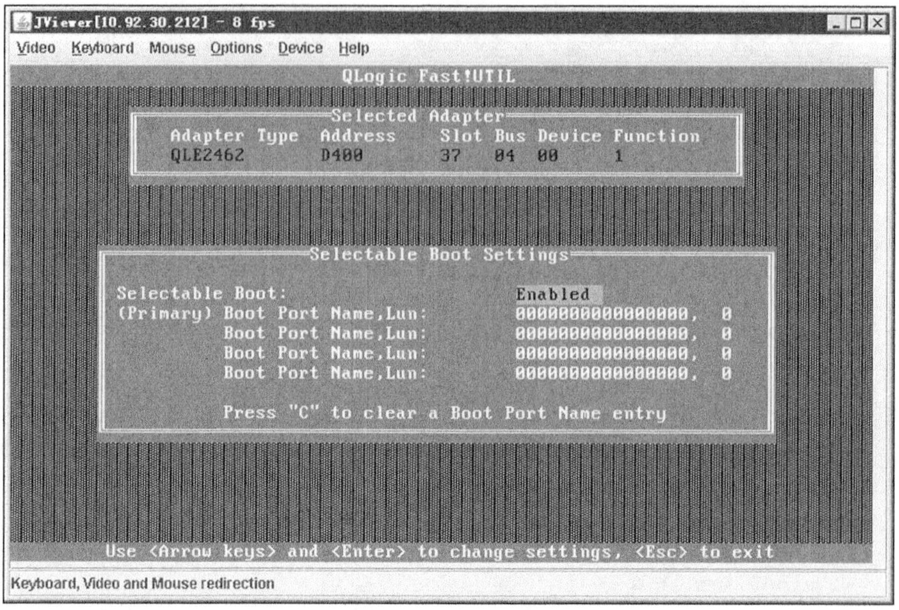

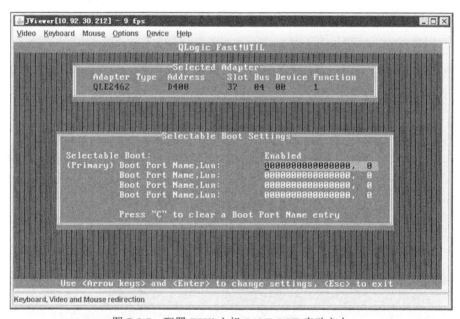

图 7-5-6 配置 ESXi 主机 SANBOOT 启动之六

图 7-5-7 配置 ESXi 主机 SANBOOT 启动之七

第 8 步，如果上述章节 DELL MD 存储以及 Cisco MDS 交换机配置没有问题，FC HBA 卡会找到 DELL MD 存储（如图 7-5-8 所示），按"回车"键。

第 9 步，选择需要使用的 LUN，在前面小节中创建了用于 ESXi 主机使用的 sanboot LUN，其容量为 10GB（如图 7-5-9 所示），按"回车"键。

7.5 配置 ESXi 主机使用 FC 存储　　353

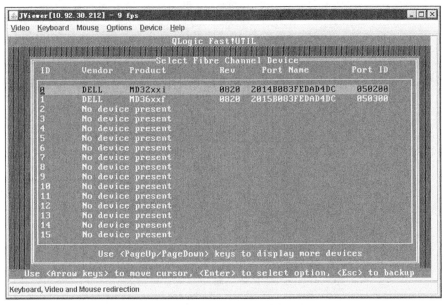

图 7-5-8　配置 ESXi 主机 SANBOOT 启动之八

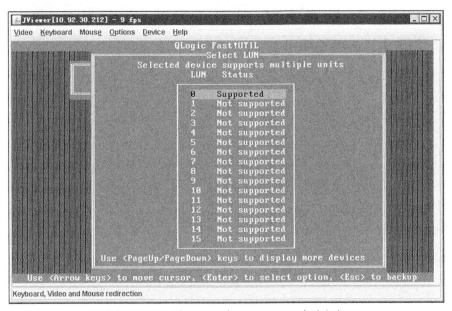

图 7-5-9　配置 ESXi 主机 SANBOOT 启动之九

第 10 步，用于 sanboot 引导的 LUN 配置完成（如图 7-5-10 所示），按"ESC"键退出配置。

第 11 步，选择"Save changes"保存相关的配置（如图 7-5-11 所示），按"回车"键。

第 12 步，重新启动服务器，可以看到 QLE 2462 的 FC HBA 卡下已存在 FC SAN 存储的 LUN，同时这台服务器未配置本地硬盘（如图 7-5-12 所示）。

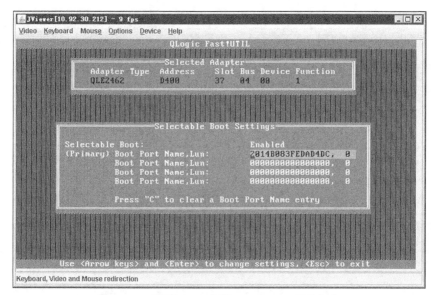

图 7-5-10 配置 ESXi 主机 SANBOOT 启动之十

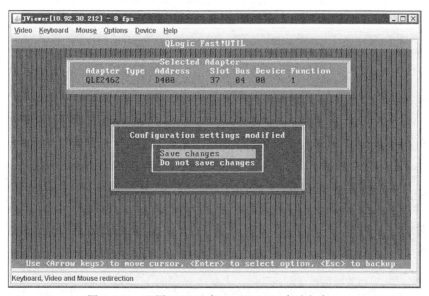

图 7-5-11 配置 ESXi 主机 SANBOOT 启动之十一

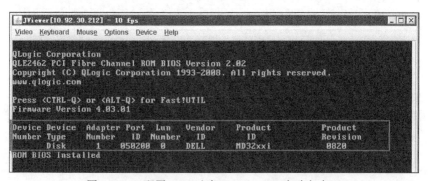

图 7-5-12 配置 ESXi 主机 SANBOOT 启动之十二

第 13 步，加载 ESXi 6.5 安装 ISO，在选择硬盘处可以看到 FC SAN 存储的 LUN（如图 7-5-13 所示），安装过程与之前介绍的完全一样。

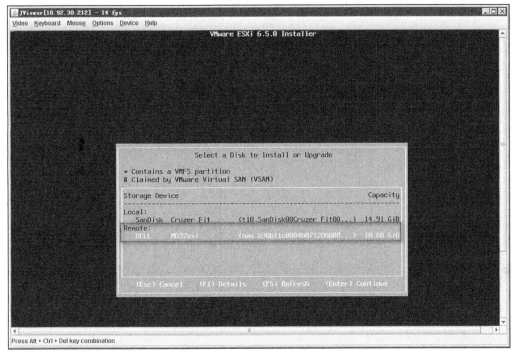

图 7-5-13　配置 ESXi 主机 SANBOOT 启动之十三

第 14 步，安装完成后使用 VMware vSphere Client 工具登录 ESXi12 主机，可以看到数据存储使用的都是 FC SAN 存储（如图 7-5-14 所示）。

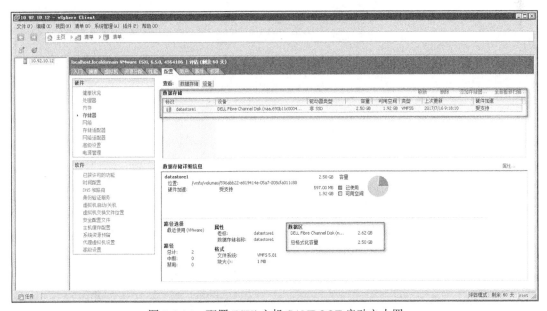

图 7-5-14　配置 ESXi 主机 SANBOOT 启动之十四

第 15 步，查看"存储适配器"，可以看到 DELL MD 存储相关信息（如图 7-5-15 所示）。

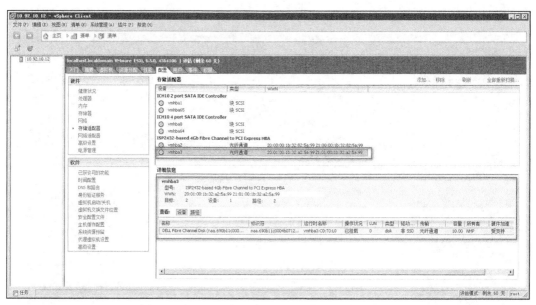

图 7-5-15　配置 ESXi 主机 SANBOOT 启动之十五

第 16 步，按照相同的操作方法配置 ESXi13 主机 sanboot 引导，完成后查看 ESXi13 主机的本地存储信息（如图 7-5-16 所示）。

图 7-5-16　配置 ESXi 主机 SANBOOT 启动之十六

第 17 步，查看 ESXi13 主机"存储适配器"，可以看到 DELL MD 存储相关信息（如图 7-5-17 所示）。

7.5 配置 ESXi 主机使用 FC 存储   357

图 7-5-17　配置 ESXi 主机 SANBOOT 启动之十七

7.5.2　配置 ESXi 主机使用 FC 共享存储

生产环境中，ESXi 主机除使用 sanboot 引导外，使用 FC 存储作为共享也是比较常见的。本小节介绍 ESXi 主机如何配置使用 FC 共享存储。关于 DELL MD 存储以及 FC 交换机配置可参考前面相关章节，此处不再演示。

第 1 步，ESXi13 主机配置有双口 FC HBA 卡，其中 vmhba3 用于 sanboot，vmhba2 用于访问 FC 共享存储（如图 7-5-18 所示）。

图 7-5-18　配置 ESXi 主机使用 FC 共享存储之一

第 2 步，查看 ESXi13 主机"存储适配器"，可以看到 DELL MD 存储相关信息，共容量为 100GB（如图 7-5-19 所示）。

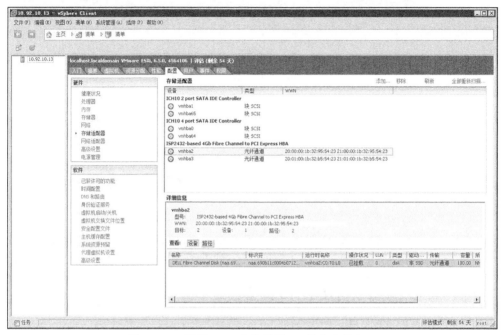

图 7-5-19　配置 ESXi 主机使用 FC 共享存储之二

第 3 步，查看 ESXi12 主机"存储器"信息，FC 共享存储未添加，因此数据存储仅有安装 ESXi 6.5 系统所使用的存储（如图 7-5-20 所示），单击"添加存储器"。

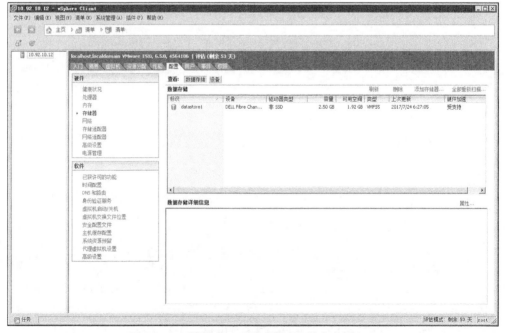

图 7-5-20　配置 ESXi 主机使用 FC 共享存储之三

第 4 步，选择添加存储器的类型（如图 7-5-21 所示），单击"下一步"按钮。

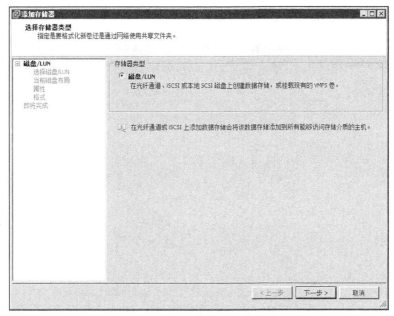

图 7-5-21　配置 ESXi 主机使用 FC 共享存储之四

第 5 步，ESXi 主机检测到未使用的 DELL MD 存储 LUN（如图 7-5-22 所示），单击"下一步"按钮。

图 7-5-22　配置 ESXi 主机使用 FC 共享存储之五

第 6 步，系统检查当前磁盘使用情况（如图 7-5-23 所示），单击"下一步"按钮。

图 7-5-23　配置 ESXi 主机使用 FC 共享存储之六

第 7 步，输入添加存储器的名称（如图 7-5-24 所示），单击 "下一步" 按钮。

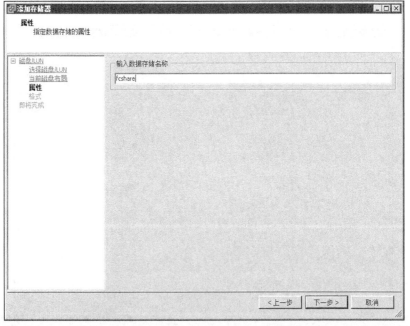

图 7-5-24　配置 ESXi 主机使用 FC 共享存储之七

第 8 步，指定数据存储使用的容量（如图 7-5-25 所示），单击 "下一步" 按钮。
第 9 步，完成数据存储的添加（如图 7-5-26 所示），单击 "完成" 按钮。

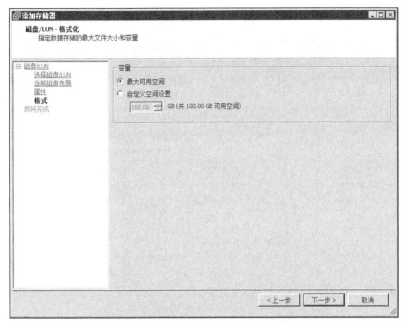

图 7-5-25　配置 ESXi 主机使用 FC 共享存储之八

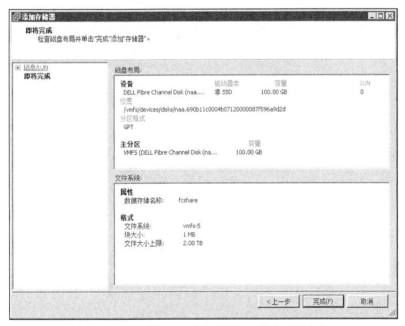

图 7-5-26　配置 ESXi 主机使用 FC 共享存储之九

第 10 步，查看 ESXi12 主机"存储器"信息，可以看到添加成功的 FC 数据存储（如图 7-5-27 所示）。

第 11 步，查看 ESXi13 主机"存储器"信息，可以看到添加成功的 FC 数据存储（如图 7-5-28 所示）。

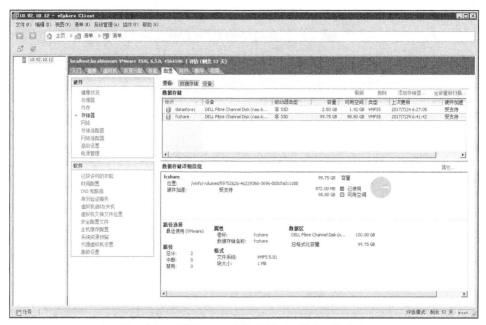

图 7-5-27　配置 ESXi 主机使用 FC 共享存储之十

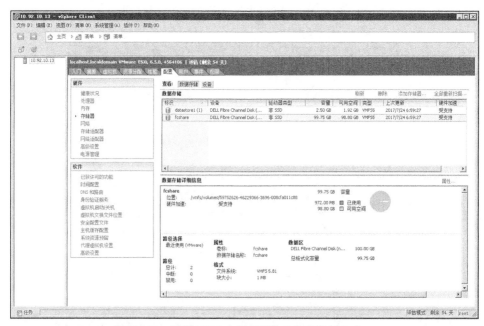

图 7-5-28　配置 ESXi 主机使用 FC 共享存储之十一

7.5.3　ESXi 主机在 FC 存储下高级特性使用

完成 FC 共享存储配置后，需要验证其高级特性的使用情况。本小节演示基于 FC 共享存储虚拟机的迁移。

第 1 步，将 ESX12、ESXi13 主机加入到 vCenter Server 进行统一管理，同时准备好虚拟机（如图 7-5-29 所示）。

7.5 配置 ESXi 主机使用 FC 存储   363

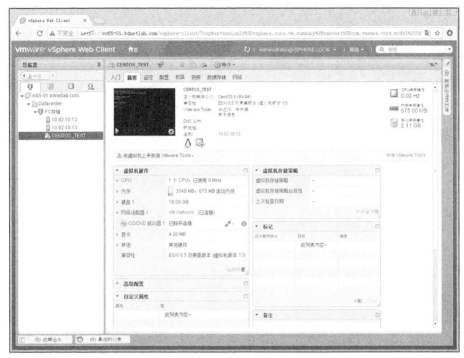

图 7-5-29　ESXi 主机在 FC 存储下高级特性的使用之一

第 2 步，名为 CENTOS_TEST 的虚拟机目前运行在 ESXi12 主机上（如图 7-5-30 所示），在虚拟机上单击右键，选择"迁移"。

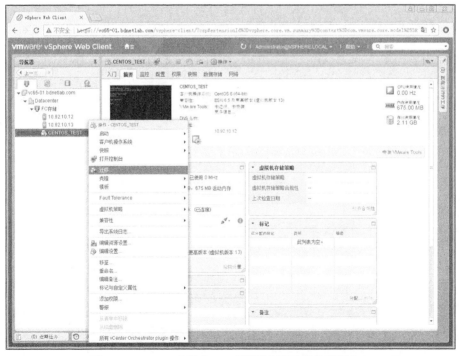

图 7-5-30　ESXi 主机在 FC 存储下高级特性的使用之二

第 3 步，选择迁移类型"仅更改计算资源"，将虚拟机迁移到另一个主机（如图 7-5-31 所示），单击"下一步"按钮。

图 7-5-31　ESXi 主机在 FC 存储下高级特性的使用之三

第 4 步，手动选择将虚拟机迁移到 ESXi13 主机上（如图 7-5-32 所示）。

图 7-5-32　ESXi 主机在 FC 存储下高级特性的使用之四

第 5 步，选择网络，根据生产环境的情况确认是否修改目标网络（如图 7-5-33 所示），单击"下一步"按钮。

图 7-5-33　ESXi 主机在 FC 存储下高级特性的使用之五

第 6 步，选择 vMotion 优先级，使用正常优先级即可（如图 7-5-34 所示），单击"下一步"按钮。

图 7-5-34　ESXi 主机在 FC 存储下高级特性的使用之六

第 7 步，确认迁移参数设置正确（如图 7-5-35 所示），单击"下一步"按钮。

366 第 7 章 部署使用 FC SAN 存储

图 7-5-35 ESXi 主机在 FC 存储下高级特性的使用之七

第 8 步，虚拟机 CENTOS_TEST 成功迁移到 ESXi13 主机（如图 7-5-36 所示），说明基于 FC 共享存储虚拟机迁移正常。

图 7-5-36 ESXi 主机在 FC 存储下高级特性的使用之八

7.6 配置 ESXi 主机使用 FCoE 存储

基于 FCoE 存储近几年来在数据中心得到大量应用，其主要因素之一是刀片服务器大量使用，FCoE 可以在不调整硬件的情况下使用 10GE 以太网传输 FC 数据。本节介绍如何配置 ESXi 主机使用 FCoE 存储。

7.6.1 配置 FCoE 存储准备工作

实验环境准备了两台 ESXi 主机用于 FCoE 存储配置，ESXi 主机均配置 1 个 10GE 以太网络适配器，连接至 Cisco Nexus 5548UP 融合交换机。

第 1 步，使用 VMware vSphere Client 客户端登录 ESXi 主机，确认主机 10GE 以太网络适配器正常工作（如图 7-6-1 所示）。

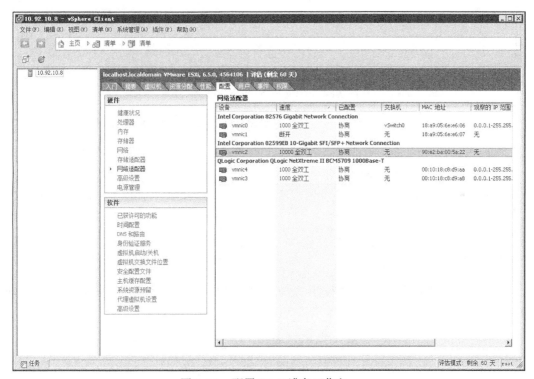

图 7-6-1　配置 FCoE 准备工作之一

第 2 步，在"存储适配器"中选择"添加"（如图 7-6-2 所示）。

第 3 步，选择"添加软件 FCoE 适配器"（如图 7-6-3 所示），单击"确定"按钮。生产环境如果预算允许，可以考虑购置硬件 FCoE 适配器。

第 4 步，将 ESXi 主机 10GE 适配器关联到软件 FCoE（如图 7-6-4 所示），单击"确定"按钮。

图 7-6-2 配置 FCoE 准备工作之二

图 7-6-3 配置 FCoE 准备工作之三

图 7-6-4 配置 FCoE 准备工作之四

第 5 步，关联软件 FCoE 适配器出现错误提示（如图 7-6-5 所示），原因是 ESXi 主机 10GE 适配器未关联虚拟交换机，单击"关闭"按钮。

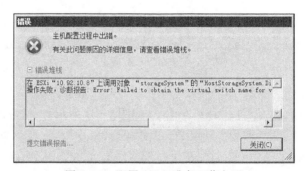

图 7-6-5 配置 FCoE 准备工作之五

第 6 步，单击"添加网络"（如图 7-6-6 所示）。

7.6 配置 ESXi 主机使用 FCoE 存储　　369

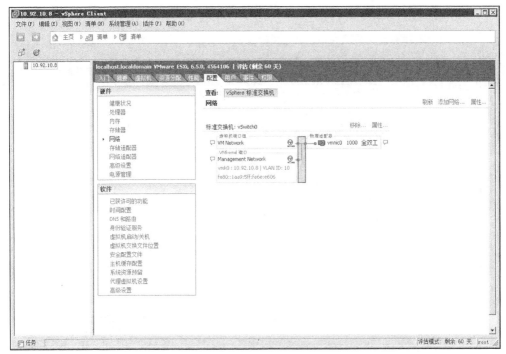

图 7-6-6　配置 FCoE 准备工作之六

第 7 步，选择基于"VMkernel"的连接类型（如图 7-6-7 所示），单击"下一步"按钮。

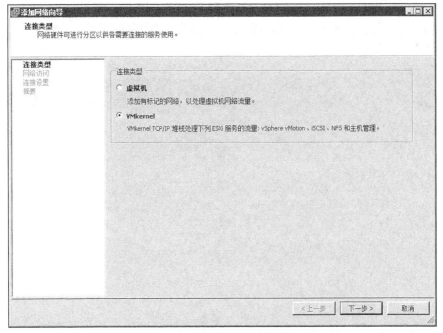

图 7-6-7　配置 FCoE 准备工作之七

第 8 步，勾选用于 FCoE 的 10GE 适配器 vmnic2（如图 7-6-8 所示），单击"下一步"按钮。

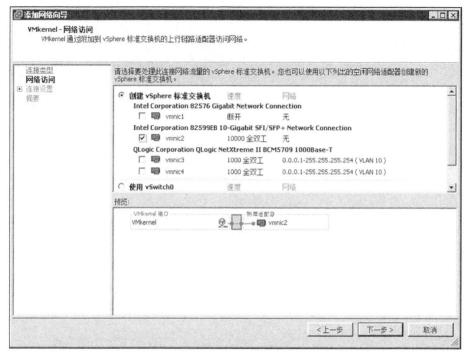

图 7-6-8　配置 FCoE 准备工作之八

第 9 步，对端口组进行命名，生产环境推荐配置便于日常管理（如图 7-6-9 所示），单击"下一步"按钮。

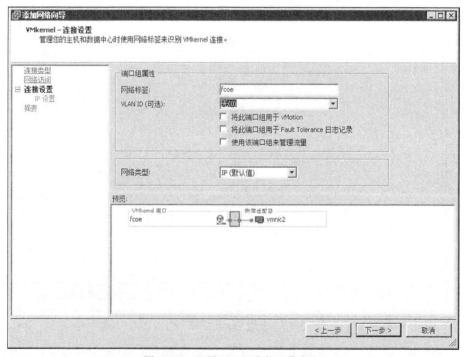

图 7-6-9　配置 FCoE 准备工作之九

第 10 步，VMkernel 需要配置 IP 地址连接，选择自动获取或者手动设置均可（如图 7-6-10 所示），单击"下一步"按钮。

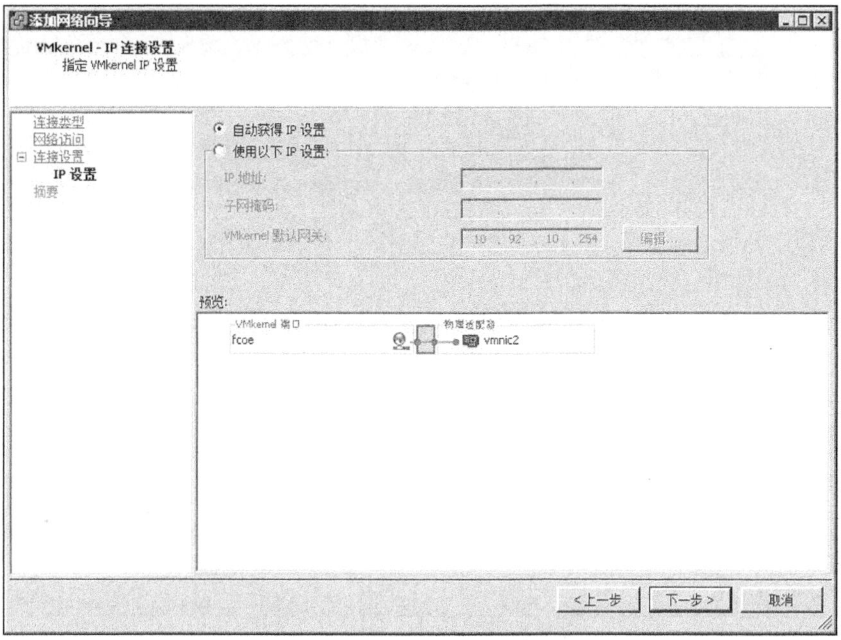

图 7-6-10　配置 FCoE 准备工作之十

第 11 步，确认标准交换机参数正确（如图 7-6-11 所示），单击"完成"按钮。

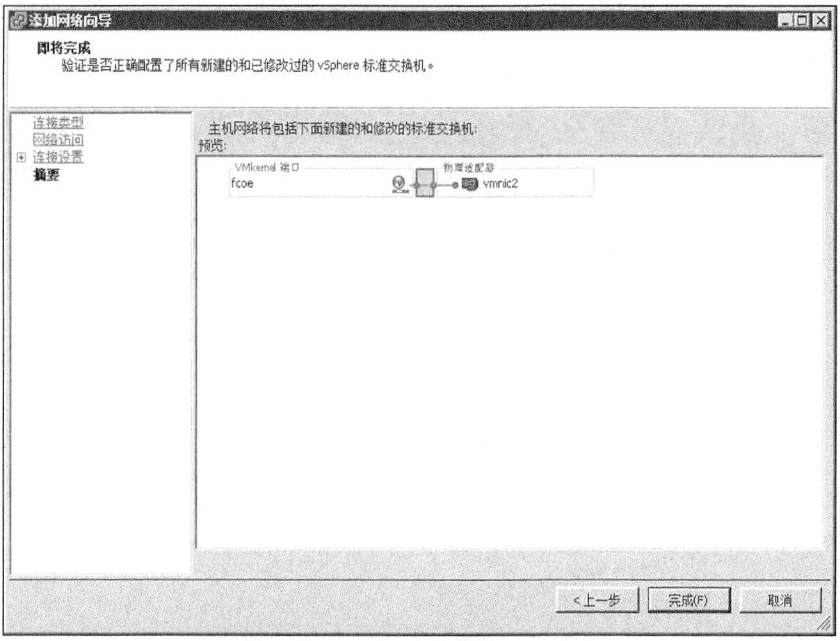

图 7-6-11　配置 FCoE 准备工作之十一

第 12 步，用于 FCoE 存储 10GE 适配器关联标准交换机完成（如图 7-6-12 所示）。

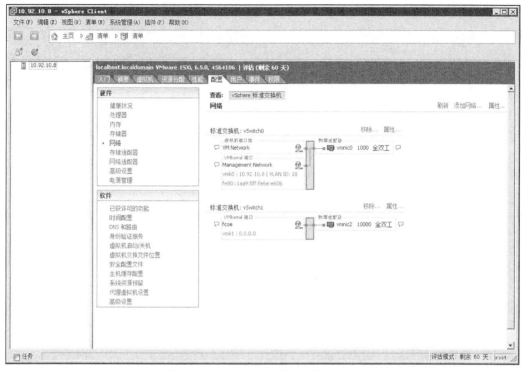

图 7-6-12　配置 FCoE 准备工作之十二

第 13 步，重新添加"软件 FCoE 适配器"成功（如图 7-6-13 所示）。

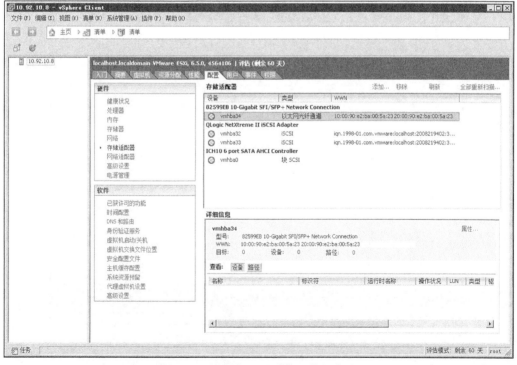

图 7-6-13　配置 FCoE 准备工作之十三

7.6.2 配置 FCoE 交换机

FCoE 存储配置中很重要的环节就是 FCoE 交换机配置，FCoE 交换机负责 FC 存储到以太网通道之间的数据转换，实验环境配置一台 Cisco Nexus 5548UP 交换机用于 FCoE 操作，该交换机可以通过命令的方式将接口切换为以太网接口或 FC 接口。本小节介绍 Cisco Nexus 5548UP 交换机 FCoE 存储配置。

第 1 步，登录 DC2-N5K-01 交换机，使用命令"show interface brief"查看接口信息，该设备一共有 32 个接口，目前均处于以太网接口模式。注意 Eth1/25-30 模块状态为"SFP validation failed"，因为该接口插入的是 SFP 存储模块，而接口配置处于传统以太网模式，所以会提示 SFP 验证失败，通过命令修改接口类型则恢复正常。

```
DC2-N5K-01# show interface brief
--------------------------------------------------------------------------
Ethernet      VLAN    Type Mode    Status  Reason                  Speed      Port
Interface                                                                      Ch #
Eth1/1        1       eth  fabric  up      none                    10G(D)   --
Eth1/2        1       eth  fabric  up      none                    10G(D)   --
Eth1/3        1       eth  access  down    Link not connected      10G(D)   --
Eth1/4        1       eth  access  down    Link not connected      10G(D)   --
Eth1/5        1       eth  access  down    SFP not inserted        10G(D)   --
Eth1/6        1       eth  access  down    SFP not inserted        10G(D)   --
Eth1/7        1       eth  access  up      none                    10G(D)   --
Eth1/8        1       eth  access  up      none                    10G(D)   --
……
Eth1/25       1       eth  access  down    SFP validation failed   10G(D)   --
Eth1/26       1       eth  access  down    SFP validation failed   10G(D)   --
Eth1/27       1       eth  access  down    SFP validation failed   10G(D)   --
Eth1/28       1       eth  access  down    SFP validation failed   10G(D)   --
Eth1/29       1       eth  access  down    SFP validation failed   10G(D)   --
Eth1/30       1       eth  access  down    SFP validation failed   10G(D)   --
Eth1/31       1       eth  access  down    SFP not inserted        10G(D)   --
Eth1/32       1       eth  access  down    SFP not inserted        10G(D)   --
--------------------------------------------------------------------------
```

第 2 步，通过命令行方式将接口修改为 FC 模式。

```
DC2-N5K-01# configure terminal      //进入全局配置模式
Enter configuration commands, one per line.   End with CNTL/Z.
DC2-N5K-01(config)# slot 1      //指定插槽位 1
DC2-N5K-01(config-slot)# port 25-32 type ?      //选择将具体端口修改为 FC
  ethernet    Ethernet Port
  fc          FC Port
DC2-N5K-01(config-slot)# port 25-32 type fc    //将 25-32 号端口修改为 FC 接口
Port type is changed. Please copy configuration and reload the switch
DC2-N5K-01(config-slot)# reload      //重新启动交换机
```

```
WARNING: There is unsaved configuration!!!
WARNING: This command will reboot the system
Do you want to continue? (y/n) [n] y
```

第 3 步，再次使用命令 "show interface brief" 查看接口信息，可以看到 24 个以太网接口，刚切换的 FC 接口不可见，其原因是未启用 FCoE 特性。

```
DC2-N5K-01# show interface bri
--------------------------------------------------------------------------------
Ethernet      VLAN   Type Mode    Status   Reason                Speed      Port
Interface                                                                   Ch #
--------------------------------------------------------------------------------
Eth1/1        1      eth  fabric  up       none                  10G(D)     --
Eth1/2        1      eth  fabric  up       none                  10G(D)     --
Eth1/3        1      eth  access  down     Link not connected    10G(D)     --
Eth1/4        1      eth  access  down     Link not connected    10G(D)     --
Eth1/5        1      eth  access  down     SFP not inserted      10G(D)     --
Eth1/6        1      eth  access  down     SFP not inserted      10G(D)     --
Eth1/7        1      eth  access  up       none                  10G(D)     --
Eth1/8        1      eth  access  up       none                  10G(D)     --
……
Eth1/19       1      eth  access  down     SFP not inserted      10G(D)     --
Eth1/20       1      eth  access  down     SFP not inserted      10G(D)     --
Eth1/21       1      eth  access  down     SFP not inserted      10G(D)     --
Eth1/22       1      eth  access  down     SFP not inserted      10G(D)     --
Eth1/23       1      eth  access  down     SFP not inserted      10G(D)     --
Eth1/24       1      eth  access  down     SFP not inserted      10G(D)     --
--------------------------------------------------------------------------------
```

第 4 步，默认情况下 Cisco Nexus 交换机未开启 FCoE 特性，使用命令 "feature fcoe" 开启。

```
DC2-N5K-01(config)# feature fcoe    //开启 FCoE 特性
FC license checked out successfully
fc_plugin extracted successfully
FC plugin loaded successfully
FCoE manager enabled successfully
FC enabled on all modules successfully
Enabled FCoE QoS policies successfully
```

第 5 步，再次使用命令 "show interface brief" 查看接口信息，可以看到 25-32 端口变为 FC 接口，其中 fc1/31-32 未插入模块。

```
DC2-N5K-01(config)# show interface brief
--------------------------------------------------------------------------------
Interface   Vsan   Admin   Admin   Status            SFP    Oper   Oper    Port
                   Mode    Trunk                            Mode   Speed   Channel
```

		Mode				(Gbps)	
fc1/25	1	auto	on	down	swl	--	--
fc1/26	1	auto	on	down	swl	--	--
fc1/27	1	auto	on	down	swl	--	--
fc1/28	1	auto	on	down	swl	--	--
fc1/29	1	auto	on	down	swl	--	--
fc1/30	1	auto	on	down	swl	--	--
fc1/31	1	auto	on	sfpAbsent	--	--	--
fc1/32	1	auto	on	sfpAbsent	--	--	--

第 6 步，配置 VSAN 数据库并将 DELL MD 存储接口划入 VSAN。

```
DC2-N5K-01(config)# vsan database
DC2-N5K-01(config-vsan-db)# vsan 21 name fcoe   //创建 VSAN 21 并命名为 fcoe
DC2-N5K-01(config-vsan-db)# vsan 21 interface fc1/25   //将 DELL MD 存储主控 0 接口加入 VSAN 21
Traffic on fc1/25 may be impacted. Do you want to continue? (y/n) [n] y
DC2-N5K-01(config-vsan-db)# vsan 21 interface fc1/26   //将 DELL MD 存储主控 1 接口加入 VSAN 21
Traffic on fc1/26 may be impacted. Do you want to continue? (y/n) [n] y
```

第 7 步，查看 VSAN 接口划入情况。

```
DC2-N5K-01(config)# show vsan membership
vsan 1 interfaces:
    fc1/27          fc1/28          fc1/29          fc1/30
    fc1/31          fc1/32
vsan 21 interfaces:
    fc1/25          fc1/26
vsan 4079(evfp_isolated_vsan) interfaces:
vsan 4094(isolated_vsan) interfaces:
```

第 8 步，ESXi 主机通过以太网访问 FC 存储，需要通过 VLAN 关联到 VSAN。

```
DC2-N5K-01(config)# vlan 21       //创建 VLAN 21
DC2-N5K-01(config-vlan)# fcoe vsan 21     //关联 VSAN 21
```

第 9 步，查看 VLAN 与 VSAN 之间的关联情况。

```
DC2-N5K-01(config)# show vlan fcoe
Original VLAN ID     Translated VSAN ID     Association State
---------------      ------------------     -----------------
      21                    21                 Operational
```

第 10 步，配置 ESXi 主机用于 FCoE 适配器。为了实现以太网与 FC SAN 数据之间的转换，需要配置虚拟的 FC 接口，然后再绑定以太网接口。

```
DC2-N5K-01(config-if)# interface e1/8    //连接 ESXi8 主机网络适配器
DC2-N5K-01(config-if)# switchport mode trunk   //配置为 TRUNK 模式
```

```
DC2-N5K-01(config-if-range)# no shutdown
DC2-N5K-01(config-if)# interface vfc 8            //创建虚拟 vfc 接口
DC2-N5K-01(config-if)# bind interface e1/8        //将物理接口绑定至 vfc 虚拟接口
DC2-N5K-01(config-if)# no shutdown
```

第 11 步，查看 vfc 虚拟接口信息。

```
DC2-N5K-01(config)# show interface vfc 8
vfc8 is trunking
    Bound interface is Ethernet1/8
    Hardware is Ethernet
    Port WWN is 20:07:54:7f:ee:99:8c:7f
    Admin port mode is F, trunk mode is on
    snmp link state traps are enabled
    Port mode is TF
    Port vsan is 1
    Trunk vsans (admin allowed and active) (1,21)
    Trunk vsans (up)                       ()
    Trunk vsans (isolated)                 ()
    Trunk vsans (initializing)             (1,21)
    1 minute input rate 0 bits/sec, 0 bytes/sec, 0 frames/sec
    1 minute output rate 0 bits/sec, 0 bytes/sec, 0 frames/sec
      0 frames input, 0 bytes
        0 discards, 0 errors
      0 frames output, 0 bytes
        0 discards, 0 errors
    last clearing of "show interface" counters Thu Apr 23 11:12:18 2009

    Interface last changed at Thu Apr 23 11:12:18 2009
```

第 12 步，将 vfc 虚拟接口划入 VSAN。

```
DC2-N5K-01(config)# vsan database
DC2-N5K-01(config-vsan-db)# vsan 21 interface vfc 8
```

第 13 步，使用命令"show vsan membership"查看接口划入情况。

```
DC2-N5K-01(config)# show vsan membership
vsan 1 interfaces:
    fc1/27          fc1/28          fc1/29          fc1/30
    fc1/31          fc1/32
vsan 21 interfaces:
    fc1/25          fc1/26          vfc8
vsan 4079(evfp_isolated_vsan) interfaces:
vsan 4094(isolated_vsan) interfaces:
```

第 14 步，使用命令"show flogi database"可以看到 DELL MD 存储以及 ESXi 主机注册信息。

7.6 配置 ESXi 主机使用 FCoE 存储

```
DC2-N5K-01# show flogi database
--------------------------------------------------------------------------------
INTERFACE        VSAN     FCID       PORT NAME                NODE NAME
--------------------------------------------------------------------------------
fc1/25           21       0xcb0000   20:44:b0:83:fe:da:d4:dc 20:04:b0:83:fe:da:d4:dc
fc1/26           21       0xcb0020   20:45:b0:83:fe:da:d4:dc 20:04:b0:83:fe:da:d4:dc
vfc8             21       0xcb0040   20:00:90:e2:ba:00:5a:23 10:00:90:e2:ba:00:5a:23
Total number of flogi = 3.
```

第 15 步，创建 ZONE 并将接口划入 ZONE。

```
DC2-N5K-01(config)# zone name fcoe vsan 21
DC2-N5K-01(config-zone)# member pwwn 20:00:90:e2:ba:00:5a:23
DC2-N5K-01(config-zone)# member interface fc1/25
DC2-N5K-01(config-zone)# member interface fc1/26
```

第 16 步，使用命令"show zone"查看 ZONE 信息。

```
DC2-N5K-01(config)# show zone
zone name fcoe vsan 21
  pwwn 20:00:90:e2:ba:00:5a:23
  interface fc1/25 swwn 20:00:54:7f:ee:99:8c:40
  interface fc1/26 swwn 20:00:54:7f:ee:99:8c:40
```

第 17 步，配置 ZONESET 并关联 ZONE。

```
DC2-N5K-01(config)# zoneset name zs_fcoe vsan 21
DC2-N5K-01(config-zoneset)# member fcoe
```

第 18 步，使用命令"show zoneset"查看 ZONESET 信息。

```
DC2-N5K-01(config)# show zoneset
zoneset name zs_fcoe vsan 21
  zone name fcoe vsan 21
    pwwn 20:00:90:e2:ba:00:5a:23
    interface fc1/25 swwn 20:00:54:7f:ee:99:8c:40
    interface fc1/26 swwn 20:00:54:7f:ee:99:8c:40
```

第 19 步，使用命令"zoneset activate"激活 ZONESET。

```
DC2-N5K-01(config)# zoneset activate name zs_fcoe vsan 21
Zoneset activation initiated. check zone status
```

第 20 步，使用命令"show zoneset active"查看激活 DELL MD 存储与 ESXi 主机之间的映射。

```
DC2-N5K-01(config)# show zoneset active
zoneset name zs_fcoe vsan 21
  zone name fcoe vsan 21
  * fcid 0xcb0040 [pwwn 20:00:90:e2:ba:00:5a:23]
```

```
* fcid 0xcb0000 [interface fc1/25 swwn 20:00:54:7f:ee:99:8c:40]
* fcid 0xcb0020 [interface fc1/26 swwn 20:00:54:7f:ee:99:8c:40]
```

7.6.3 使用 FCoE 存储

完成 FCoE 交换机相关配置后 ESXi 主机就可以使用 FCoE 存储。如果配置完成后不能看到存储相关信息，应对配置进行检查。

第 1 步，查看 ESXi 主机"存储适配器"，可以看到 FCoE 存储相关信息，共容量为 50GB（如图 7-6-14 所示）。

图 7-6-14　配置 ESXi 主机使用 FCoE 存储之一

第 2 步，查看 ESXi 主机"存储器"信息，FCoE 存储未添加，因此数据存储仅有安装 ESXi 6.5 系统所使用的存储（如图 7-6-15 所示），单击"添加存储器"。

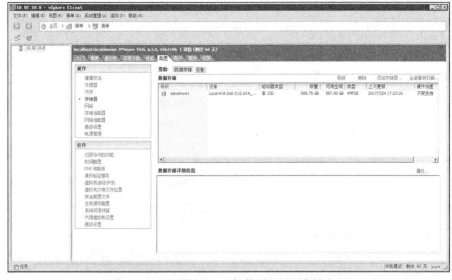

图 7-6-15　配置 ESXi 主机使用 FCoE 存储之二

7.6 配置 ESXi 主机使用 FCoE 存储　　379

第 3 步，选择添加存储器的类型（如图 7-6-16 所示），单击"下一步"按钮。

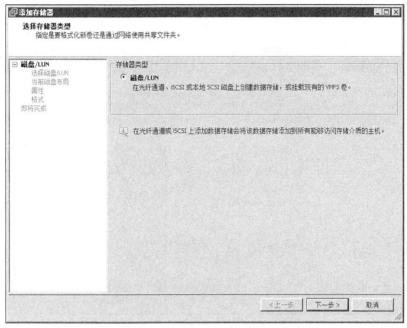

图 7-6-16　配置 ESXi 主机使用 FCoE 存储之三

第 4 步，ESXi 主机检测到未使用的 DELL MD 存储 LUN（如图 7-6-17 所示），单击"下一步"按钮。

图 7-6-17　配置 ESXi 主机使用 FCoE 存储之四

第 5 步，系统检查当前磁盘使用情况（如图 7-6-18 所示），单击"下一步"按钮。

图 7-6-18　配置 ESXi 主机使用 FCoE 存储之五

第 6 步，输入添加存储器的名称（如图 7-6-19 所示），单击"下一步"按钮。

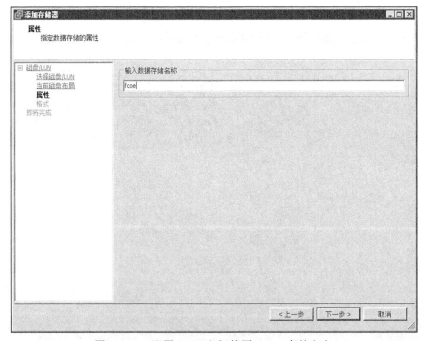

图 7-6-19　配置 ESXi 主机使用 FCoE 存储之六

第 7 步，指定数据存储使用的容量（如图 7-6-20 所示），单击"下一步"按钮。

图 7-6-20　配置 ESXi 主机使用 FCoE 存储之七

第 8 步，完成数据存储的添加（如图 7-6-21 所示），单击"完成"按钮。

图 7-6-21　配置 ESXi 主机使用 FCoE 存储之八

第 9 步，查看 ESXi 主机"存储器"信息，可以看到添加成功的 FCoE 数据存储（如图 7-6-22 所示）。

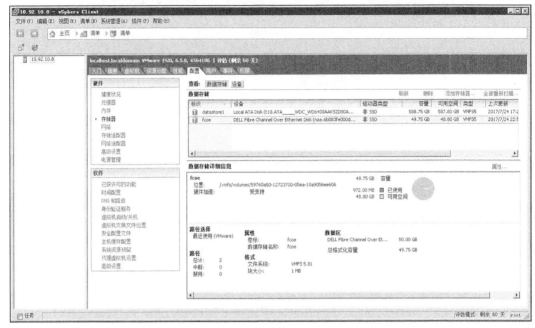

图 7-6-22　配置 ESXi 主机使用 FCoE 存储之九

第 10 步，按照上述方式配置其他 ESXi 主机 FCoE 存储（如图 7-6-23 所示）。

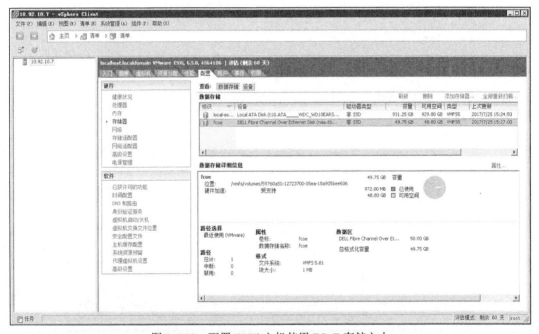

图 7-6-23　配置 ESXi 主机使用 FCoE 存储之十

ESXi 主机基于 FCoE 存储高级特性的使用与 FC 存储相同，均支持各种高级特性，读者可参考 FC 存储部分内容。

7.7 实验 FC 设备配置信息

在本章的最后部分，作者给出所涉及 FC 交换机以及 FCoE 交换机的完整配置信息供读者参考。

7.7.1 Cisco MDS 交换机配置信息

1. 基于 FC 存储 DC1-MDS-01 交换机完整配置

```
DC1-MDS-01# show running-config
!Command: show running-config
!Time: Tue Jul 25 06:04:33 2017
version 5.2(2)
feature npiv
feature telnet
role name default-role
  description This is a system defined role and applies to all users.
  rule 5 permit show feature environment
  rule 4 permit show feature hardware
  rule 3 permit show feature module
  rule 2 permit show feature snmp
  rule 1 permit show feature system
username admin password 5 $1$gjRu.j67$ZlO/qeVPiW5Kiz9FnNuZY0  role network-admin
banner motd #
Welcome to BDNETLAB DataCenter Lab
You are using now is "DC1-MDS-01"
#
ip domain-lookup
ip host DC1-MDS-01 10.92.30.249
aaa group server radius radius
snmp-server user admin network-admin auth md5 0x2ca7cd69de58ba6481deef6501d212bd priv 0x2ca7cd69d
e58ba6481deef6501d212bd localizedkey
rmon event 1 log trap public description FATAL(1) owner PMON@FATAL
rmon event 2 log trap public description CRITICAL(2) owner PMON@CRITICAL
rmon event 3 log trap public description ERROR(3) owner PMON@ERROR
rmon event 4 log trap public description WARNING(4) owner PMON@WARNING
rmon event 5 log trap public description INFORMATION(5) owner PMON@INFO
vsan database
  vsan 10 name "sanboot"
  vsan 20 name "sanshare"
cfs ipv4 distribute
device-alias database
  device-alias name esx12-share pwwn 21:00:00:1b:32:82:5a:99
```

```
    device-alias name esx13-share pwwn 21:00:00:1b:32:95:54:23

device-alias commit

fcdomain fcid database
    vsan 1 wwn 21:00:00:1b:32:82:01:9b fcid 0x320000 dynamic
    vsan 1 wwn 21:00:00:e0:8b:9a:67:13 fcid 0x320100 dynamic
    vsan 1 wwn 20:14:b0:83:fe:da:d4:dc fcid 0x320200 dynamic
    vsan 1 wwn 20:0b:00:05:9b:76:93:80 fcid 0x320300 dynamic
    vsan 1 wwn 21:01:00:e0:8b:ba:67:13 fcid 0x320301 dynamic
    vsan 1 wwn 21:01:00:1b:32:a2:01:9b fcid 0x320401 dynamic
    vsan 1 wwn 21:00:00:1b:32:04:63:b8 fcid 0x320600 dynamic
    vsan 1 wwn 20:0c:00:05:9b:76:93:80 fcid 0x320400 dynamic
    vsan 1 wwn 21:01:00:1b:32:a2:5a:99 fcid 0x320700 dynamic
    vsan 1 wwn 24:0b:00:05:9b:76:93:80 fcid 0x320500 dynamic
    vsan 1 wwn 21:01:00:1b:32:b5:54:23 fcid 0x320800 dynamic
    vsan 1 wwn 20:15:b0:83:fe:da:d4:dc fcid 0x320900 dynamic
    vsan 10 wwn 21:01:00:1b:32:a2:5a:99 fcid 0x050000 dynamic
    vsan 10 wwn 21:01:00:1b:32:b5:54:23 fcid 0x050100 dynamic
    vsan 10 wwn 20:14:b0:83:fe:da:d4:dc fcid 0x050200 dynamic
    vsan 10 wwn 20:15:b0:83:fe:da:d4:dc fcid 0x050300 dynamic
    vsan 10 wwn 20:0b:00:05:9b:76:93:80 fcid 0x050400 dynamic
    vsan 10 wwn 20:0c:00:05:9b:76:93:80 fcid 0x050500 dynamic
    vsan 20 wwn 20:0b:00:05:9b:76:93:80 fcid 0xe50000 dynamic
    vsan 20 wwn 20:0c:00:05:9b:76:93:80 fcid 0xe50100 dynamic
    vsan 20 wwn 21:00:00:1b:32:95:54:23 fcid 0xe50001 dynamic
!             [esx13-share]
    vsan 20 wwn 20:24:b0:83:fe:da:d4:dc fcid 0xe50101 dynamic
    vsan 20 wwn 20:25:b0:83:fe:da:d4:dc fcid 0xe50002 dynamic
    vsan 20 wwn 21:00:00:1b:32:82:5a:99 fcid 0xe50102 dynamic
!             [esx12-share]

vsan database
    vsan 10 interface fc1/5
    vsan 10 interface fc1/6
    vsan 20 interface fc1/11
    vsan 20 interface fc1/12
    vsan 10 interface fc1/15
    vsan 10 interface fc1/16
ip default-gateway 10.92.30.254
switchname DC1-MDS-01
line console
line vty
boot kickstart bootflash:/m9100-s2ek9-kickstart-mz.5.2.2.bin
boot system bootflash:/m9100-s2ek9-mz.5.2.2.bin
interface fc1/1
……        #省略接口
```

```
interface fc1/11
    switchport mode F
interface fc1/12
    switchport mode F
……
interface fc1/24
zone mode enhanced vsan 1
!Full Zone Database Section for vsan 10
zone name sanboot vsan 10
    member pwwn 21:01:00:1b:32:a2:5a:99
    member pwwn 21:01:00:1b:32:b5:54:23
    member interface fc1/15 swwn 20:00:00:0d:ec:90:e0:00
    member interface fc1/16 swwn 20:00:00:0d:ec:90:e0:00

zoneset name zs_sanboot vsan 10
    member sanboot

zoneset activate name zs_sanboot vsan 10
!Full Zone Database Section for vsan 20
zone name fcshare vsan 20
    member pwwn 21:00:00:1b:32:82:5a:99
!           [esx12-share]
    member pwwn 21:00:00:1b:32:95:54:23
!           [esx13-share]
    member pwwn 20:24:b0:83:fe:da:d4:dc
    member pwwn 20:25:b0:83:fe:da:d4:dc

zoneset name zs_fcshare vsan 20
    member fcshare

zoneset activate name zs_fcshare vsan 20

interface fc1/1
    port-license acquire
    no shutdown
……    #省略接口
interface fc1/11
    switchport trunk allowed vsan 10-20
    port-license acquire
    no shutdown

interface fc1/12
    switchport trunk allowed vsan 10-20
    port-license acquire
    no shutdown
……
interface mgmt0
```

```
    ip address 10.92.30.249 255.255.255.0
    switchport speed 100
```

2. 基于 FC 存储 DC1-MDS-02 交换机完整配置，启用 NPV 模式

```
DC1-MDS-02# show running-config
!Command: show running-config
!Time: Sun Sep  2 06:49:35 2001
version 5.0(4d)
feature telnet

role name default-role
    description This is a system defined role and applies to all users.
    rule 5 permit show feature environment
    rule 4 permit show feature hardware
    rule 3 permit show feature module
    rule 2 permit show feature snmp
    rule 1 permit show feature system
username admin password 5 $1$eiXE56q4$zO/1tiEJxF2sNhCJGmjuV/    role network-admin
ip domain-lookup
ip host DC1-MDS-02 10.92.30.248
aaa group server radius radius
rmon event 1 log trap public description FATAL(1) owner PMON@FATAL
rmon event 2 log trap public description CRITICAL(2) owner PMON@CRITICAL
rmon event 3 log trap public description ERROR(3) owner PMON@ERROR
rmon event 4 log trap public description WARNING(4) owner PMON@WARNING
rmon event 5 log trap public description INFORMATION(5) owner PMON@INFO

vsan database
    vsan 10
    vsan 20
cfs ipv4 distribute

vsan database
    vsan 20 interface fc1/5
    vsan 20 interface fc1/6
    vsan 20 interface fc1/11
    vsan 20 interface fc1/12
    vsan 20 interface fc1/15
    vsan 20 interface fc1/16

ip default-gateway 10.92.30.254
switchname DC1-MDS-02
line console
boot kickstart bootflash:/m9100-s2ek9-kickstart-mz.5.0.4d.bin
boot system bootflash:/m9100-s2ek9-mz.5.0.4d.bin
feature npv
```

```
interface fc1/1
   switchport mode F
……    #省略接口
interface fc1/11
   switchport mode NP
interface fc1/12
   switchport mode NP
……    #省略接口
interface fc1/1
   port-license acquire
……    #省略接口
interface fc1/11
   port-license acquire
   no shutdown

interface fc1/12
   port-license acquire
   no shutdown
interface mgmt0
   ip address 10.92.30.248 255.255.255.0
   switchport speed 100
```

7.7.2 Cisco Nexus 交换机配置信息

基于 FCoE 存储 DC2-N5K-01 交换机完整配置如下。

```
DC2-N5K-01# show running-config
!Command: show running-config
!Time: Fri Apr 24 03:59:18 2009
version 7.3(0)N1(1)
feature fcoe
hostname DC2-N5K-01
feature telnet
feature lacp
feature lldp
feature fex
username admin password 5 $1$XZkOxUET$S5djE6W2TUbJ86gcRIeLM1  role network-admin
no password strength-check
banner motd #
Welcome to BDNETLAB DataCenter Lab
You are using now is "DC2-N5K-01"
#
ip domain-lookup
control-plane
   service-policy input copp-system-policy-customized
policy-map type control-plane copp-system-policy-customized
   class copp-system-class-default
```

```
      police cir 2048 kbps bc 6400000 bytes
  fex 105
    pinning max-links 1
    description "FEX0105"
  slot 1
    port 1-24 type ethernet
    port 25-32 type fc
  snmp-server user admin network-admin auth md5 0x2e7a7da98ecd5a860d4510e8a094ec66 priv
0x2e7a7da98
    ecd5a860d4510e8a094ec66 localizedkey
  rmon event 1 log description FATAL(1) owner PMON@FATAL
  rmon event 2 log description CRITICAL(2) owner PMON@CRITICAL
  rmon event 3 log description ERROR(3) owner PMON@ERROR
  rmon event 4 log description WARNING(4) owner PMON@WARNING
  rmon event 5 log description INFORMATION(5) owner PMON@INFO
  system qos
    service-policy type queuing input fcoe-default-in-policy
    service-policy type queuing output fcoe-default-out-policy
    service-policy type qos input fcoe-default-in-policy
    service-policy type network-qos fcoe-default-nq-policy


  vlan 1, 10, 20
  vlan 21
    fcoe vsan 21
  vrf context management
    ip route 0.0.0.0/0 10.92.30.254
  vsan database
    vsan 21 name "fcoe"
  fcdomain fcid database
    vsan 1 wwn 20:44:b0:83:fe:da:d4:dc fcid 0x00ec0000 dynamic
    vsan 1 wwn 20:45:b0:83:fe:da:d4:dc fcid 0x00ec0020 dynamic
    vsan 21 wwn 20:44:b0:83:fe:da:d4:dc fcid 0x00cb0000 dynamic
    vsan 21 wwn 20:45:b0:83:fe:da:d4:dc fcid 0x00cb0020 dynamic
    vsan 21 wwn 20:00:90:e2:ba:00:5a:23 fcid 0x00cb0040 dynamic
    vsan 21 wwn 20:00:90:e2:ba:09:e4:eb fcid 0x00cb0060 dynamic

  interface port-channel201
    switchport mode trunk
    speed 10000


  interface vfc7
    bind interface Ethernet1/7
    no shutdown


  interface vfc8
    bind interface Ethernet1/8
    no shutdown
```

```
vsan database
  vsan 21 interface vfc7
  vsan 21 interface vfc8
  vsan 21 interface fc1/25
  vsan 21 interface fc1/26

interface fc1/25
  no shutdown

interface fc1/26
  no shutdown
……    #省略接口

interface Ethernet1/1
  switchport mode fex-fabric
  fex associate 105

interface Ethernet1/2
  switchport mode fex-fabric
  fex associate 105
……    #省略接口
interface Ethernet1/17
  switchport mode trunk
  channel-group 201 mode active

interface Ethernet1/18
  switchport mode trunk
  channel-group 201 mode active
……    #省略接口
interface mgmt0
  vrf member management
  ip address 10.92.30.244/24

interface Ethernet105/1/1
  switchport mode trunk
……    #省略接口
interface Ethernet105/1/48
  switchport mode trunk
line console
line vty
boot kickstart bootflash:/n5000-uk9-kickstart.7.3.0.N1.1.bin
boot system bootflash:/n5000-uk9.7.3.0.N1.1.bin
interface fc1/25
interface fc1/26
interface fc1/27
interface fc1/28
interface fc1/29
```

```
interface fc1/30
interface fc1/31
interface fc1/32
zone name fcoe vsan 21
   member interface fc1/25 swwn 20:00:54:7f:ee:99:8c:40
   member interface fc1/26 swwn 20:00:54:7f:ee:99:8c:40
   member pwwn 20:00:90:e2:ba:09:e4:eb
   member pwwn 20:00:90:e2:ba:00:5a:23
zoneset name zs_fcoe vsan 21
   member fcoe
zoneset activate name zs_fcoe vsan 21
```

7.8 本章小结

本章对 FC 存储进行了详细的介绍，包括 FC 存储的基本概念以及配置。为保证操作的有效性，特别使用了全物理设备进行操作。

对于 FC 存储来说，其重点在于 FC 存储本身以及 FC 交换机的配置，ESXi 主机仅作为客户端连接。

对于 FCoE 存储来说，其重点在于 FC 存储本身以及 FCoE 交换机配置。需要注意的是，FCoE 需要使用 10GE 以太网络。

第 8 章　部署使用 iSCSI 存储

2003 年 2 月，互联网工程任务组 IETF 批准通过 iSCSI 协议，这项由 IBM、Cisco 公司共同发起的技术标准，经过三年 20 个版本的不断完善，终于得到了 IETF 认可。这吸引了很多的厂商参与到相关产品的开发中来，也推动了更多的用户采用 iSCSI 的解决方案。本章介绍 iSCSI 的基本概念以及如何在 ESXi 主机上配置使用 iSCSI 存储。

本章要点
- iSCSI 协议介绍
- 配置 Open-E 存储服务器
- 配置 ESXi 主机使用 iSCSI 存储

8.1　iSCSI 协议介绍

iSCSI（Internet Small Computer System Interface）是通过 TCP/IP 网络传输 SCSI 指令的协议。iSCSI 协议参照 SAM-3（SCSI Architecture Model-3）制定。在 SAM-3 的体系结构中，iSCSI 属于传输层协议，在 TCP/IP 模型中属于应用层协议。

8.1.1　SCSI 协议介绍

在了解 iSCSI 协议前，需要了解 SCSI。SCSI 全称是 Small Computer System Interface（小型计算机接口）。SCSI 是 1979 年由美国的施加特公司（希捷的前身）研发并制定，由美国国家标准协会（ANSI）公布的接口标准。SCSI Architecture Model（SAM-3）用一种较松散的方式定义了 SCSI 的体系架构。

SCSI Architecture Model-3，是 SCSI 体系模型的标准规范，它自底向上分为 4 个层次。

物理连接层（Physical Interconnects）。如 Fibre Channel Arbitrated Loop、Fibre Channel Physical Interfaces；

SCSI 传输协议层（SCSI Transport Protocols）。如 SCSI Fibre Channel Protocol、Serial Bus Protocol、Internet SCSI；

共享指令集（SCSI Primary Command）。适用于所有设备类型；

专用指令集（Device-Type Specific Command Sets）。如块设备指令集 SBC（SCSI Block Commands）、流设备指令集 SSC（SCSI Stream Commands）、多媒体指令集 MMC（SCSI-3 Multimedia Command Set）。

简单地说，SCSI 定义了一系列规则提供给 I/O 设备，用以请求相互之间的服务。每个

I/O 设备称为"逻辑单元"（LU），每个逻辑单元都有唯一的地址来区分它们，这个地址称为"逻辑单元号"（LUN）。SCSI 模型采用客户端/服务器（C/S，Client/Server）模式，客户端称为 Initiator，服务器称为 Target。数据传输时，Initiator 向 Target 发送 request，Target 回应 response，在 iSCSI 协议中也沿用了这套思路。

8.1.2　iSCSI 协议基本概念

iSCSI 协议是集成了 SCSI 协议和 TCP/IP 协议的新协议。它在 SCSI 基础上扩展了网络功能，也就是可以让 SCSI 命令通过网络传送到远程 SCSI 设备上，而 SCSI 协议只能访问本地的 SCSI 设备。iSCSI 是传输层之上的协议，使用 TCP 连接建立会话。在 Initiator 端的 TCP 端口号随机选取，Target 的端口号默认是 3260。iSCSI 使用客户/服务器模型。客户端称为 Initiator，服务器端称为 Target。

Initiator：通常指用户主机系统，用户产生 SCSI 请求，并将 SCSI 命令和数据封装到 TCP/IP 包中发送到 IP 网络中。

Target：通常存在于存储设备上，用于转换 TCP/IP 包中的 SCSI 命令和数据。

8.1.3　iSCSI 协议名字规范

在 iSCSI 协议中，Initiator 和 Target 是通过名字进行通信的，因此，每一个 iSCSI 节点（即 Initiator）必须拥有一个 iSCSI 名字。iSCSI 协议定义了 3 类名称结构。

1. iqn（iSCSI Qualified Name）

格式为"iqn"+"年月"+"."+"域名的颠倒"+"："+"设备的具体名称"。之所以颠倒域名是为了避免可能的冲突。

2. eui（Extend Unique Identifier）

eui 来源于 IEEE 中的 EUI，格式是为"eui"+"64bits 的唯一标识（16 个字母）"。64bits 中，前 24bits（6 个字母）是公司的唯一标识，后面 40bits（10 个字母）是设备的标识。

3. naa（Network Address Authority）

由于 SAS 协议和 FC 协议都支持 naa，因此 iSCSI 协议定义也支持这种名字结构。naa 格式为"naa"+"64bits（16 个字母）或者 128bits（32 个字母）的唯一标识"。

8.2　配置 Open-E 存储服务器

第 6 章简要介绍了 Open-E 存储服务器的安装配置，本节介绍 Open-E 存储服务器磁盘创建、映射以及其他常用的配置。

8.2.1　配置 Open-E 存储磁盘

Open-E 存储提供多种存储服务支持，比较常见的是 iSCSI、NFS、FC 等，在使用前都需要对磁盘进行配置。本小节介绍 Open-E 存储磁盘的基本配置。

第 1 步，Open-E 支持软件以及硬件阵列，如果需要使用硬件阵列可参考官方兼容列表。实验环境使用软件阵列操作，单击"SETUP"菜单中"Software RAID"（如图 8-2-1 所示）。

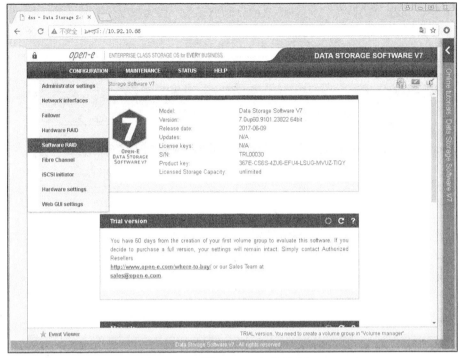

图 8-2-1　配置 Open-E 存储磁盘之一

第 2 步，实验使用的 Open-E 存储服务器配置 4 块 50GB 的磁盘，未进行任何阵列配置（如图 8-2-2 所示），Open-E 支持 RAID0、RAID1、RAID5、RAID6 阵列模式。

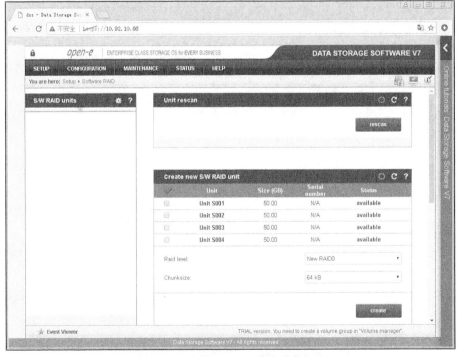

图 8-2-2　配置 Open-E 存储磁盘之二

第 3 步，勾选 4 块磁盘，同时将 RAID 模式选择为 RAID5（如图 8-2-3 所示），单击"create"按钮。

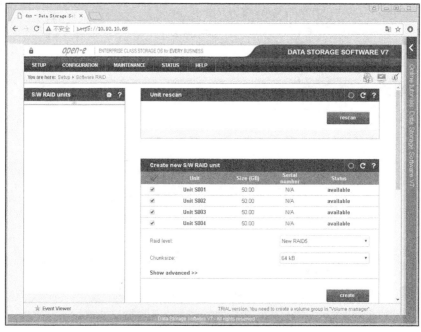

图 8-2-3　配置 Open-E 存储磁盘之三

第 4 步，软件 RAID5 阵列创建完成，阵列容量为 150GB（如图 8-2-4 所示）。

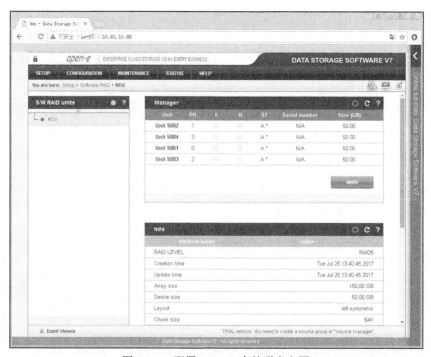

图 8-2-4　配置 Open-E 存储磁盘之四

第 5 步，单击"CONFIGURATION"菜单中"Volume manager"→"Volume groups"创建卷组（如图 8-2-5 所示）。

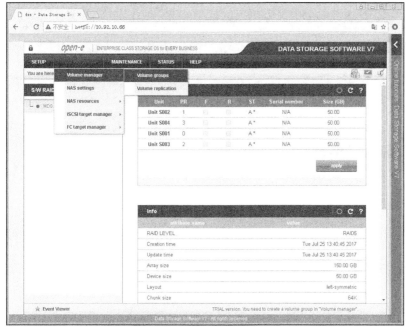

图 8-2-5　配置 Open-E 存储磁盘之五

第 6 步，卷组还未创建，可以使用的单元为刚创建的软件 RAID5 阵列，容量为 150GB（如图 8-2-6 所示）。

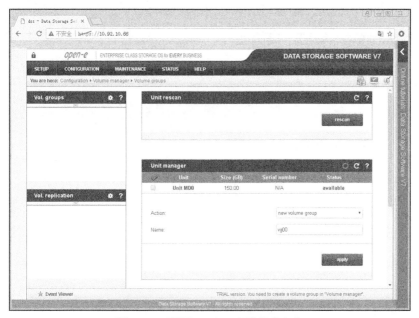

图 8-2-6　配置 Open-E 存储磁盘之六

第 7 步，勾选 Unit MD0，同时将卷组命名为 vgiscsi（如图 8-2-7 所示），单击"apply"按钮。

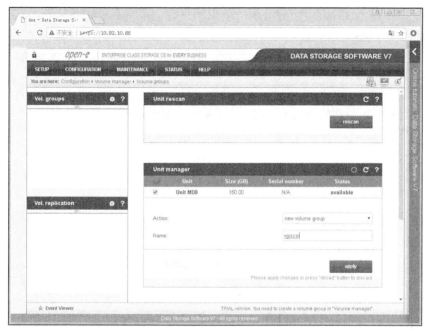

图 8-2-7　配置 Open-E 存储磁盘之七

第 8 步，名为 vgiscsi 的卷组创建完成，其状态变为 in use（如图 8-2-8 所示）。

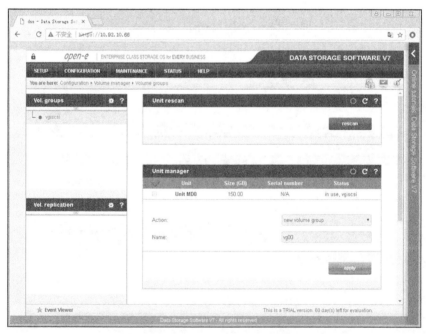

图 8-2-8　配置 Open-E 存储磁盘之八

8.2.2　配置 Open-E 存储 iSCSI 选项

Open-E 存储服务器磁盘配置以及卷组创建完成后，即可进行 iSCSI 以及其他配置。本小节介绍如何配置 iSCSI 选项。

第 1 步，选择 vgiscsi 卷组，卷组支持多种卷模式（如图 8-2-9 所示），可以创建 NAS 卷。

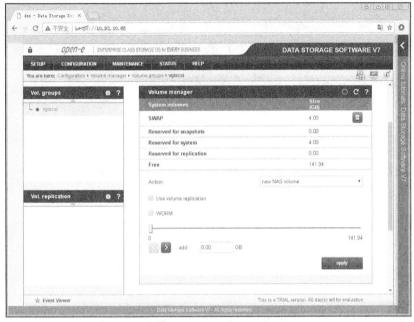

图 8-2-9　配置 Open-E 存储 iSCSI 选项之一

第 2 步，在"Action"选项中选择"new iSCSI volume"创建新的 iSCSI 卷，在"Options"选项中选择"Create new target automatically"自动创建新的 iSCSI target，iSCSI 存储的容量根据实际情况设置（如图 8-2-10 所示），单击"apply"按钮。

图 8-2-10　配置 Open-E 存储 iSCSI 选项之二

第 3 步，容量为 100GB 的 iscsi 卷创建完成（如图 8-2-11 所示）。

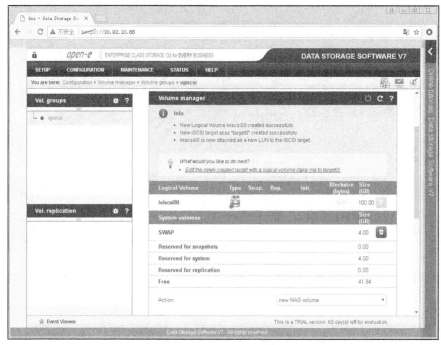

图 8-2-11　配置 Open-E 存储 iSCSI 选项之三

第 4 步，按照上述方法再创建一个容量为 20GB 的 iscsi 卷（如图 8-2-12 所示）。

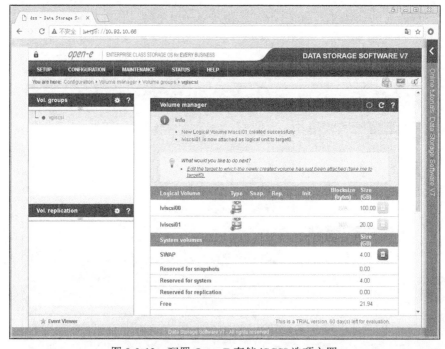

图 8-2-12　配置 Open-E 存储 iSCSI 选项之四

8.2.3 配置 Open-E 存储负载均衡

生产环境中使用 Open-E 存储，负载均衡以及容错非常重要，Open-E 存储基于 Linux 内核进行开发，使用 Linux 端口绑定提供负载均衡以及容错（如图 8-2-13 所示）。

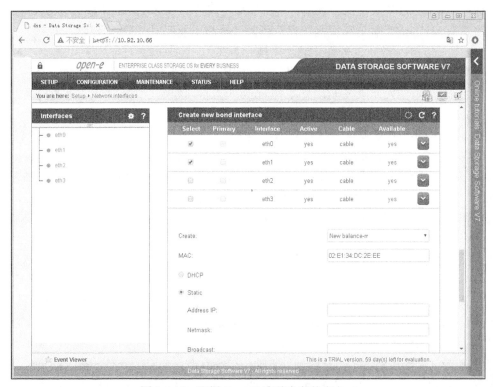

图 8-2-13　配置 Open-E 存储负载均衡之一

Open-E 存储负载均衡模式解释如下。

1. balance-rr（轮询负载模式）

标准链路负载均衡，增加带宽同时支持容错，当一条链路出现故障时会自动切换正常链路，物理交换机需要配置端口聚合。传输数据包顺序是依次传输（比如第 1 个数据包走 eth0，下一个包就走 eth1，一直循环下去，直到最后一个传输完毕），此模式提供负载平衡和容错能力；但是如果一个连接或者会话的数据包从不同的接口发出，中途再经过不同的链路，在客户端很有可能出现数据包无序到达的问题，而无序到达的数据包需要重新被发送，这样网络的吞吐量就会下降。

2. active-backup （主备模式）

主备模式下只有一块适配器处于 active 状态，另一块是备用的 standby 状态，所有流量都在 active 链路上处理，主备模式不能在物理交换机上配置端口聚合。此模式只提供容错能力，它的资源利用率较低，只有一个适配器处于工作状态。

3. balance-xor（基于 XOR HASH 模式）

基于 XOR Hash 负载均衡，基于指定的传输 HASH 策略传输数据包。缺省的策略是（源

MAC 地址 XOR 目标 MAC 地址）% slave 数量，其他的传输策略可以通过 xmit_hash_policy 选项指定。此模式提供负载平衡和容错能力。

4. Broadcast（广播模式）

所有包从所有网络接口发出，不提供负载均衡，只提供容错功能。

5. 802.3ad（IEEE 802.3ad 动态链接聚合）

标准的 802.3ad 协议，配合物理交换机的端口聚合 LACP 使用。该模式与 balance-rr 模式外的其他 bonding 负载均衡模式一样，任何连接都不能使用多于一个接口的带宽。

6. balance-tlb（适配器传输负载均衡模式）

根据每个 slave 的负载情况选择 slave 进行发送，接收时使用当前轮到的 slave。该模式要求 slave 接口的网络设备驱动有某种 ethtool 支持，该模式不需要物理交换机配置端口聚合。

7. alance-alb（适配器适应性负载均衡模式）

该模式包含 balance-tlb 模式，同时加上针对 IPV4 流量的接收负载均衡（receive load balance，rlb），而且不需要任何物理交换机端口聚合支持，接收负载均衡是通过 ARP 协商实现的。bonding 驱动截获本机发送的 ARP 应答，并把源硬件地址改写为 bond 中某个 slave 的唯一硬件地址，从而使得不同的对端使用不同的硬件地址进行通信。来自服务器端的接收流量也会被均衡。当本机发送 ARP 请求时，bonding 驱动把对端的 IP 信息从 ARP 包中复制并保存下来。当 ARP 应答从对端到达时，bonding 驱动把它的硬件地址提取出来，并发起一个 ARP 应答给 bond 中的某个 slave。使用 ARP 协商进行负载均衡的一个问题是：每次广播 ARP 请求时都会使用 bond 的硬件地址，因此对端学习到这个硬件地址后，接收流量将会全部流向当前的 slave。这个问题可以通过给所有的对端发送更新（ARP 应答）来解决，应答中包含它们独一无二的硬件地址，从而导致流量重新分布。当新的 slave 加入到 bond 中或者某个未激活的 slave 重新激活时，接收流量也要重新分布。接收的负载被顺序地分布（round robin）在 bond 中最高速的 slave 上，当某个链路被重新接上，或者一个新的 slave 加入到 bond 中，接收流量在所有当前激活的 slave 中全部重新分配，通过使用指定的 MAC 地址给每个 client 发起 ARP 应答。下面介绍的 updelay 参数必须被设置为某个大于等于 switch（交换机）转发延时的值，从而保证发往对端的 ARP 应答不会被 switch（交换机）阻截。

注意：balance-tlb 和 balance-alb 模式不需要配置物理交换机端口聚合，适配器自动形成聚合；802.3ad 模式需要物理交换机支持 802.3ad；balance-rr、balance-xor 和 broadcast 需要物理交换机配置端口聚合。

了解 Open-E 存储负载均衡模式后，下面配置主备模式来验证其负载均衡效果。

第 1 步，勾选 eth0 以及 eth1 进行端口绑定，所有的以太网适配器线路均连接，在 Create 选项处选择 "New active-backup"，输入 IP 地址信息（如图 8-2-14 所示），单击 "create" 按钮。

第 2 步，端口绑定配置完成，在 Interfaces 处新增了 bond0 接口（如图 8-2-15 所示）。

8.2 配置 Open-E 存储服务器　　401

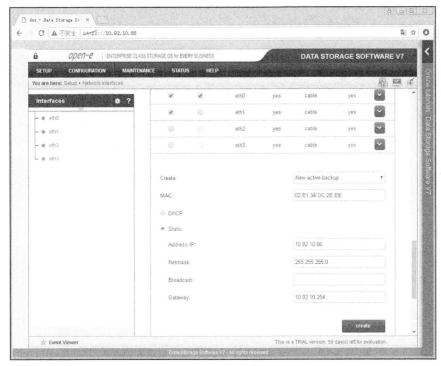

图 8-2-14　配置 Open-E 存储负载均衡之二

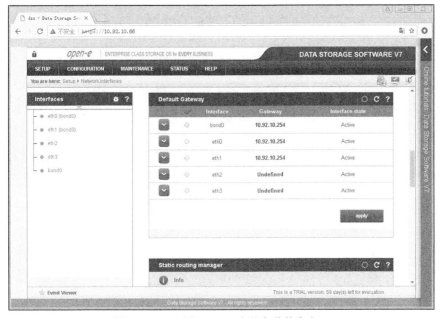

图 8-2-15　配置 Open-E 存储负载均衡之三

第 3 步，模拟故障，断开 Open-E 存储服务器 eth0 网络适配器，eth0 状态为 no cable（如图 8-2-16 所示）。

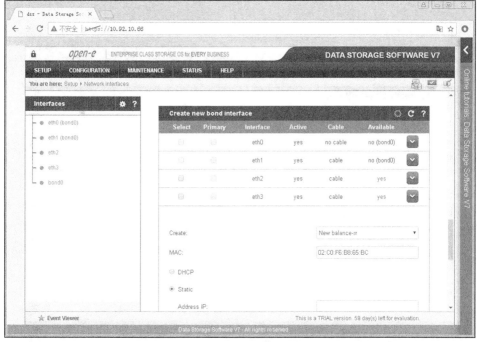

图 8-2-16　配置 Open-E 存储负载均衡之四

第 4 步，查看 ESXi 主机软件 iSCSI 适配器状态，可以看到 Open-E 存储服务器连接正常（如图 8-2-17 所示），说明已切换到 eth1 网络适配器提供 iSCSI 存储访问，容错配置生效。

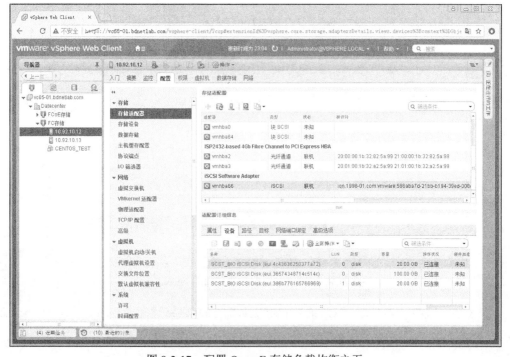

图 8-2-17　配置 Open-E 存储负载均衡之五

第 5 步，模拟故障，再断开 Open-E 存储服务器 eth1 网络适配器（如图 8-2-18 所示），iSCSI 适配器已经无法访问 Open-E 存储服务器，虚拟机 CENTOS_TEST 处于 inaccessible 状态。

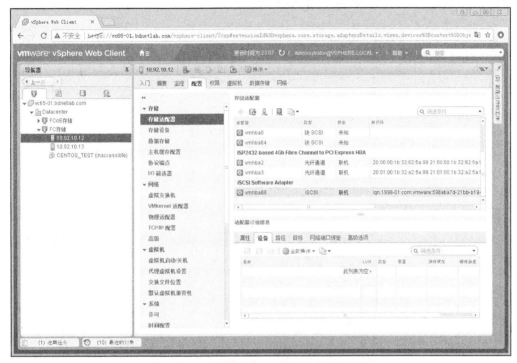

图 8-2-18　配置 Open-E 存储负载均衡之六

8.3　配置 ESXi 主机使用 iSCSI 存储

Open-E 存储 iSCSI 卷创建完成后，ESXi 主机就可以配置使用。本节介绍 ESXi 主机如何配置使用 iSCSI 存储以及需要注意的事项。

8.3.1　配置 ESXi 主机启用 iSCSI 存储

一般生产环境未配置专用的 iSCSI HBA 卡，通常使用普通以太网适配器承载 iSCSI 流量，也就是软件 iSCSI 适配器。

第 1 步，使用浏览器登录 vCenter Server，确定需要 ESXi 主机，选择"管理"→"存储适配器"，单击"添加软件 iSCSI 适配器"（如图 8-3-1 所示）。

第 2 步，系统提示将向列表中添加软件 iSCSI 适配器（如图 8-3-2 所示），单击"确定"按钮。

第 3 步，ESXi12 主机软件 iSCSI 适配器添加成功，软件 iSCSI 适配器名为 vmhba66（如图 8-3-3 所示）。

第 8 章 部署使用 iSCSI 存储

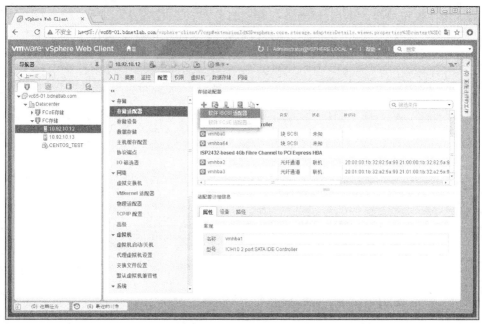

图 8-3-1 配置 ESXi 主机使用 iSCSI 存储之一

图 8-3-2 配置 ESXi 主机使用 iSCSI 存储之二

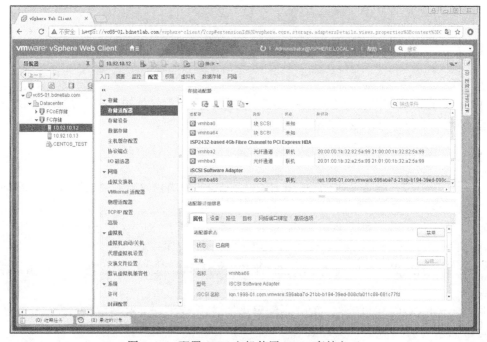

图 8-3-3 配置 ESXi 主机使用 iSCSI 存储之三

8.3 配置 ESXi 主机使用 iSCSI 存储

第 4 步，单击软件 iSCSI 适配器，在"目标"→"动态发现"处选择"添加"（如图 8-3-4 所示）。

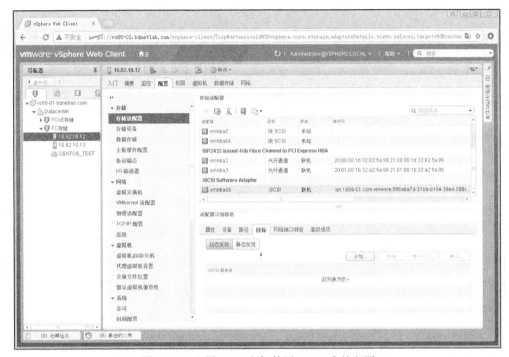

图 8-3-4 配置 ESXi 主机使用 iSCSI 存储之四

第 5 步，输入 iSCSI 服务器 IP 地址，端口默认为 3260（如图 8-3-5 所示），单击"确定"按钮。

图 8-3-5 配置 ESXi 主机使用 iSCSI 存储之五

第 6 步，添加完成后系统会提示存储发生了更改，需要重新扫描存储适配器（如图 8-3-6 所示），单击"重新扫描"按钮。

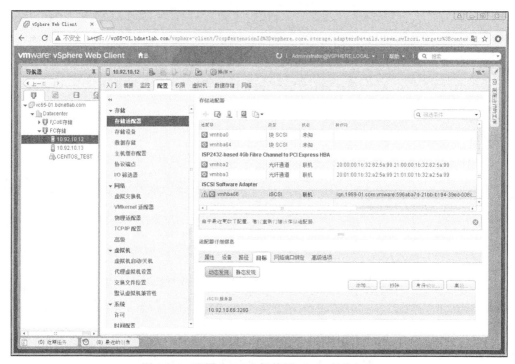

图 8-3-6　配置 ESXi 主机使用 iSCSI 存储之六

第 7 步，勾选"扫描新的存储设备"以及"扫描新的 VMFS 卷"重新扫描存储（如图 8-3-7 所示），单击"确定"按钮。

图 8-3-7　配置 ESXi 主机使用 iSCSI 存储之七

第 8 步，扫描存储适配器完成，在设备选项中可以看到刚创建的容量为 100GB 以及 20GB 的 iSCSI 卷（如图 8-3-8 所示）。

第 9 步，查看 ESXi 主机"数据存储"信息，目前 ESXi 主机有安装 ESXi 系统使用的存储以及 FC 存储，单击"新建数据存储"（如图 8-3-9 所示）。

8.3 配置 ESXi 主机使用 iSCSI 存储　407

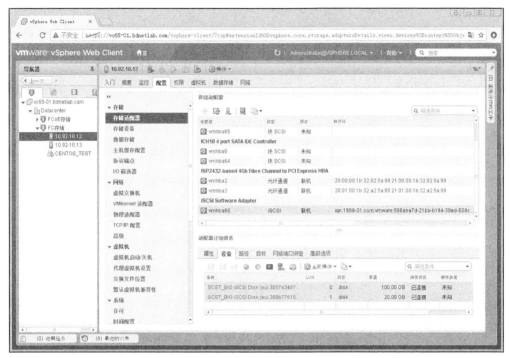

图 8-3-8　配置 ESXi 主机使用 iSCSI 存储之八

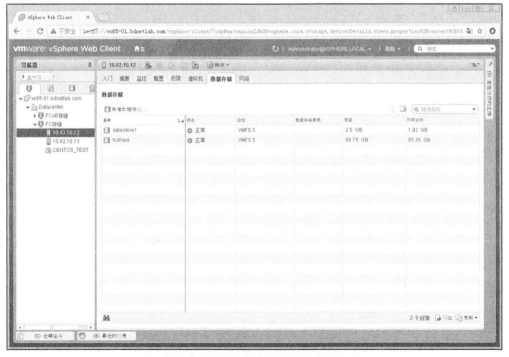

图 8-3-9　配置 ESXi 主机使用 iSCSI 存储之九

第 10 步，选择新建数据存储类型为 **VMFS**（如图 8-3-10 所示），单击 "下一步" 按钮。

图 8-3-10　配置 ESXi 主机使用 iSCSI 存储之十

第 11 步，输入数据存储名称并选择容量为 100GB 的卷（如图 8-3-11 所示），单击"下一步"按钮。

图 8-3-11　配置 ESXi 主机使用 iSCSI 存储之十一

第 12 步，选择新建数据存储 VMFS 版本，ESXi 6.5 系统 VMFS 版本已经升级到 VMFS 6（如图 8-3-12 所示），单击"下一步"按钮。

图 8-3-12　配置 ESXi 主机使用 iSCSI 存储之十二

第 13 步，配置新建数据存储的分区（如图 8-3-13 所示），单击"下一步"按钮。

图 8-3-13　配置 ESXi 主机使用 iSCSI 存储之十三

第 14 步，确认新建数据存储参数设置正确（如图 8-3-14 所示），单击"完成"按钮。

图 8-3-14　配置 ESXi 主机使用 iSCSI 存储之十四

第 15 步，名为 iscsi 的数据存储创建完成（如图 8-3-15 所示）。

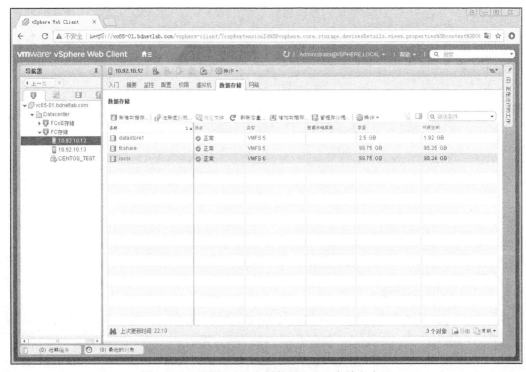

图 8-3-15　配置 ESXi 主机使用 iSCSI 存储之十五

第 16 步，经过配置，ESXi13 主机也可以看到该存储（如图 8-3-16 所示）。

图 8-3-16　配置 ESXi 主机使用 iSCSI 存储之十六

第 17 步，在 FC 存储集群上单击"数据存储"，iscsi 数据存储容量为 99.75GB（如图 8-3-17 所示），将未配置的 20GB 卷添加到该存储，单击"增加数据存储容量"。

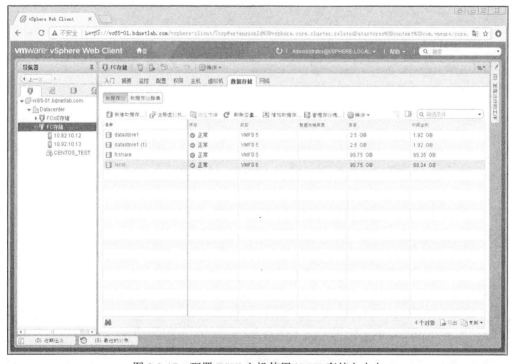

图 8-3-17　配置 ESXi 主机使用 iSCSI 存储之十七

第 18 步，选择未配置的 20GB 存储卷（如图 8-3-18 所示），单击"下一步"按钮。

图 8-3-18　配置 ESXi 主机使用 iSCSI 存储之十八

第 19 步，配置增加数据存储的分区（如图 8-3-19 所示），单击"下一步"按钮。

图 8-3-19　配置 ESXi 主机使用 iSCSI 存储之十九

第 20 步，确认增加数据存储参数设置正确（如图 8-3-20 所示），单击"完成"按钮。

8.3　配置 ESXi 主机使用 iSCSI 存储　　413

图 8-3-20　配置 ESXi 主机使用 iSCSI 存储之二十

第 21 步，iscsi 数据存储增加容量完成，目前容量为 119.5GB（如图 8-3-21 所示）。

图 8-3-21　配置 ESXi 主机使用 iSCSI 存储之二十一

8.3.2　配置 ESXi 主机绑定 iSCSI 流量

生产环境中使用 iSCSI 存储，大多数是采用物理适配器运行 iSCSI 存储流量。为了保证其传输的效率不受影响，VMware vSphere 提供了 iSCSI 存储网络绑定的特性，也就是可以将 iSCSI 存储流量指定在某个物理适配器独立运行。本小节介绍如何绑定 iSCSI 存储网络。

第 1 步，生产环境推荐使用独立虚拟交换机 iSCSI 存储流量，确定配置的 ESXi 主机，选择"配置"→"网络"→"VMkernel 适配器"（如图 8-3-22 所示），单击"添加主机网络"。

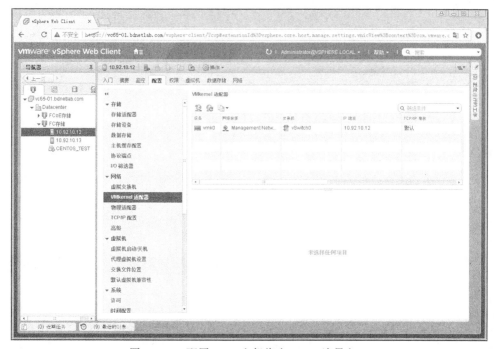

图 8-3-22　配置 ESXi 主机绑定 iSCSI 流量之一

第 2 步，iSCSI 需要使用 VMkernel 网络适配器，选择"VMkernel 网络适配器"（如图 8-3-23 所示），单击"下一步"按钮。

图 8-3-23　配置 ESXi 主机绑定 iSCSI 流量之二

第 3 步，选择新建标准交换机绑定 iSCSI 存储流量（如图 8-3-24 所示），单击"下一步"按钮。

图 8-3-24　配置 ESXi 主机绑定 iSCSI 流量之三

第 4 步，为新创建标准交换机分配适配器（如图 8-3-25 所示），单击"添加适配器"按钮。

图 8-3-25　配置 ESXi 主机绑定 iSCSI 流量之四

第 5 步，选择 vmnic1 作为新创建标准交换机的上行链路适配器（如图 8-3-26 所示），单击"确定"按钮。

图 8-3-26　配置 ESXi 主机绑定 iSCSI 流量之五

第 6 步，确认将 vmnic1 适配器添加到新创建的标准交换机（如图 8-3-27 所示），单击"下一步"按钮。

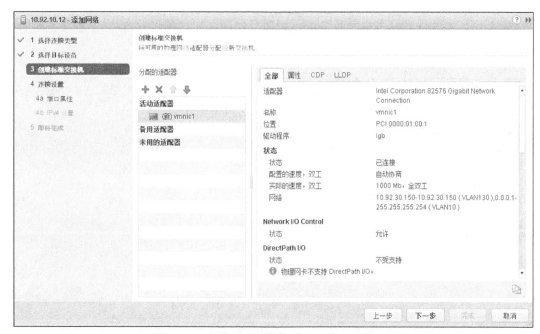

图 8-3-27　配置 ESXi 主机绑定 iSCSI 流量之六

第 7 步，配置 VMkernel 端口属性（如图 8-3-28 所示），单击"下一步"按钮。

图 8-3-28　配置 ESXi 主机绑定 iSCSI 流量之七

第 8 步，配置 VMkernel 端口 IP 地址（如图 8-3-29 所示），单击"下一步"按钮。

图 8-3-29　配置 ESXi 主机绑定 iSCSI 流量之八

第 9 步，确认参数配置正确（如图 8-3-30 所示），单击"完成"按钮。

图 8-3-30 配置 ESXi 主机绑定 iSCSI 流量之九

第 10 步，名为 vSwitch1 的标准交换机创建完成（如图 8-3-31 所示）。

图 8-3-31 配置 ESXi 主机绑定 iSCSI 流量之十

第 11 步，查看 ESXi 主机 iSCSI 软件适配器网络端口绑定，目前没有任何 VMkernel 网络适配器绑定到此 iSCSI 主机总线适配器（如图 8-3-32 所示），单击"添加"。

8.3 配置 ESXi 主机使用 iSCSI 存储 419

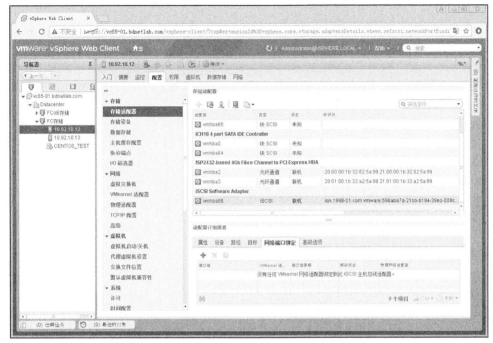

图 8-3-32 配置 ESXi 主机绑定 iSCSI 流量之十一

第 12 步，勾选需要绑定的 VMkernel 网络适配器，系统会检测端口组策略是否为"合规"状态（如图 8-3-33 所示），单击"确定"按钮。

图 8-3-33 配置 ESXi 主机绑定 iSCSI 流量之十二

第 13 步，添加完成后系统会提示存储发生了更改，需要重新扫描存储适配器（如图 8-3-34 所示），单击"重新扫描"按钮。

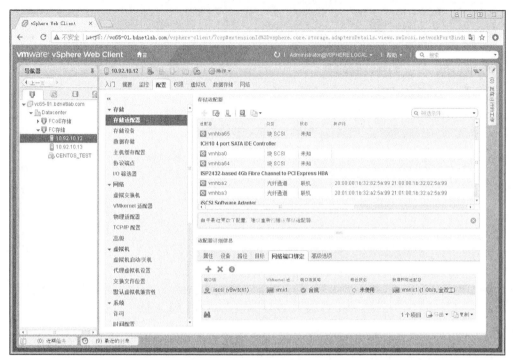

图 8-3-34　配置 ESXi 主机绑定 iSCSI 流量之十三

第 14 步，勾选"扫描新的存储设备"以及"扫描新的 VMFS 卷"重新扫描存储（如图 8-3-35 所示），单击"确定"按钮。

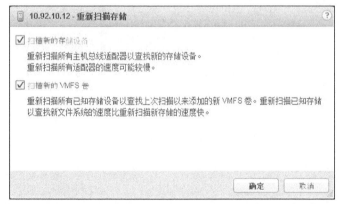

图 8-3-35　配置 ESXi 主机绑定 iSCSI 流量之十四

第 15 步，重新扫描存储完成后，ESXi 主机 iSCSI 软件适配器网络端口绑定完成且处于活动状态（如图 8-3-36 所示）。

8.3 配置 ESXi 主机使用 iSCSI 存储

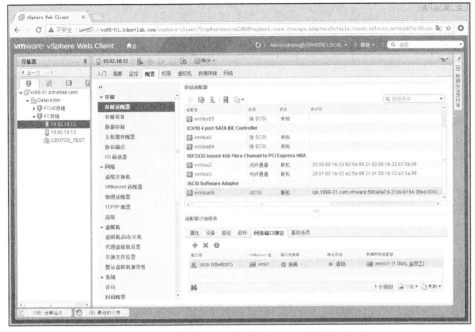

图 8-3-36 配置 ESXi 主机绑定 iSCSI 流量之十五

8.3.3 配置使用 ESXi 主机高级特性

完成 ESXi 主机 iSCSI 基本应用与 iSCSI 流量绑定后，可以测试 ESXi 主机在 iSCSI 存储下虚拟机迁移是否正常。

第 1 步，虚拟机 CENTOS_TEST 目前使用的存储为 FC 存储（如图 8-3-37 所示）。

图 8-3-37 配置使用 ESXi 主机高级特性之一

第 2 步，确定虚拟机迁移类型，选择更改存储（如图 8-3-38 所示），单击"下一步"按钮。

图 8-3-38　配置使用 ESXi 主机高级特性之二

第 3 步，选择将虚拟机存储迁移到 iscsi 存储，虚拟磁盘格式以及虚拟机存储策略使用现有的设置（如图 8-3-39 所示），单击"下一步"按钮。

图 8-3-39　配置使用 ESXi 主机高级特性之三

第 4 步，确认配置参数正确（如图 8-3-40 所示），单击"完成"按钮。

图 8-3-40　配置使用 ESXi 主机高级特性之四

第 5 步，虚拟机 CENTOS_TEST 使用的存储从 FC 存储顺利迁移到 iSCSI 存储（如图 8-3-41 所示）。

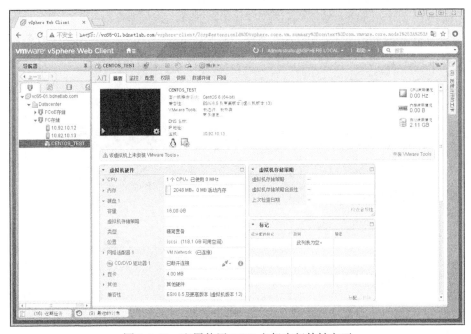

图 8-3-41　配置使用 ESXi 主机高级特性之五

8.3.4　配置 ESXi 主机启用 iSCSI 安全特性

生产环境中使用 iSCSI 存储，为保证存储安全性，推荐启用其安全特性。无论是 Open-E 存储还是 ESXi 主机，均支持基于 iSCSI CHAP 认证。

第 1 步，参考前面章节，在 Open-E 存储上再创建一个容量为 20GB 的 iSCSI 卷（如图 8-3-42 所示）。

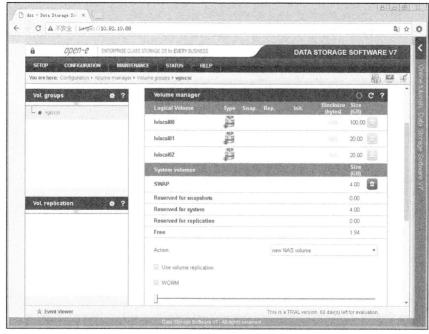

图 8-3-42　配置 ESXi 主机使用 iSCSI 安全特性之一

第 2 步，单击"CONFIGURATION"菜单中的"iSCSI target manager"，选择"CHAP users"（如图 8-3-43 所示）。

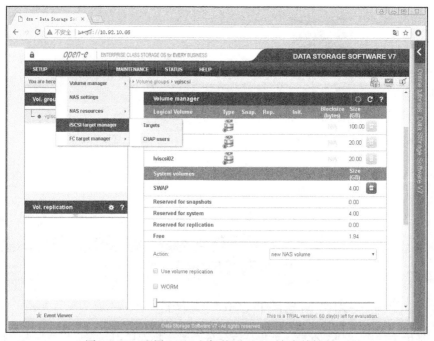

图 8-3-43　配置 ESXi 主机使用 iSCSI 安全特性之二

第 3 步，创建一个新的 CHAP user（如图 8-3-44 所示），单击"create"按钮。

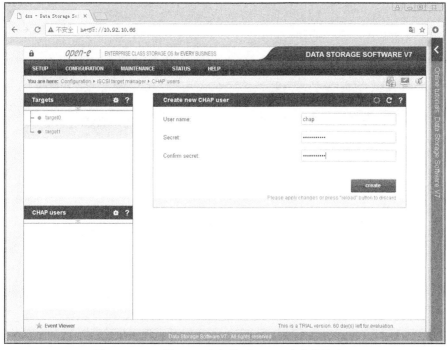

图 8-3-44　配置 ESXi 主机使用 iSCSI 安全特性之三

第 4 步，CHAP user 用户创建完成（如图 8-3-45 所示）。

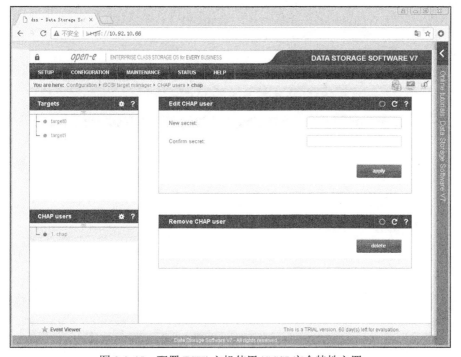

图 8-3-45　配置 ESXi 主机使用 iSCSI 安全特性之四

第 5 步，选择新创建的 iSCSI 卷关联的 Targets（如图 8-3-46 所示），CHAP 认证处于未启用状态。

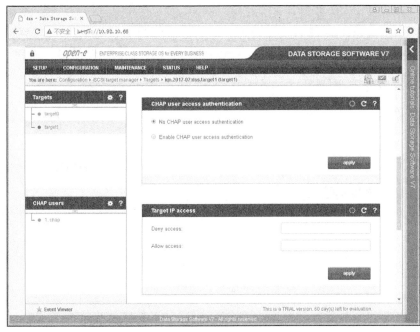

图 8-3-46　配置 ESXi 主机使用 iSCSI 安全特性之五

第 6 步，单击 "Enable CHAP user access authentication" 启用 CHAP 认证，并将创建的 chap 用户添加到允许访问（如图 8-3-47 所示）。

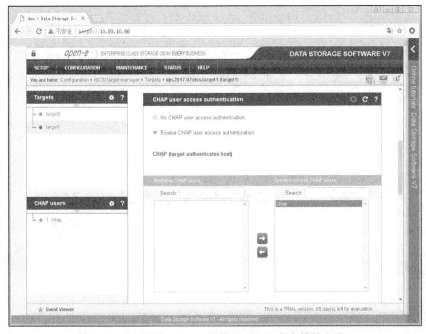

图 8-3-47　配置 ESXi 主机使用 iSCSI 安全特性之六

8.3 配置 ESXi 主机使用 iSCSI 存储　427

第 7 步，选择 ESXi 主机 iSCSI 软件适配器配置身份验证（如图 8-3-48 所示），单击"身份验证"。

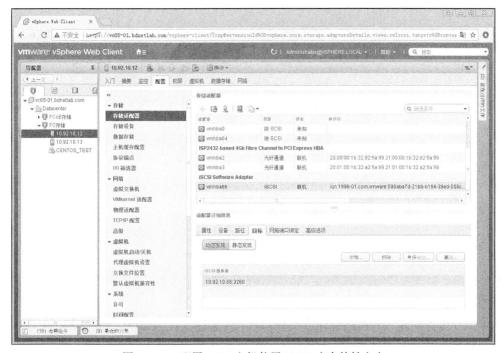

图 8-3-48　配置 ESXi 主机使用 iSCSI 安全特性之七

第 8 步，默认情况下，CHAP 身份验证处于未启用状态（如图 8-3-49 所示）。

图 8-3-49　配置 ESXi 主机使用 iSCSI 安全特性之八

第 9 步，取消勾选"从父项继承设置-vmhba66"，配置身份验证方法为"使用单向 CHAP（如果目标需要）"，输入 CHAP 凭据信息（如图 8-3-50 所示），单击"确定"按钮。

图 8-3-50 配置 ESXi 主机使用 iSCSI 安全特性之九

第 10 步，添加完成后系统会提示存储发生了更改，需要重新扫描存储适配器（如图 8-3-51 所示），单击"重新扫描"按钮。

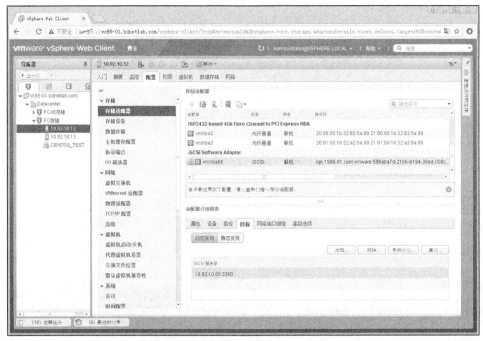

图 8-3-51 配置 ESXi 主机使用 iSCSI 安全特性之十

第 11 步，勾选"扫描新的存储设备"以及"扫描新的 VMFS 卷"重新扫描存储（如

图 8-3-52 所示），单击"确定"按钮。

图 8-3-52　配置 ESXi 主机使用 iSCSI 安全特性之十一

第 12 步，查看 ESXi 主机 iSCSI 软件适配器，将增加容量为 20GB 的 iSCSI 卷（如图 8-3-53 所示）。

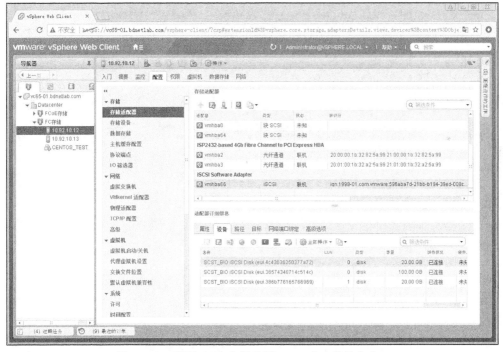

图 8-3-53　配置 ESXi 主机使用 iSCSI 安全特性之十二

第 13 步，查看未配置 CHAP 验证信息的 ESXi 主机 iSCSI 软件适配器，不能看到新增加的 iSCSI 卷信息（如图 8-3-54 所示）。

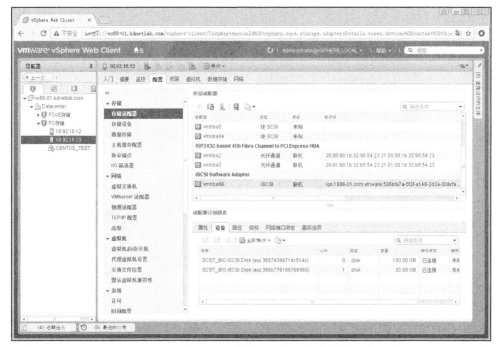

图 8-3-54　配置 ESXi 主机使用 iSCSI 安全特性之十三

第 14 步，在 ESXi 主机"数据存储"单击"新建数据存储"（如图 8-3-55 所示）。

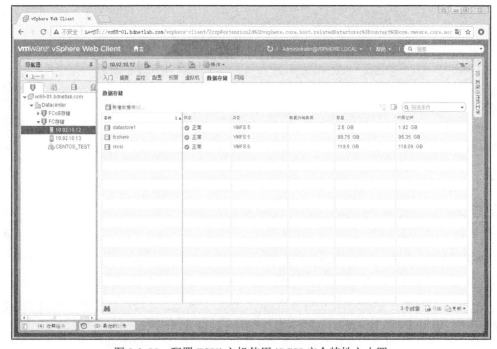

图 8-3-55　配置 ESXi 主机使用 iSCSI 安全特性之十四

第 15 步，选择新建数据存储类型为 VMFS（如图 8-3-56 所示），单击"下一步"按钮。

图 8-3-56　配置 ESXi 主机使用 iSCSI 安全特性之十五

第 16 步，输入数据存储名称并选择容量为 20GB 的卷（如图 8-3-57 所示），单击"下一步"按钮。

图 8-3-57　配置 ESXi 主机使用 iSCSI 安全特性之十六

第 17 步，选择新建数据存储 VMFS 版本（如图 8-3-58 所示），单击"下一步"按钮。

图 8-3-58 配置 ESXi 主机使用 iSCSI 安全特性之十七

第 18 步，配置新建数据存储的分区（如图 8-3-59 所示），单击"下一步"按钮。

图 8-3-59 配置 ESXi 主机使用 iSCSI 安全特性之十八

第 19 步，确认新建数据存储参数设置正确（如图 8-3-60 所示），单击"完成"按钮。

8.3 配置 ESXi 主机使用 iSCSI 存储　　433

图 8-3-60　配置 ESXi 主机使用 iSCSI 安全特性之十九

第 20 步，名为 iscsi-chap 的数据存储创建完成（如图 8-3-61 所示）。

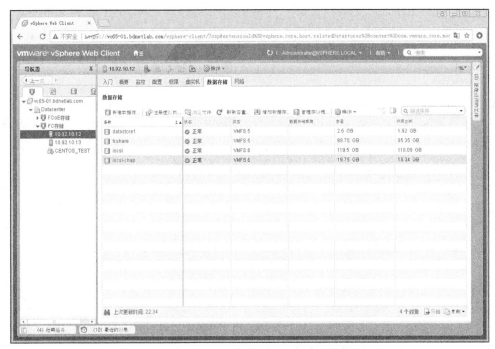

图 8-3-61　配置 ESXi 主机使用 iSCSI 安全特性之二十

第 21 步，查看未配置 CHAP 身份验证的 ESXi 主机，无法访问新创建的 iscsi-chap 存储，说明 CHAP 身份验证生效（如图 8-3-62 所示）。

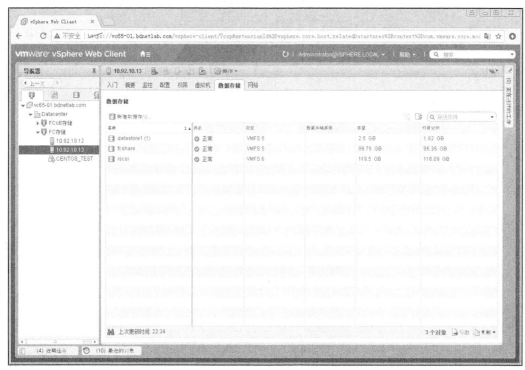

图 8-3-62 配置 ESXi 主机使用 iSCSI 安全特性之二十一

8.4 生产环境使用 iSCSI 存储讨论

VMware 官方将 iSCSI 存储定义为虚拟化平台中性价比最好的存储，可以在不改变当前生产环境架构的情况下配置使用，其构建以及维护也相对简单。那么，在生产环境中使用 iSCSI 存储需要注意什么呢？本节将讨论生产环境使用 iSCSI 存储需要考虑的事项。

8.4.1 生产环境选择 iSCSI 存储还是 FC 存储

近几年来，生产环境虚拟化选择 iSCSI 存储还是 FC 存储一直是人们议论的热点。不可否认，FC 有自己独特的优势，不失为虚拟化一项选择技术，对于 iSCSI 来说也有自己独特的优势。那么究竟应该如何选择呢？下面从几个方面进行分析。

1．从当前架构考虑

一些企业在进行虚拟化建设前，生产环境已经配置有 FC 存储，对于这样的环境来说，没有必要再构建一套 iSCSI 存储。

2．从成本考虑

对于中小企业来说，如果预算有限，那么 iSCSI 存储应当是首选，因为从投入成本来说，iSCSI 存储不需要购买专业的 FC 交换机、FC HBA 卡等硬件。

iSCSI 存储有大量开源免费软件或直接使用 Linux 搭建，几乎不需要考虑存储软件成本；而 FC 存储市面上基本没有免费，需要花费资金购置 FC 存储软件或 FC 存储服务器。

3. 从传输速率考虑

FC 存储在生产环境的主要应用有 4GB、8GB 以及 16GB，其传输速率较高；对 iSCSI 存储来说，多数环境使用 1GE 网络传输，部分环境使用 10GE 网络传输。

4. 从负载考虑

iSCSI 存储通过以太网进行传输，由于网络本身的局限性，存在延时等问题，如果使用普通物理适配器承载 iSCSI 存储流量，则需要使用 CPU 来处理数据信息。使用 TOE 芯片可以减少 CPU 的工作量，提高效率，但带 TOE 芯片的物理适配器价格相对较贵。

综合以上几个方面，对于中小企业来说，如果在预算有限的情况下，推荐使用 iSCSI 存储，尽可能考虑使用 10GE iSCSI 存储。

8.4.2 生产环境 iSCSI 存储网络设计

iSCSI 存储对于大多数中小企业来说成本低是它的最大优点，几乎所有的以太网都是基于 TCP/IP 的，它易于在中小企业内部获得、安装、配置和升级。

在生产环境中使用 iSCSI 存储，网络设计需要考虑以下方面。

1. 1GE 还是 10GE 网络用于 iSCSI 存储

对于 iSCSI 存储在生产环境中使用，最低的要求是使用 1GE 网络，如果预算充足，推荐使用 10GE 网络用于 iSCSI 存储。

2. iSCSI 存储使用独立网段

无论是从稳定性考虑还是从安全性考虑，生产环境使用 iSCSI 存储都推荐规划一个独立的网段，与其他网段进行隔离。

3. 使用 iSCSI HBA 卡或带 TOE 功能的网络适配器

在生产环境中使用 iSCSI 存储，为了减少软件 iSCSI 适配器对 CPU 资源的占用，推荐使用 iSCSI HBA 卡，如果预算不足使用普通以太网络适配器，推荐使用带 TOE 功能的网络适配器，因为 TOE 功能可以卸载需要处理的 TCP/IP 数据，减少对服务器 CPU 资源的占用，从而提高传输效率。

4. 对于 iSCSI 存储流量启用巨型帧

巨型帧也称为 Jumbo 帧，可以将 MTU 从 1500 字节增加到 9000 字节，这样可以解决 iSCSI 存储流量拥塞的问题。但需要注意的是，启用巨型帧需要承载 iSCSI 存储流量的以太网适配器以及物理交换机支持。

8.5 本章小结

本章对 iSCSI 协议以及如何使用 Open-E 存储构建 iSCSI 存储为 ESXi 主机提供服务进行了详细介绍。

VMware 官方把 iSCSI 存储定义为性价比最高的存储，对于预算有限的生产环境来说，iSCSI 存储是首选，可以通过开源免费的存储软件或 Linux 系统进行搭建，其搭建过程与后期维护相对简单。随着以太网技术的发展，生产环境推荐使用 10GE 网络构建 iSCSI 存储服务器，高带宽能进一步发挥存储的速率。

第 9 章 部署使用 Virtual SAN

Virtual SAN 是 VMware 软件定义存储的实现方式，适用于 VMware 超融合软件解决方案的企业级存储，Virtual SAN 使用内嵌的方式集成于 VMware vSphere 虚拟化平台，可以为虚拟机应用提供经过闪存优化的超融合存储。2013 年 10 月的 VMware vForum 大会发布了 Virtual SAN BETA 版本，2014 年 3 月推出正式版本 Virtual SAN 5.5（内部版本号 1.0），集成于 VMware vSphere 5.5 Update 1 版本中，2015 年发布的 VMware vSphere 6.0 版本中，Virtual SAN 的内部版本由原来的 1.0 升级到 2.0（也称为 Virtual SAN 6.0），2016 年陆续发布 Virtual SAN 6.2，2017 年发布的 VMware vSphere 6.5 版本中，Virtual SAN 升级为 Virtual SAN 6.5。本书写作的时候，Virtual SAN 版本已发布 6.6。本章介绍生产环境如何部署使用 Virtual SAN 存储。

本章要点
- Virtual SAN 存储介绍
- 部署使用 Virtual SAN 存储
- 生产环境使用 Virtual SAN 存储讨论

9.1 Virtual SAN 存储介绍

Virtual SAN 存储是一种基于软件的分布式存储解决方案，集成于 VMware vSphere 虚拟化平台，简化了存储配置，直接使用 ESXi 主机本地存储为虚拟机提供高性能、高可靠的存储服务。

9.1.1 软件定义存储介绍

软件定义存储是一种数据存储方式，所有存储相关的控制工作都仅在相对于物理存储硬件的外部软件中。这个软件不是作为存储设备中的固件，而是在一个服务器上或者作为操作系统（OS）或者作为 hypervisor 的一部分。软件定义存储是一个较大的行业发展趋势，这个行业还包括软件定义网络（SDN）和软件定义数据中心（SDDC）。与 SDN 情况类似，软件定义存储可以保证系统的存储访问能在一个精准的水平上更灵活地管理。软件定义存储是从硬件存储中抽象出来的，这也意味着它可以变成一个不受物理系统限制的共享池，以便于最有效地利用资源。它还可以通过软件和管理进行部署和供应，也可以通过基于策略的自动化管理来进一步简化。

1. 软件定义存储困难

使用软件定义存储进行长期存储充满了诸多风险，例如数据位错误、硬盘故障、网络攻击、人为失误以及自然灾害等。然而对数据进行长期存储除了要面对以上风险外，还要考虑到其他问题，例如硬件架构、软件平台、应用以及数据格式的变化等。图 9-1-1 显示了客户在存储上目前面临的挑战。与此同时，对于数据可访问性、协同性以及大数据分析与日俱增的需求，也使得问题不仅仅停留在储存时限层面，还要考虑到其可用性。

图 9-1-1　客户面临的挑战

2. 软件定义存储优势

软件定义存储允许客户将存储服务集成到服务器的软件层。软件定义存储将软件从原有的存储控制器中抽离出来，使得它们的功能得以进一步地发挥而不仅仅局限在单一的设备中。软件定义存储将数据去重功能或者精简配置局限在单一的硬盘上与把其扩展到全部存储平台层面相比，实在是没有什么太大意义。在这一点上，软件定义存储的从业者们并没有拖大家的后腿。

3. 软件定义存储隐形成本

业内鲜有提及 SDS 相关的潜在隐性成本的。SDS 可能带来的硬件混合以及匹配将反转集成软硬件到终端用户的成本或风险。在选择 SDS 产品时，评估厂商是否能够提供 SDS 的利好十分关键。时代和技术都发生了某种程度的改变，有些人说驱动器的标准已经改进。但作者认为，新的硬件技术，比如固态存储每一天都在演进。如果我们延伸存储软件抽象化的能力到最大，它应该能够协调任何硬件。如果这将作为 SDS 部署的理想状态，那么一个系统中可能的技术组合将是无限的。在这种设想之下，验证和集成新硬件技术的职责和成本都将归于 IT。

9.1.2　什么是 Virtual SAN

在 VMware 软件定义数据中心（SDDC）潮流的带动下，软件定义网络（SDN）与软件定义存储（SDS)等概念纷纷涌现。任何软件定义的基础设施组件无非是存储、计算或网络，这其中的关键是建立一种抽取组件的方式，并以编程的方式对其进行控制。与传统的虚拟存储对数据层进行了抽象不同，软件定义的存储产品主要是进行通信的编排。这意味着软件定义存储在应用程序和物理存储资源之间建立一个独立于硬件并与工作负载无关的存储应用层。

1. VMware Virtual SAN 基本概念

Virtual SAN 对存储进行了虚拟化，在提供访问共享存储目标与路径的同时具备数据层控制功能，并能够基于服务器硬件创建策略驱动的存储。实际上，Virtual SAN 就是一种数据存储方式，而所有存储相关的控制工作放在相对于物理存储硬件的外部软件中，这个软件不是作为存储设备中的固件，而是在一个服务器上或者作为操作系统（OS）或 hypervisor 的一部分。Virtual SAN 被集成到 VMware vSphere 5.5 U1 之后版本中，并且与 VMware vSphere 高可用、分布式资源调度以及 vMotion 深度集成在一起，通过 Web Client 客户端进行管理。Virtual SAN 最大的好处在于即使底层物理架构存储乱七八糟，面目全非，但是 Virtual SAN 是透明的，上面的应用、中间件与数据库等部署方式仍然不会发生变化，而且其上代码与业务逻辑也不会发生变化。

2. VMware Virtual SAN 的未来

无论是存储还是网络，除了物理连接无法通过软件来替代、盘片是不能模拟外，其他部件与要素都可以通过软件实现，在一个通用服务器上像应用程序那样运行。所有工作负载，也就是控制的逻辑全部运行在服务器上。这样看来，传统的硬件厂商退出存储业务，或者被没有将其 API 与存储虚拟机管理程序的厂商共享的硬件厂商收购。单独的工作负载可能会都转移到或者一部分转移到或者分步转移到通用的 CPU 上来运行，在价值链中体现真正价值的会是什么？控制厂商，也就是控制逻辑的这些厂商。控制逻辑厂商能够放在 CPU 上来运行，不是直接部署在 CPU 上。VMware 有虚拟化数据中心的能力，将所有资源池化，变成一个可以管理与移动的容器，以服务器的形式交付，并让数据中心完全由软件自动控制。

9.1.3 Virtual SAN 功能介绍

从 Virtual SAN 存储第一个版本到本书写作使用 Virtual SAN 6.5（写作本书的时候 Virtual SAN 已发布 6.6 版本），其发展非常迅速，新发布的版本不仅是修改老版本的 BUG，而且增加了非常多的特性，在开始部署使用 Virtual SAN 之前，需要了解各个 Virtual SAN 版本所具有功能特性。

1. Virtual SAN 5.5

Virtual SAN 5.5 称为第一代 Virtual SAN，集成于 VMware vSphere 5.5 U1 版本。该版本具有软件定义存储的基本功能，VMware vSphere 的一些高级特性无法在 Virtual SAN 5.5 上使用。从生产环境使用上看，Virtual SAN 5.5 基本用于测试使用。

2. Virtual SAN 6.0

Virtual SAN 6.0 为第二代 Virtual SAN，集成于 VMware vSphere 6.0 版本。该版本不仅修复了 Virtual SAN 5.5 存在的一些 BUG，而且增加了大量新的功能，其新增主要功能如下。

支持混合架构以及全闪存架构。

支持通过配置故障域（机架感知）解决 Virtual SAN 集群免于机架故障。

支持在删除 Virtual SAN 存储前将 Virtual SAN 数据迁移。

支持硬件层面的数据校验，检测并解决磁盘问题，从而提供更高的数据完整性。

支持运行状态服务监控，可以监控 Virtual SAN 以及集群、网络、物理磁盘的状况。

3. Virtual SAN 6.1

Virtual SAN 6.1 为第三代 Virtual SAN，集成于 VMware vSphere 6.0 U1 版本。该版本在 Virtual SAN 6.0 的基础上再次增加新的功能，其新增主要功能如下。

支持延伸集群，也就是使用 Virtual SAN 构建双活数据中心，延伸集群支持横跨两个地理位置的集群，这样可以最大程度地保护数据不受 Virtual SAN 站点故障或网络故障出现选择。

支持 ROBO，支持两节点方式部署 Virtual SAN，可以通过延伸集群功能，把见证主机放在总部数据中心，简化 Virtual SAN 部署。

支持统一的磁盘组声明，在创建 Virtual SAN 时，统一声明磁盘组的容量层与缓存层。

支持 Virtual SAN 磁盘在线升级，可以通过管理端在线将 Virtual SAN 磁盘格式升级到 2.0。

4. Virtual SAN 6.2

Virtual SAN 6.2 为第四代 Virtual SAN。该版本在 Virtual SAN 6.1 基础上增加了更多更实用的特性，其新增主要功能如下。

支持对全闪存架构的 Virtual SAN 数据去重，并采用 LZ4 算法对容量层数据进行压缩。

支持通过纠删码对 Virtual SAN 数据进行跨网络的 RAID 5/6。

支持对不同虚拟机设置不同的 IOPS。

支持纯 IPv6 运行模式。

支持软件层面的数据校验，检测并解决磁盘问题，从而提供更高的数据完整性。

5. Virtual SAN 6.5

Virtual SAN 6.5 为第五代 Virtual SAN。该版本在 Virtual SAN 6.2 基础上再次增加了新的特性，其新增主要功能如下。

支持将 Virtual SAN 配置为 iSCSI Target，通过 iSCSI 支持来连接非虚拟化工作负载。

通过直接使用交叉电缆连接两个节点来消除路由器/交换机成本，降低 ROBO 成本。

扩展了对容器和 CNA 的支持，可以使用 Docker、Swarm、Kubernetes 等随时开展工作。

Virtual SAN 6.5 标准版本提供对全闪存硬件的支持，降低构建成本。

6. Virtual SAN 6.6

Virtual SAN 6.6 为第六代 Virtual SAN。该版本是业界首个原生 HCI 安全功能、高度可用的延伸集群，同时将关键业务和新一代工作负载的全闪存性能提高了 50%，其新增主要功能如下。

针对静态数据的原生 HCI 加密解决方案，可以保护关键数据免遭不利访问。Virtual SAN 加密具有硬件独立性并简化了密钥管理，因而可降低成本并提高灵活性。不再要求部署特定的自加密驱动器 (SED)。Virtual SAN 加密还支持双因素身份验证（SecurID 和 CAC），因而能够很好地保证合规性。另外，它还是首个采用 DISA 批准的 STIG 的 HCI 解决方案。

支持单播网络连接，以帮助简化初始 Virtual SAN 设置，可以为 Virtual SAN 网络连接使用单播，不再需要设置多播。这使得 Virtual SAN 可以在更广泛的本地和云环境中部署而无需更改网络。

优化的数据服务进一步扩大了 Virtual SAN 的性能优势，具体就是，与以前的 Virtual SAN 版本相比，它可将每台全闪存主机的 IOPS 提升 50%之多，从而使每台主机的 IOPS 超过 150K。提升的性能有助于加快关键任务应用的速度，并提供更高的工作负载整合率。

借助对最新闪存技术（包括新的 Intel Optane 3D XPoint NVMe SSD 等解决方案）的现成支持，客户可加快新硬件的采用。此外，Virtual SAN 现在还提供更多的缓存驱动器选择（包括 1.6 TB 闪存），方便客户利用更大容量的最新闪存。

经验证的全新体系结构为部署 Splunk、Big Data 和 Citrix XenApp 等新一代应用提供了一条行之有效的途径。此外，现在 Photon Platform 1.1 中提供了适用于 Photon 的 Virtual SAN，而新的 Docker Volume Driver 则提供了对多租户、基于策略的管理、快照和克隆的支持。

借助新增的永不停机保护功能，Virtual SAN 可确保用户的应用正常运行和使用，而不会受潜在硬件难题的影响。新的降级设备处理 (DDH) 功能可智能地监控驱动器的运行状况，并在发生故障前主动撤出数据。新的智能驱动器重建和部分重建功能可在硬件发生故障时提供更快速的恢复，并降低集群流量以提高性能。

9.1.4 Virtual SAN 常用术语

Virtual SAN 集成于 VMware vSphere，但也可以把它看成是一个独立的组件。了解完 Virtual SAN 的基本概念后，需要对其常用术语进一步了解。

1. 对象

Virtual SAN 中一个重要的概念就是对象，Virtual SAN 是基于对象的存储，虚拟机由大量不同的存储对象组成，与之前不同的是由一组文件，而对象是一个独立的存储块设备。存储块包括虚拟机主页名字空间、虚拟机交换文件、VMDK 等。

2. 组件

从另外一个方面来看，Virtual SAN 可以理解为网络 RAID 1，Virtual SAN 在 ESXi 主机之间使用 RAID 1 阵列来实现存储对象的高可用。每个存储对象都是一个组件，组件的具体数量与存储策略有直接的关系。

3. 副本

Virtual SAN 使用 RAID 方式来实现高可用，那么一个对象就存在多个副本又避免单点故障，副本的数量与存储策略有直接的关系。

4. 见证（Witness）

见证可以理解为仲裁，在 VMware vSphere 中翻译为"证明"。见证属于比较特殊的组件，不包括元数据，仅用于当 Virtual SAN 发生故障后进行仲裁时用来确定如何恢复。

5. 磁盘组（Disk Group）

Virtual SAN 的核心之一，由 SSD 硬盘和其他硬盘（SATAT、SAS）组成，用于缓存和存储数据，是构建 Virtual SAN 的基础。

6. 基于存储策略的管理（Storage Policy-Based Management）

基于存储策略的管理 Virtual SAN 的核心，所有部署在 Virtual SAN 上的虚拟机都必须使用一种存储策略。如果没有创建新的存储策略，虚拟机将使用默认策略。本书 9.1.5 小节将详细介绍存储策略。

9.1.5 Virtual SAN 存储策略介绍

Virtual SAN 使用 Storage Policy-Based Management（简称 SPBM，中文翻译为"基于存储策略的管理"）部署虚拟机。通过使用基于存储策略的管理，虚拟机可以根据生产环境的需求并且在不关机的情况应用不同的策略。所有部署在 vSAN 上的虚拟机都必须使用一种存储策略，如果没有创建新的存储策略，虚拟机将使用默认策略。Virtual SAN 存储策略主要有以下几种类型。

1. Number of Failures to Tolerate

Number of Failures to Tolerate，简称为 FTT，中文翻译为"允许的故障数"。该策略定义在集群中存储对象针对主机数量、磁盘或网络故障的同时发生故障的数量，默认情况下 FTT 值为 1，FTT 的值决定了 Virtual SAN 群集需要的 ESXi 主机数量，假设 FTT 的值设置为 n，则将会有 n+1 份拷贝，要求 2n+1 台主机，FTT 值对应 ESXi 主机列表参考表 9-1-1。如果使用双节点 Virtual SAN，则配置额外的见证主机，表 9-1-1 不适用于双节点 Virtual SAN 配置。

表 9-1-1　　　　　　　　　　FTT 值对应 ESXi 主机列表

FTT	副本	见证	ESXi 主机数
0	1	0	1
1	2	1	3
2	3	2	5
3	4	3	7

2. Number of Disk Stripes per Object

Number of Disk Stripes per Object，简称为 Stripes，中文翻译为"每个对象的磁盘带数"，表示存储对象的磁盘跨越主机的拷贝数。Stripes 值相当于 RAID0 的环境，分布在多个物理磁盘上。一般来说，Stripes 默认值为 1，最大值为 12。如果将该参数值设置为大于 1 时，虚拟机可以获取更好的 IOPS 性能，但会占用更多的系统资源。默认值 1 可以满足大多数虚拟机负载使用，对于磁盘 I/O 密集型运算可以调整 Stripes 值。当一个对象大小超过 255GB 时，即使 Stripes 默认为 1，系统还是会对对象进行强行分割。

需要说明的是，在 Virtual SAN 环境中，所有的写操作都是先写入 SSD 硬盘，增加条带对性能可能没有增加，因为系统无法保证新增加的条带会使用不同的 SSD 硬盘，新的条带可能会放置在位于同一个磁盘组的硬盘上。当然，如果新的条带被放置在不同的磁盘组中，就会使用到新的 SSD，这种情况下会带来性能上的提升。

3. Flash Read Cache Reservation

Flash Read Cache Reservation，中文翻译为"闪存读取缓存预留"。默认为 0，这个参数结合虚拟机磁盘大小来设定 Read Cache 大小，计算方式为百分比，可以精确到小数点后 4 位，如果虚拟机磁盘大小为 100GB，闪存读取缓存预留设置为 10%，闪存读取缓存预留值会使用 10GB 的 SSD 容量，当虚拟机磁盘越大的时候，会占用大量的闪存空间。在生产环境中，一般不配置闪存读取缓存预留，因为为虚拟机预留的闪存读取缓存不能用于其他对象，而未预留的闪存可以共享给所有对象使用。需要注意的是，Read Cache 在全闪存环境下失效。

4. Force Provisioning

Force Provisioning，中文翻译为"强制置备"。通过强制置备，可以强行配置具体的存储策略。启用强制置备后，Virtual SAN 会监控存储策略应用，在存储策略无法满足需求时，如果选择了强制置备，则策略将被强行设置为：

FTT=0

Stripe=1

Object Space Reservation=0

警告：这个参数只能在绝对必要时作为例外情况使用，一般不作为默认参数使用，有可能将虚拟机以及关联的数据置于危险环境。

5. Object Space Reservation

Object Space Reservation，简称为 OSR，中文翻译为"对象空间预留"。默认为 0，也就是说虚拟机的硬盘模式为 Thin Provisioning（精简置备），意味着虚拟机部署的时候不会预留任何空间，只有当虚拟机存储增长时空间才会被使用。对象空间预留值如果设置为 100%，虚拟机存储对容量的要求会被预先保留，也就是 Thick Provisioning（厚置备）。需要注意的是，vSAN 中 Thick Provisioning，只存在 Lazy Zeroed Thick（厚置备延迟置零，LZT），不存在 Eager Zeroed Thick（厚置备置零，EZT），也就是说在 vSAN 环境下将无法使用 vSphere 高级特性中的 Failures Tolerate 技术。

6. 容错

容错是从 Virtual SAN 6.2 版本引入的新的虚拟机存储策略，其主要是为了解决老版本 Virtual SAN 使用 RAID 1 技术占用大量的磁盘空间问题。Virtual SAN 6.2 版本开始提供 RAID 5/6 纠删码技术，这样可以减少虚拟机对磁盘空间的占用，提供更多的 Virtual SAN 存储空间。

7. 对象 IOPS 限制

对象 IOPS 限制是从 Virtual SAN 6.2 版本完善的虚拟机存储策略，可以对虚拟机按应用需求进行不同的 IOPS 限制，提高 I/O 效率。

8. 禁用对象校验和

禁用对象校验和是为了保证 Virtual SAN 数据的完整性，系统在读写操作时会检查检验数据，如果数据有问题，则会对数据进行修复操作。禁用对象校验和设置为 NO，系统会对问题数据进行修复；设置为 YES，系统不会对问题数据进行修复。

9.2 部署 Virtual SAN 6.5

Virtual SAN 集成于 VMware vSphere 内核中，只需满足条件，启用 Virtual SAN 即可使用，其重点在于各种特性的配置使用。本节使用全闪存架构配置 Virtual SAN。

9.2.1 使用 Virtual SAN 要求

在部署 Virtual SAN 之前，为保证生产环境的稳定性，需要了解其软硬件要求，否则可能导致生产环境的 Virtual SAN 出现严重问题。

1. 物理服务器以及硬件

生产环境一般使用大厂品牌服务器，而这些主流服务器一般都会通过 VMware 官方认证。需要注意的是，VMware 官方针对 Virtual SAN 专门发布了硬件兼容性列表，主要是针对存储控制器、SSD 等进行兼容性要求。

生产环境使用 Virtual SAN，对物理服务器内存也提出了要求。VMware 官方推荐使用 Virtual SAN 的物理服务器最少配置 6GB 内存，如果物理服务器配置多个磁盘组，推荐使用 32GB 以上的内存。

生产环境使用 Virtual SAN，在混合环境下，至少需要配置 1GE 网络承载 Virtual SAN 流量；中大型环境或全闪存环境使用 Virtual SAN，推荐使用 10GE 网络承载 Virtual SAN 流量。

2. Virtual SAN 集群中 ESXi 主机数量

表 9-1-1 显示了根据不同的副本数量集群中需要配置的 ESXi 主机数量，生产环境中强烈不推荐使用最低要求，比如 FTT=1 时，ESXi 主机数量要求为 3，这是最低要求，不适合于生产环境，因为可能由于组件数以及其他原因导致 Virtual SAN 故障。FTT=1 时，推荐配置使用 4 台以上的 ESXi 主机。对于生产环境其他需求，推荐 ESXi 主机数量大于最低要求数量，双节点 Virtual SAN 集群例外。

3. Virtual SAN 软件版本

Virtual SAN 版本已发布至第六代，生产环境应根据其需求进行选择，选择好 Virtual SAN 版本还需要确定是使用该版本的标准版、高级版还是企业版等，这些版本所具有功能是不一样的，比如标准版不支持去重、纠删码、延伸集群等功能。本书写作的时候 VMware 已发布 Virtual SAN 6.6，属于刚发布的产品，不推荐立即应用于生产环境。

9.2.2 配置 Virtual SAN 所需网络

Virtual SAN 需要使用 VMkernel 承载其流量，支持标准交换机以及分布式交换机配置，但如果需要进行后续的网络 I/O 配置等操作，推荐使用分布式交换机承载 Virtual SAN 流量。本小节介绍如何配置分布式交换机承载 Virtual SAN 网络。

第 1 步，准备好运行 Virtual SAN 集群的 ESXi 主机（如图 9-2-1 所示）。

图 9-2-1 配置 Virtual SAN 所需网络之一

第 2 步，选择 ESXi 主机进行配置，实验环境使用的是全闪存磁盘，因此使用 10GE 网络（如图 9-2-2 所示）。

图 9-2-2　配置 Virtual SAN 所需网络之二

第 3 步，选择导航器中的"网络"，单击"新建 Distributed Switch"（如图 9-2-3 所示）。

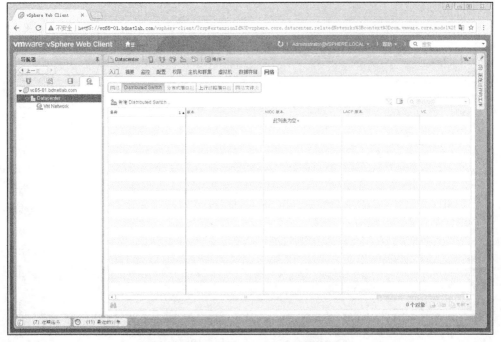

图 9-2-3　配置 Virtual SAN 所需网络之三

第 4 步，输入新建 Distributed Switch 名称（如图 9-2-4 所示），单击"下一步"按钮。

图 9-2-4　配置 Virtual SAN 所需网络之四

第 5 步，选择新建 Distributed Switch 的版本（如图 9-2-5 所示），单击"下一步"按钮。

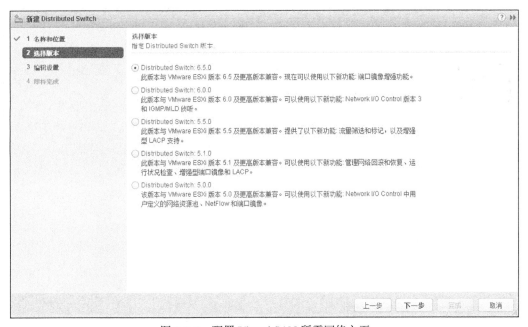

图 9-2-5　配置 Virtual SAN 所需网络之五

第 6 步，编辑新建 Distributed Switch 上行链路端口数量以及创建默认端口组，实验环境中每台 ESXi 主机配置 1 个 10GE 网络，所以上行链路配置为 1（如图 9-2-6 所示），单击"下一步"按钮。

第 9 章 部署使用 Virtual SAN

图 9-2-6 配置 Virtual SAN 所需网络之六

第 7 步，确认参数配置正确（如图 9-2-7 所示），单击 "下一步" 按钮。

图 9-2-7 配置 Virtual SAN 所需网络之七

第 8 步，新建名为 vds-vsan65 的分布式交换机完成（如图 9-2-8 所示）。

9.2 部署 Virtual SAN 6.5　　447

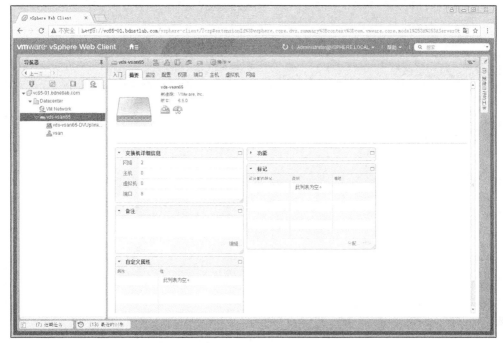

图 9-2-8　配置 Virtual SAN 所需网络之八

第 9 步，在新建的 vds-vsan65 分布式交换机上单击右键，选择"添加和管理主机"（如图 9-2-9 所示）。

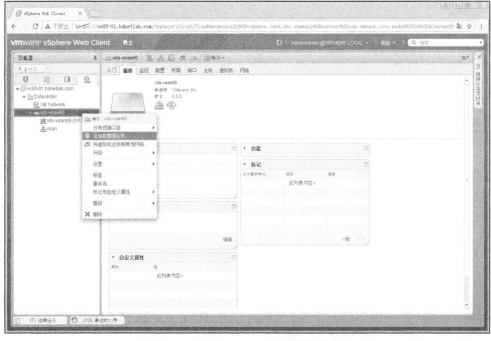

图 9-2-9　配置 Virtual SAN 所需网络之九

第 10 步，选择"添加主机"向分布式交换机添加 ESXi 主机（如图 9-2-10 所示），单击"下一步"按钮。

图 9-2-10　配置 Virtual SAN 所需网络之十

第 11 步，因为创建的 vds-vsan65 分布式交换机还未添加 ESXi 主机，所以列表为空（如图 9-2-11 所示），单击"+新主机"。

图 9-2-11　配置 Virtual SAN 所需网络之十一

第 12 步，勾选需要加入 vds-vsan65 的 ESXi 主机（如图 9-2-12 所示），单击"确定"按钮。

图 9-2-12　配置 Virtual SAN 所需网络之十二

第 13 步，ESXi 主机添加到 vds-vsan65 分布式交换机（如图 9-2-13 所示），单击"下一步"按钮。

图 9-2-13　配置 Virtual SAN 所需网络之十三

第 14 步，勾选"管理物理适配器"对 ESXi 主机连接到分布式交换机的上行链路进行配置（如图 9-2-14 所示），单击"下一步"按钮。

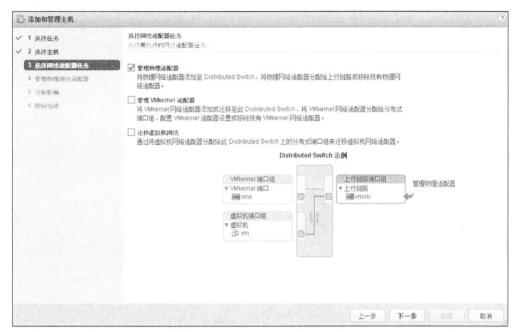

图 9-2-14　配置 Virtual SAN 所需网络之十四

第 15 步，选择 ESXi 主机物理适配器（如图 9-2-15 所示），单击"分配上行链路"。

图 9-2-15　配置 Virtual SAN 所需网络之十五

第 16 步，为物理适配器选择分配上行链路（如图 9-2-16 所示），单击"确定"按钮。

9.2 部署 Virtual SAN 6.5　　451

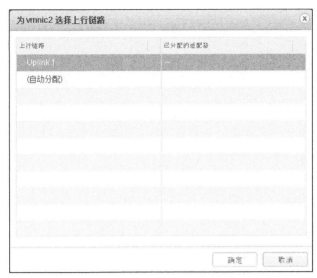

图 9-2-16　配置 Virtual SAN 所需网络之十六

第 17 步，为 ESXi 主机分配到 vds-vsan65 分布式交换机上行链路完成（如图 9-2-17 所示），按照相同的方法将添加其他 ESXi 主机上行链路，完成后单击"下一步"按钮。

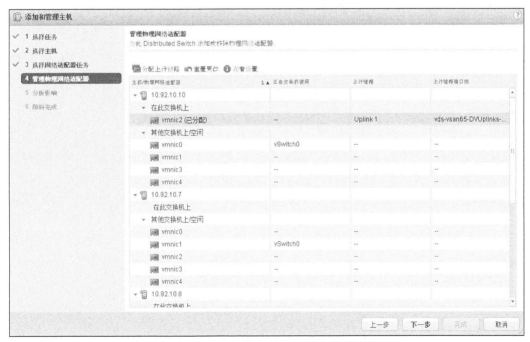

图 9-2-17　配置 Virtual SAN 所需网络之十七

第 18 步，系统对分配的上行链路进行校验，确定是否影响其他网络（如图 9-2-18 所示），单击"下一步"按钮。

图 9-2-18 配置 Virtual SAN 所需网络之十八

第 19 步，确认参数配置正确（如图 9-2-19 所示），单击"完成"按钮。

图 9-2-19 配置 Virtual SAN 所需网络之十九

第 20 步，ESXi 主机加入到 vds-vsan65 分布式交换机（如图 9-2-20 所示）。

9.2 部署 Virtual SAN 6.5   453

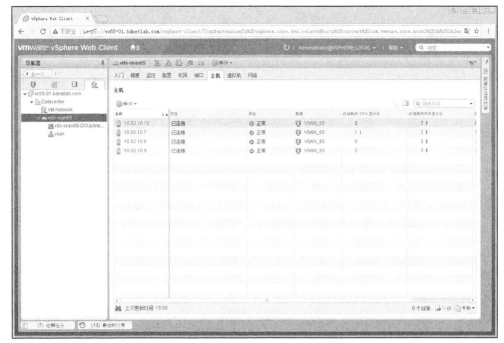

图 9-2-20　配置 Virtual SAN 所需网络之二十

第 21 步，选择 ESXi 主机创建 VMkernel 承载 Virtual SAN 流量（如图 9-2-21 所示），单击"添加网络"。

图 9-2-21　配置 Virtual SAN 所需网络之二十一

第 22 步，选择"VMkernel 网络适配器"（如图 9-2-22 所示），单击"下一步"按钮。

图 9-2-22　配置 Virtual SAN 所需网络之二十二

第 23 步，选择现有网络（如图 9-2-23 所示），单击"浏览"按钮。

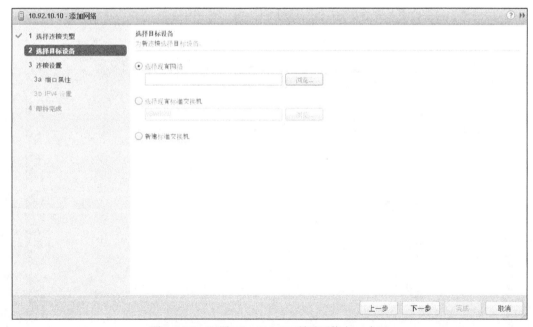

图 9-2-23　配置 Virtual SAN 所需网络之二十三

第 24 步，选择 vds-vsan65 分布式交换机上的 vsan 端口组（如图 9-2-24 所示）。

9.2 部署 Virtual SAN 6.5　　455

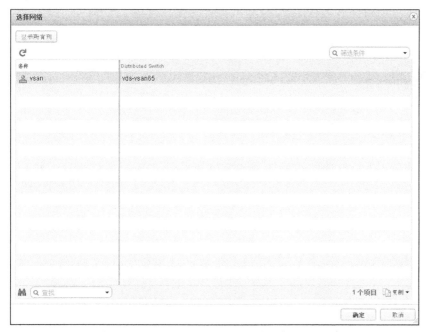

图 9-2-24　配置 Virtual SAN 所需网络之二十四

第 25 步，确认选择现有网络为 vsan（如图 9-2-25 所示），单击"下一步"按钮。

图 9-2-25　配置 Virtual SAN 所需网络之二十五

第 26 步，勾选可用服务中的"Virtual SAN"（如图 9-2-26 所示），单击"下一步"按钮。

图 9-2-26　配置 Virtual SAN 所需网络之二十六

第 27 步，选择自动获取 IP 地址（如图 9-2-27 所示），单击"下一步"按钮。

图 9-2-27　配置 Virtual SAN 所需网络之二十七

第 28 步，确认配置正确（如图 9-2-28 所示），单击"完成"按钮。

图 9-2-28　配置 Virtual SAN 所需网络之二十八

第 29 步，承载 Virtual SAN 流量的 VMkernel 创建完成（如图 9-2-29 所示）。

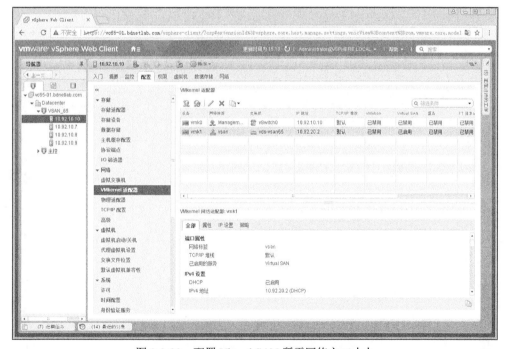

图 9-2-29　配置 Virtual SAN 所需网络之二十九

第 30 步，按照相同的方法创建其他 ESXi 主机承载 Virtual SAN 流量的 VMkernel，完成后查看整体拓扑情况（如图 9-2-30 所示）。

图 9-2-30　配置 Virtual SAN 所需网络之三十

9.2.3　启用 Virtual SAN

在启用 Virtual SAN 之前再次确认集群中 ESXi 主机是否已经准备好需要的磁盘以及网络。另外需要注意的是，启用前必须关闭 HA 特性。本小节介绍如何启用 Virtual SAN 以及创建磁盘组。

第 1 步，默认情况下 Virtual SAN 处于禁用状态（如图 9-2-31 所示），单击"配置"按钮。

图 9-2-31　启用 Virtual SAN 之一

第 2 步，开始配置 Virtual SAN 功能，磁盘声明有自动以及手动两种模式，去重和压缩功能、故障域和延伸集群后续章节再进行配置（如图 9-2-32 所示），单击"下一步"按钮。

图 9-2-32　启用 Virtual SAN 之二

第 3 步，配置向导对 Virtual SAN 使用的网络进行验证（如图 9-2-33 所示），单击"下一步"按钮。

图 9-2-33　启用 Virtual SAN 之三

第 4 步，选择不声明磁盘，待启用 Virtual SAN 后手动指定（如图 9-2-34 所示），单击"下一步"按钮。

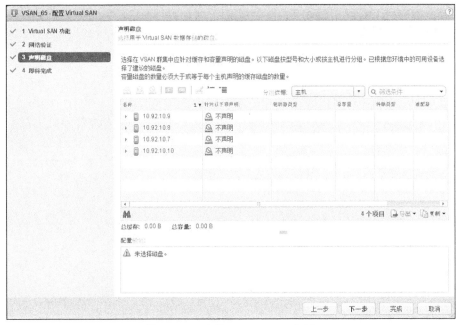

图 9-2-34　启用 Virtual SAN 之四

第 5 步，确认参数配置正确。由于未配置磁盘组，所有 Virtual SAN 容量为 0（如图 9-2-35 所示），单击"完成"按钮。

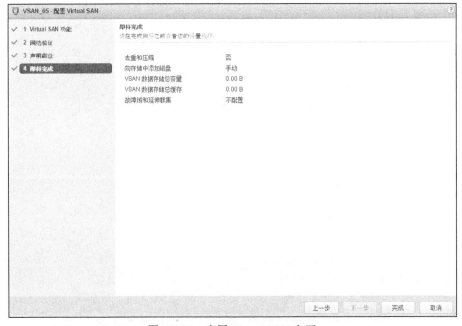

图 9-2-35　启用 Virtual SAN 之五

第 6 步，集群 Virtual SAN 启用完成（如图 9-2-36 所示）。

图 9-2-36　启用 Virtual SAN 之六

第 7 步，选择"磁盘管理"手动添加 ESXi 主机磁盘组（如图 9-2-37 所示），选择 ESXi 主机后单击"创建磁盘组"。

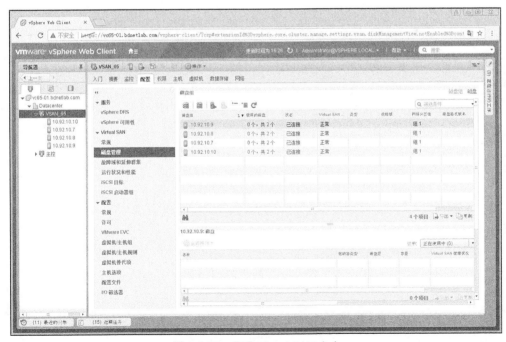

图 9-2-37　启用 Virtual SAN 之七

第 8 步，选择缓存层以及容量层磁盘（如图 9-2-38 所示），单击"下一步"按钮。

图 9-2-38　启用 Virtual SAN 之八

第 9 步，ESXi 主机磁盘组创建完成（如图 9-2-39 所示）。

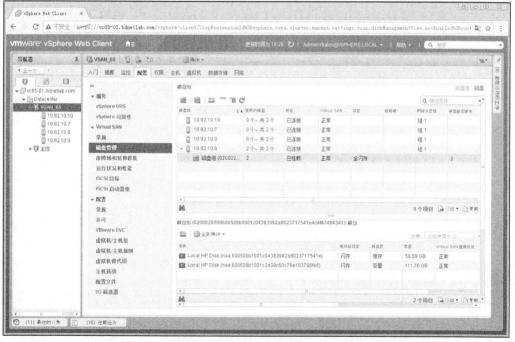

图 9-2-39　启用 Virtual SAN 之九

第 10 步，按照相同的方式在其他 ESXi 主机上创建磁盘组（如图 9-2-40 所示）。

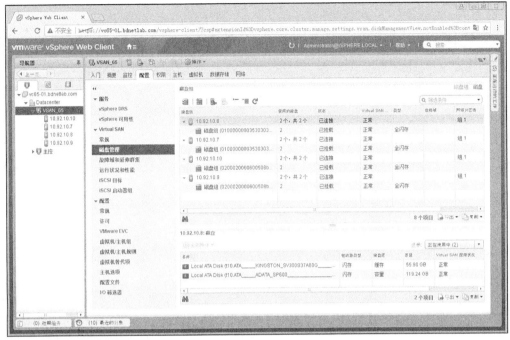

图 9-2-40　启用 Virtual SAN 之十

第 11 步，查看"常规"信息，可以看到 Virtual SAN 使用了 8 个磁盘，格式版本为 3.0（如图 9-2-41 所示）。

图 9-2-41　启用 Virtual SAN 之十一

第 12 步，选择 ESXi 主机查看 vsanDatastore 信息，可以看到格式为 vsan，容量为 464.71GB（如图 9-2-42 所示）。

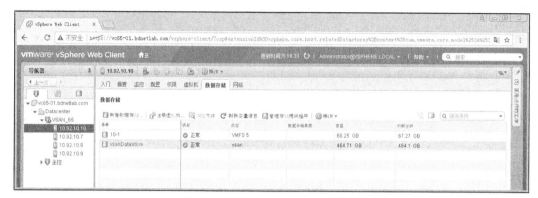

图 9-2-42　启用 Virtual SAN 之十二

第 13 步，在集群"监控"选项查看 Virtual SAN 容量，可以看到详细的 Virtual SAN 信息（如图 9-2-43 所示）。

图 9-2-43　启用 Virtual SAN 之十三

第 14 步，在集群"监控"选项查看运行状况，可以看到 Virtual SAN 相关的状态。需要特别注意的是，测试结果为"失败"状态（如图 9-2-44 所示）。实验环境使用的 HP 服务器 SCSI 控制器不在 VMware 硬件兼容列表中，会出现失败以及警告信息。Virtual SAN HCL 数据库版本可以通过在线或离线方式进行更新。

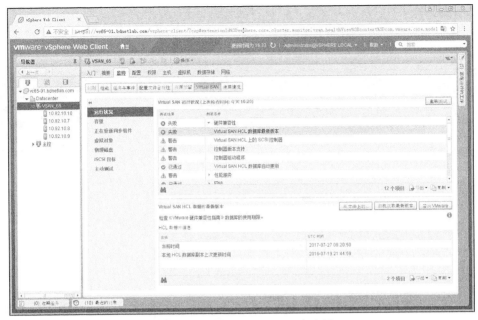

图 9-2-44　启用 Virtual SAN 之十四

9.2.4　配置 Virtual SAN 存储策略

Virtual SAN 存储策略配置影响到虚拟机的容错以及正常运行，错误的配置可能导致虚拟机运行速度缓慢，更为严重的是可能导致虚拟机数据丢失。本小节介绍 Virtual SAN 存储策略配置。

第 1 步，使用浏览器登录 vCenter Server，选择"主页"中的"虚拟机存储策略"（如图 9-2-45 所示）。

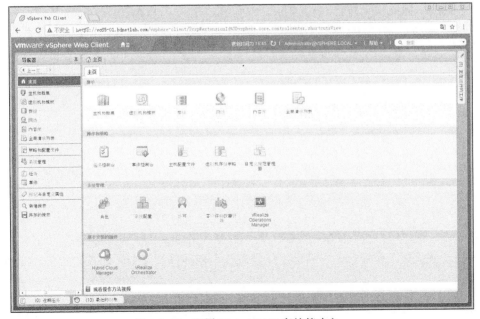

图 9-2-45　配置 Virtual SAN 存储策略之一

第 2 步，系统会默认创建 3 条虚拟机存储策略（如图 9-2-46 所示），选择 "Virtual SAN Default Storage Policy"，单击 "编辑设置"。

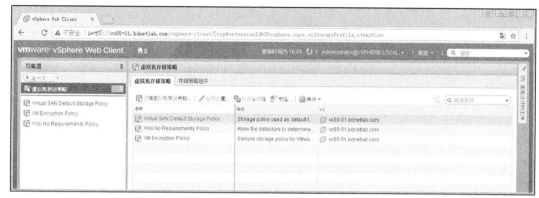

图 9-2-46　配置 Virtual SAN 存储策略之二

第 3 步，查看 Virtual SAN Default Storage Policy 默认的规则集配置（如图 9-2-47 所示），可以根据 9.1.5 小节介绍调整参数。

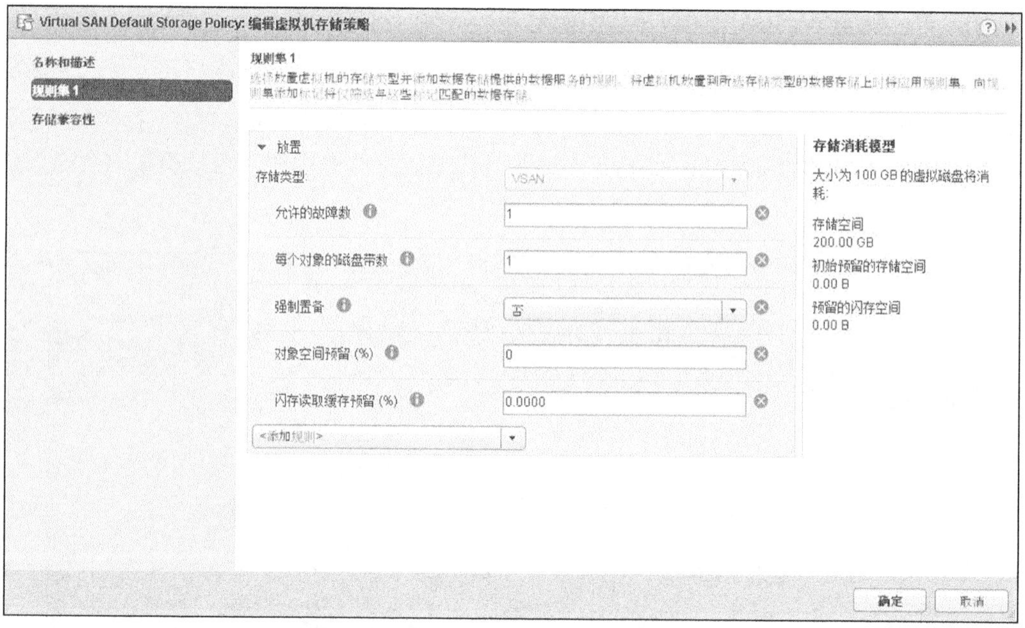

图 9-2-47　配置 Virtual SAN 存储策略之三

第 4 步，创建虚拟机 VM01-VSAN，使用默认 Virtual SAN 存储策略，可以看到虚拟机文件夹以及硬盘处于合规状态，其硬盘 1 具有 2 个副本以及 1 个见证（如图 9-2-48 所示）。

9.2 部署 Virtual SAN 6.5   467

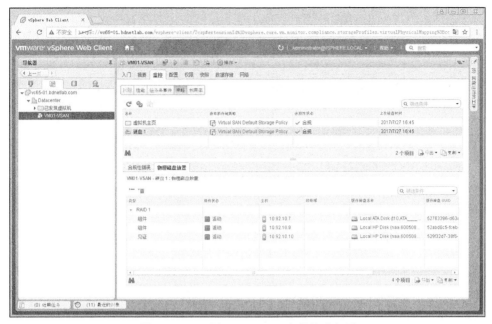

图 9-2-48　配置 Virtual SAN 存储策略之四

9.2.5　配置 Virtual SAN 去重和压缩

Virtual SAN 支持使用块级别的去重和压缩技术来节省 Virtual SAN 存储空间，当启用后去重和压缩后会减少磁盘组中的冗余数据。目前仅在全闪存环境支持去重和压缩，混合架构不能使用。

第 1 步，默认情况下去重和压缩处于已禁用状态（如图 9-2-49 所示），单击"编辑"按钮。

图 9-2-49　配置 Virtual SAN 去重和压缩之一

第 2 步，编辑 Virtual SAN 设置，启用去重和压缩（如图 9-2-50 所示），单击"确定"按钮。

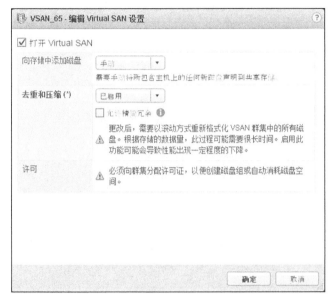

图 9-2-50　配置 Virtual SAN 去重和压缩之二

第 3 步，启用去重和压缩后，系统会对 Virtual SAN 磁盘进行重新处理（如图 9-2-51 所示）。

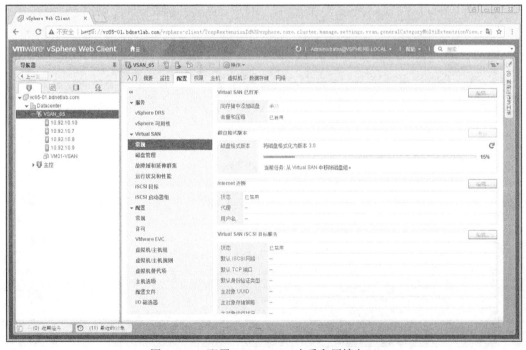

图 9-2-51　配置 Virtual SAN 去重和压缩之三

第 4 步，执行去重和压缩操作后，查看 Virtual SAN 容量信息，使用去重和压缩后节省 736MB 空间，比率为 1.12 倍（如图 9-2-52 所示）。

图 9-2-52　配置 Virtual SAN 去重和压缩之四

9.2.6　配置 Virtual SAN 纠删码

由于去重和压缩技术目前只能在全闪存架构下使用，对于混合架构，可以使用 Virtual SAN 提供的 RAID 5/6 纠删码技术来提高容量使用效率。表 9-2-1 显示了使用纠删码空间消耗情况对比。

表 9-2-1　　　　　　　　　　　纠删码空间消耗情况对比

RAID	FTT	数据大小	空间需求
RAID 1	1	100GB	200GB
RAID 1	2	100GB	300GB
RAID 5/6	1	100GB	133GB
RAID 5/6	2	100GB	150GB

如果在存储策略中启用 RAID 5/6 纠删码技术，不支持将 FTT 值设置为 3；当 FTT 值设置为 1 时，为 RAID 5 模式；当 FTT 值设置为 2 时，为 RAID 6 模式。本小节介绍 RAID 5 纠删码配置。

第 1 步，创建新的虚拟机存储策略（如图 9-2-53 所示），单击"下一步"按钮。

图 9-2-53　配置 Virtual SAN 纠删码之一

第 2 步，进入虚拟机存储策略配置向导（如图 9-2-54 所示），单击"下一步"按钮。

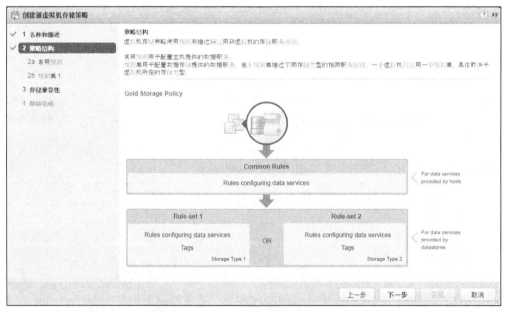

图 9-2-54　配置 Virtual SAN 纠删码之二

第 3 步，选择是否配置提供的数据服务的常用规则（如图 9-2-55 所示），单击"下一步"按钮。

图 9-2-55　配置 Virtual SAN 纠删码之三

第 4 步，配置规则集，容错方式选择"RAID-5/6（Erasure Coding）-Capacity"（如图 9-2-56 所示），单击"下一步"按钮。

图 9-2-56　配置 Virtual SAN 纠删码之四

第 5 步，对存储兼容性进行校验（如图 9-2-57 所示），单击"下一步"按钮。

图 9-2-57　配置 Virtual SAN 纠删码之五

第 6 步，确认参数配置正确（如图 9-2-58 所示），单击"完成"按钮。

图 9-2-58　配置 Virtual SAN 纠删码之六

第 7 步，新创建名为 RAID 5 的纠删码存储策略完成（如图 9-2-59 所示）。

9.2 部署 Virtual SAN 6.5 473

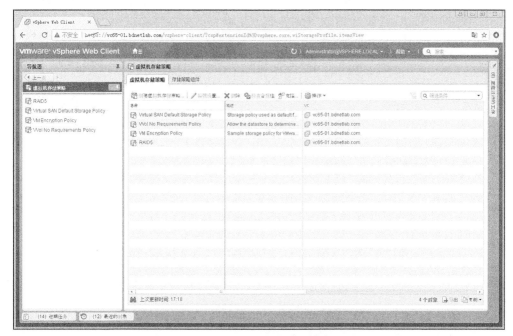

图 9-2-59 配置 Virtual SAN 纠删码之七

第 8 步，在虚拟机 VM01-VSAN 上单击右键，编辑虚拟机存储策略（如图 9-2-60 所示）。

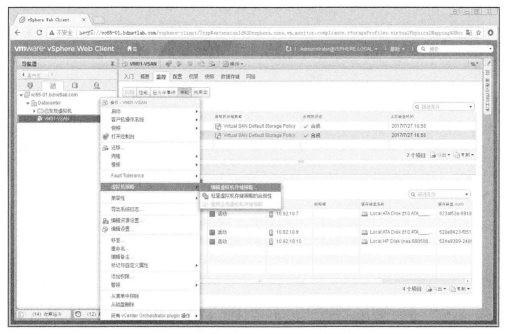

图 9-2-60 配置 Virtual SAN 纠删码之八

第 9 步，调整虚拟机存储策略为新创建的 RAID 5（如图 9-2-61 所示），单击"应用于全部"按钮。

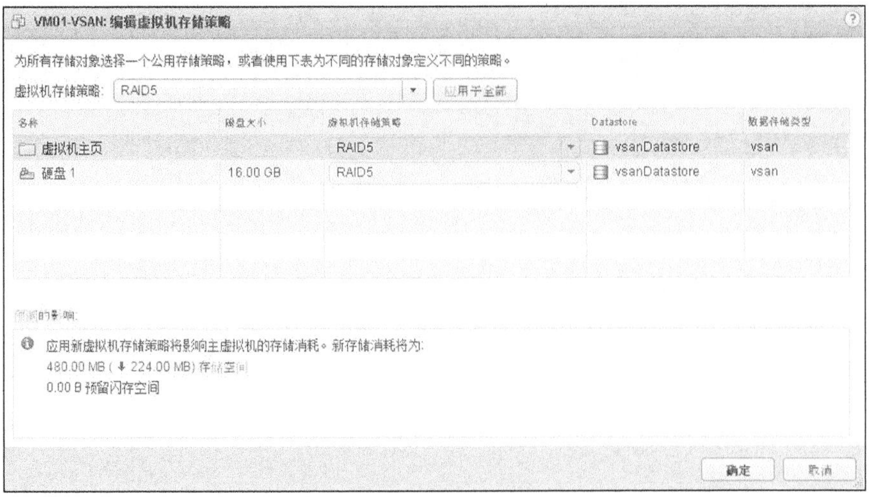

图 9-2-61　配置 Virtual SAN 纠删码之九

第 10 步，虚拟机 VM01-VSAN 存储策略调完成，查看硬盘策略处于合规状态，且使用 RAID 5 模式（如图 9-2-62 所示）。

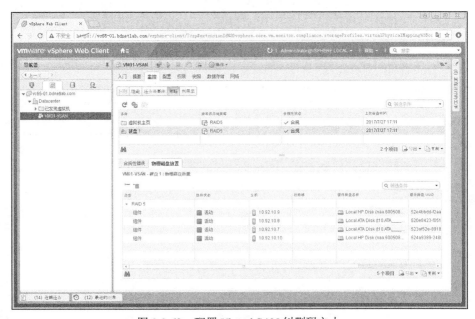

图 9-2-62　配置 Virtual SAN 纠删码之十

9.2.7　配置 Virtual SAN 故障域

Virtual SAN 故障域是从 6.0 版本新增加的功能，在生产环境中，Virtual SAN 可能使用多个机架的普通服务器或刀片服务器。以刀片服务器为例，假设虚拟机有 3 个副本，且这 3 个副本分布在这台刀片服务器的 3 个刀片上，如果这台刀片服务器发生故障，即使有再多的副本也会导致虚拟机发生故障。故障域是为了解决这些问题而出现的，当配置故障域后，副本可以分布在其他主机上，避免故障的发生。在 Virtual SAN 中启用故障域时，至少需要个 3 个故障域，每个故障域至少包含 1 台 ESXi 主机，这样可以确保 Virtual SAN 的正常运行。VMware 官方推荐至少使用 4 个故障域以支持数据迁出以及数据保护配置等。本小节介绍如何配置 Virtual SAN 故障域。

第 1 步，默认情况下，Virtual SAN 故障域未配置，但可以假设 ESXi 主机处于同一个故障域（如图 9-2-63 所示），单击"+"新建故障域。

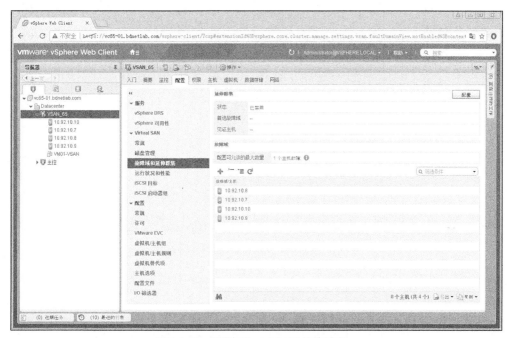

图 9-2-63　配置 Virtual SAN 故障域之一

第 2 步，输入新建故障域的名称，勾选 1 台 ESXi 主机（如图 9-2-64 所示），单击"确定"按钮。

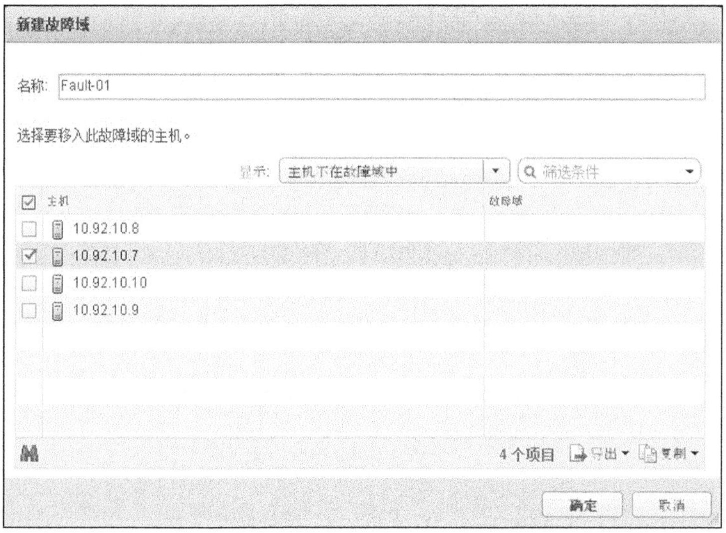

图 9-2-64　配置 Virtual SAN 故障域之二

第 3 步，创建好 1 个故障域且这个故障域有 1 台 ESXi 主机（如图 9-2-65 所示）。

图 9-2-65　配置 Virtual SAN 故障域之三

第 4 步，按照相同的方式创建两个故障域，其中 Fault-03 故障域配置 2 台 ESXi 主机（如图 9-2-66 所示）。

图 9-2-66　配置 Virtual SAN 故障域之四

第 5 步，新建虚拟机 VM02-VSAN，查看存储策略，硬盘 1 物理磁盘分布在 3 个故障域（如图 9-2-67 所示）。

9.2 部署 Virtual SAN 6.5   477

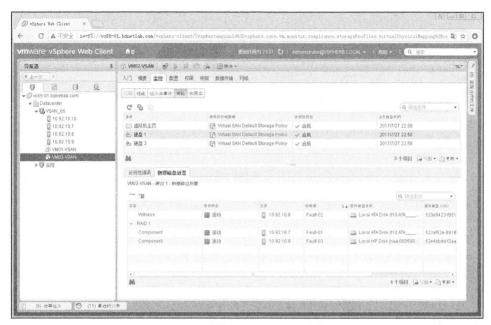

图 9-2-67　配置 Virtual SAN 故障域之五

第 6 步，查看虚拟机 VM01-VSAN 存储策略，存储策略处于不合规状态，其中两个组件位于同一个故障域 Fault-03（如图 9-2-68 所示）。

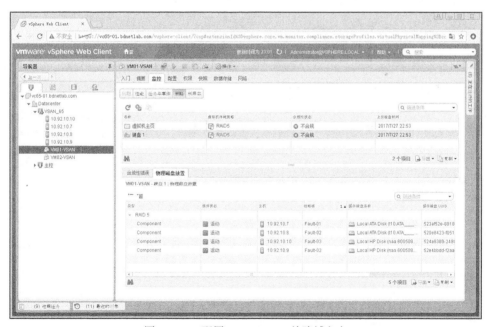

图 9-2-68　配置 Virtual SAN 故障域之六

第 7 步，添加一台新的 ESXi 主机进入 Virtual SAN 集群，刷新存储策略，虚拟机 VM01-VSAN 存储策略处于合规状态，磁盘 1 物理磁盘分布在不同的故障域中，这样可以避免同一故障域出现问题虚拟机故障的情况（如图 9-2-69 所示）。

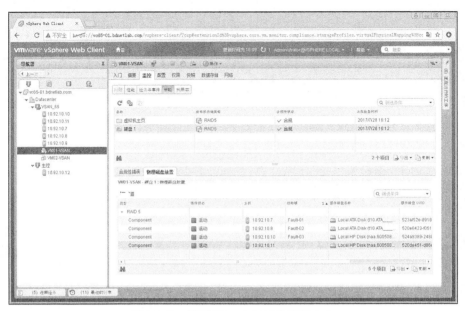

图 9-2-69　配置 Virtual SAN 故障域之七

9.2.8　配置 Virtual SAN 延伸集群

Virtual SAN 延伸集群功能可以理解为双活数据中心，每个延伸集群包括两个站点和一个见证主机，通过配置使用延伸集群，两个站点均为活动站点，当其中一个站点出现故障时，可以使用另外一个站到点，这样可以避免某个站点出现而故障影响到集群的正常运行。本小节介绍延伸集群的基本配置。

第 1 步，选择集群中的"故障域和延伸群集"（如图 9-2-70 所示），单击"配置"按钮。

图 9-2-70　配置 Virtual SAN 延伸集群之一

第 2 步，配置首选故障域和辅助故障域（如图 9-2-71 所示）。

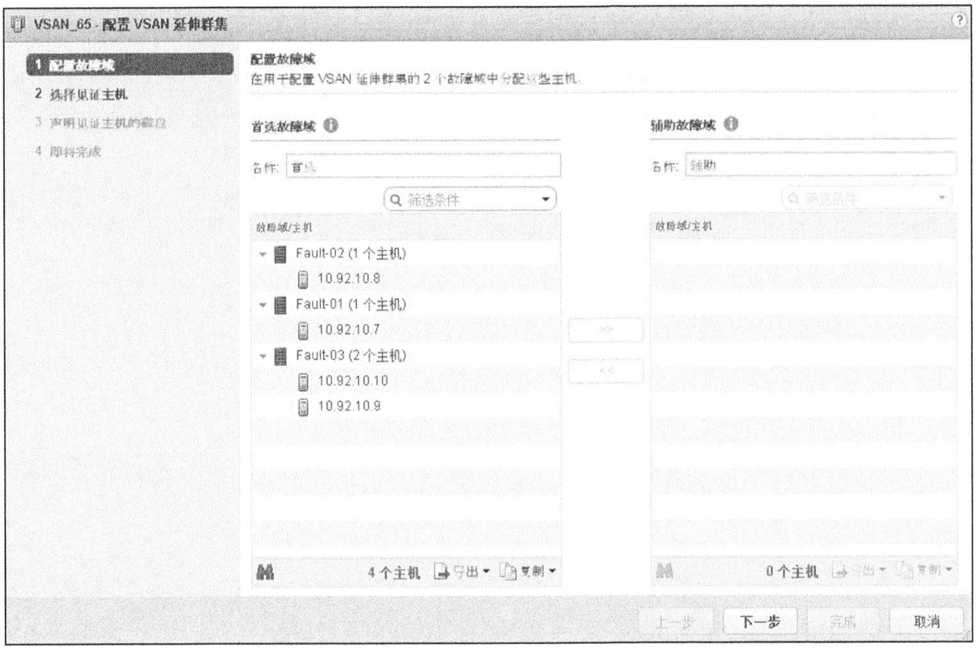

图 9-2-71　配置 Virtual SAN 延伸集群之二

第 3 步，首选故障域和辅助故障域分别选择两台 ESXi 主机（如图 9-2-72 所示），单击"下一步"按钮。

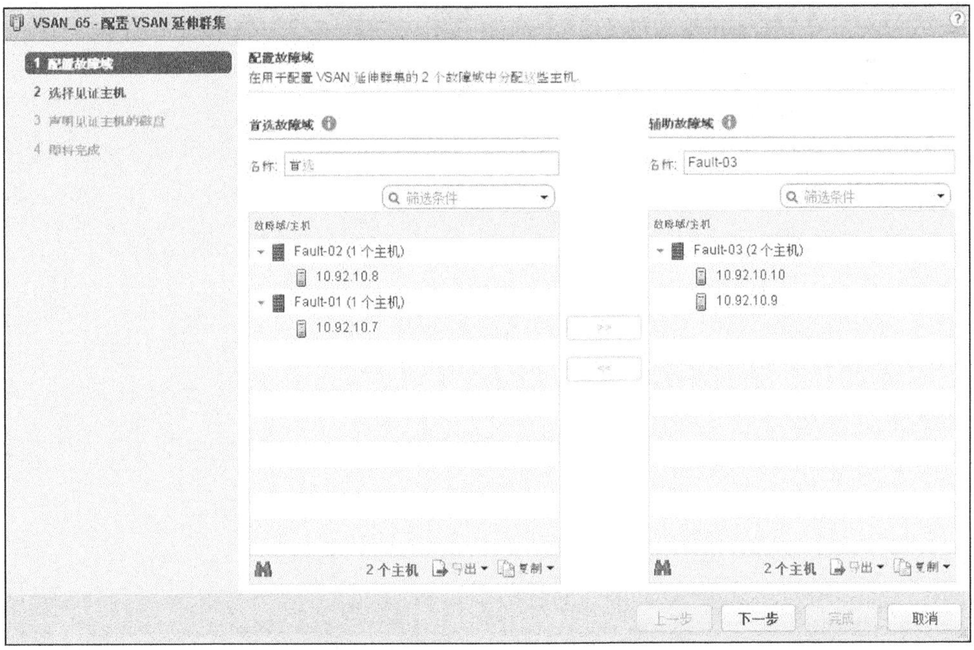

图 9-2-72　配置 Virtual SAN 延伸集群之三

第 4 步，选择见证主机（如图 9-2-73 所示），单击"下一步"按钮。

图 9-2-73　配置 Virtual SAN 延伸集群之四

第 5 步，声明见证主机使用的缓存层磁盘和容量层磁盘（如图 9-2-74 所示），单击"下一步"按钮。

图 9-2-74　配置 Virtual SAN 延伸集群之五

第 6 步，确认参数配置正确（如图 9-2-75 所示），单击"完成"按钮。

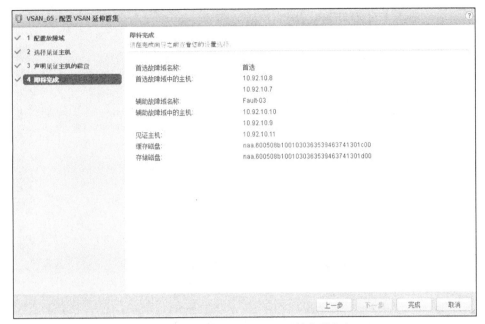

图 9-2-75　配置 Virtual SAN 延伸集群之六

第 7 步，延伸集群配置完成（如图 9-2-76 所示）。

图 9-2-76　配置 Virtual SAN 延伸集群之七

第 8 步，查看虚拟机 VM02-VSAN 存储策略，硬盘 1 具有 2 个副本，其中 1 个副本位于首选故障域，另外一个副本位于辅助故障域 Fault-03 上（如图 9-2-77 所示）。当首选故障域出现故障时，将使用辅助故障域上的副本。

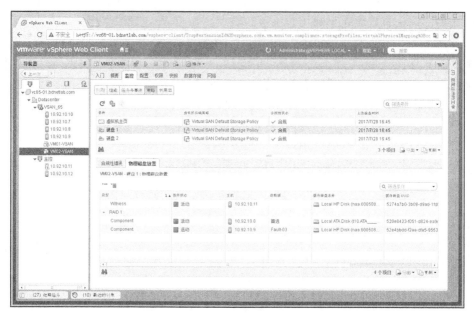

图 9-2-77　配置 Virtual SAN 延伸集群之八

9.2.9　配置 Virtual SAN 为 iSCSI 目标服务器

Virtual SAN 6.5 版本增加一个新的功能，即可以将 Virtual SAN 容量通过网络 iSCS 接口的存储块服务，从而变身成为一个专业的存储设备，其他物理服务器或非 vSphere 上的虚拟机可以使用 Virtual SAN 所提供的存储服务。本小节介绍如何配置 Virtual SAN 为 iSCSI 目标服务器。

第 1 步，选择集群中的"iSCSI 目标"，默认情况下为禁用状态（如图 9-2-78 所示），单击"编辑"按钮。

图 9-2-78　配置 Virtual SAN 为 iSCSI 目标服务器之一

第 2 步，勾选"启用 Virtual SAN iSCSI 目标服务"，配置网络、TCP 端口、身份验证以及存储策略（如图 9-2-79 所示），单击"确定"按钮。

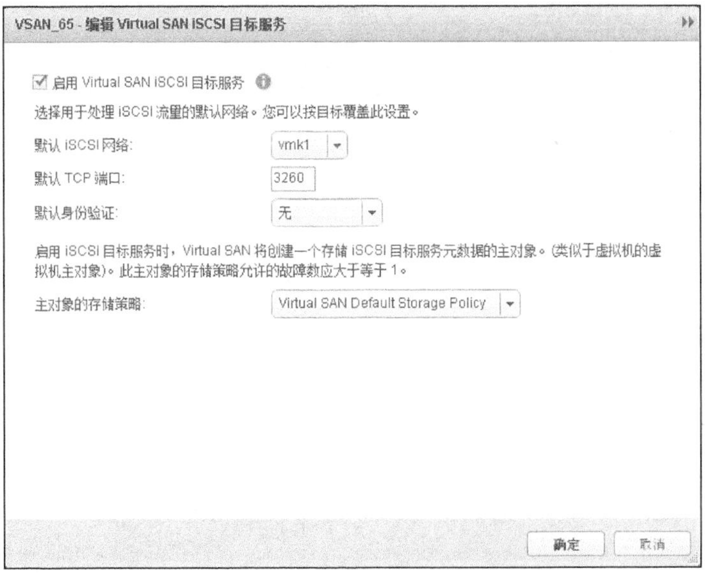

图 9-2-79　配置 Virtual SAN 为 iSCSI 目标服务器之二

第 3 步，Virtual SAN iSCSI 目标服务启用，目标的详细信息以及 LUN 还未配置，因此列表为空（如图 9-2-80 所示），单击"+"新建 iSCSI 目标。

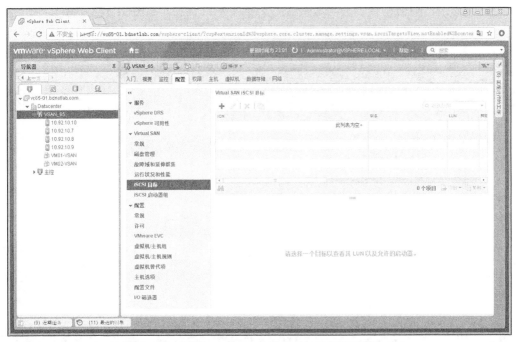

图 9-2-80　配置 Virtual SAN 为 iSCSI 目标服务器之三

第 4 步，配置目标详细信息，其中目标 IQN 会自动生成，需要配置目标别名以及其他参数,同时勾选"将第一个 LUN 添加到 iSCSI 目标(可选)",配置 LUN 相关信息(如图 9-2-81 所示),单击"确定"按钮。

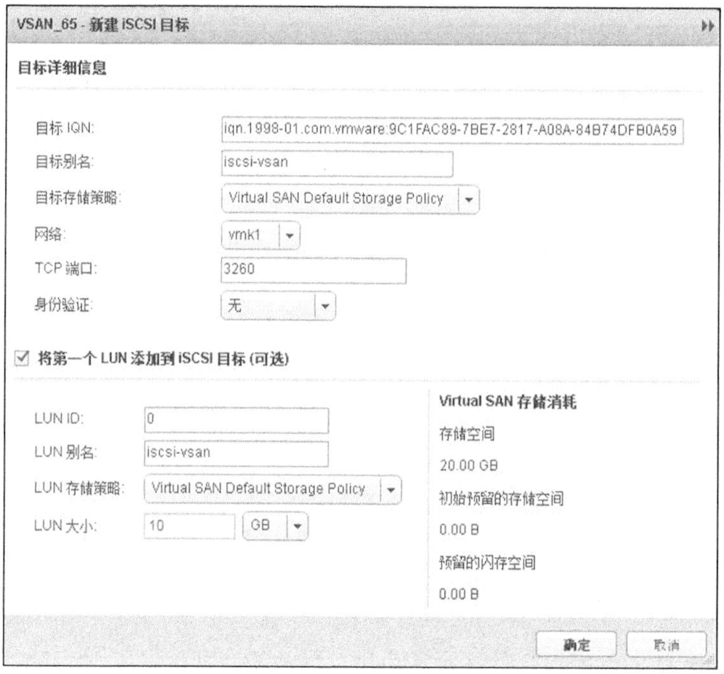

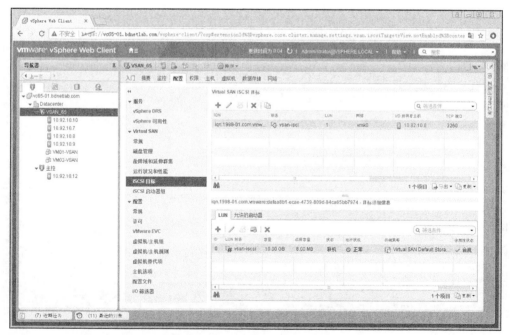

图 9-2-81　配置 Virtual SAN 为 iSCSI 目标服务器之四

第 5 步,新建 iSCSI 目标信息配置完成,处于正常状态(如图 9-2-82 所示)。

图 9-2-82　配置 Virtual SAN 为 iSCSI 目标服务器之五

第6步，添加允许的启动器（如图9-2-83所示），单击"+"。

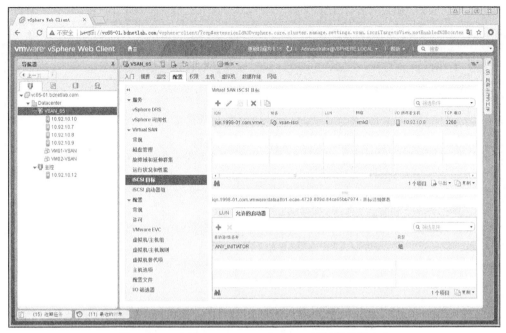

图 9-2-83　配置 Virtual SAN 为 iSCSI 目标服务器之六

第7步，输入允许访问的启动器 IQN（如图9-2-84所示），单击"确定"按钮。

图 9-2-84　配置 Virtual SAN 为 iSCSI 目标服务器之七

第8步，允许的启动器添加完成（如图9-2-85所示）。

第9步，使用 Windows 系统自带的 iSCSI 发起程序连接 Virtual SAN 存储 iSCSI 目标（如图9-2-86所示），状态已连接。

486　第 9 章　部署使用 Virtual SAN

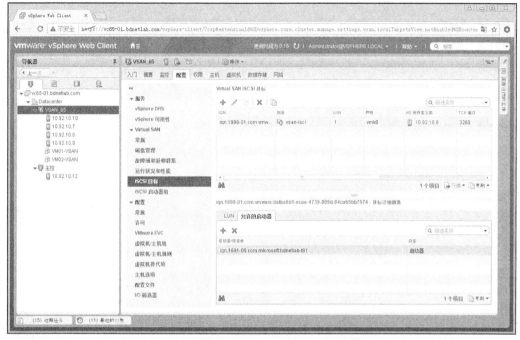

图 9-2-85　配置 Virtual SAN 为 iSCSI 目标服务器之八

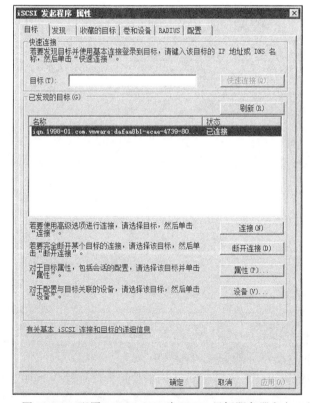

图 9-2-86　配置 Virtual SAN 为 iSCSI 目标服务器之九

第 10 步，打开 Windows 系统磁盘管理，未分配的硬盘 2 容量为 10GB（如图 9-2-87 所示），该 10GB 硬盘容量为 Virtual SAN 存储 iSCSI 目标。

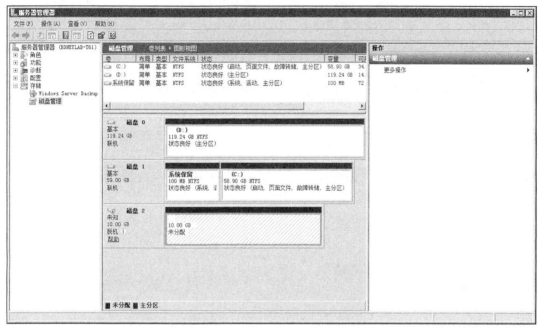

图 9-2-87　配置 Virtual SAN 为 iSCSI 目标服务器之十

第 11 步，查看集群 Virtual SAN iSCSI 目标信息，运行状态正常（如图 9-2-88 所示）。

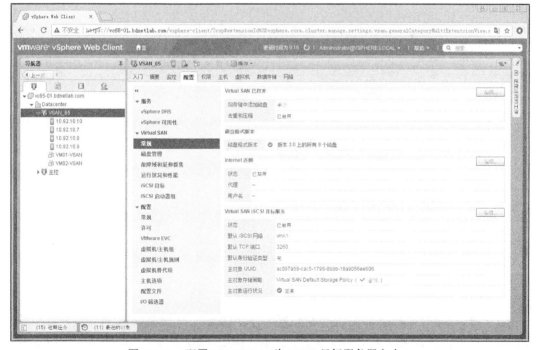

图 9-2-88　配置 Virtual SAN 为 iSCSI 目标服务器之十一

9.2.10 配置 Virtual SAN 性能服务

Virtual SAN 性能服务是从 Virtual SAN 6.2 版本开始提供的新功能，可以通过 Web Client 直接对 Virtual SAN 进行监控，性能服务不仅可以查看 Virtual SAN 集群的相关性能数据，而且可以对主机、虚拟机以及磁盘、磁盘组性能进行查看。本小节介绍如何配置 Virtual SAN 性能服务。

第 1 步，选择集群中的"监控"，查看性能，默认情况下 Virtual SAN 性能服务为禁用状态（如图 9-2-89 所示），单击"立即打开"。

图 9-2-89　配置 Virtual SAN 性能监控之一

第 2 步，勾选"打开 Virtual SAN 性能服务"（如图 9-2-90 所示），单击"确定"按钮。

图 9-2-90　配置 Virtual SAN 性能监控之二

第 3 步，单击"Virtual SAN-虚拟机消耗"，可以看到 Virtual SAN 集群中所有虚拟机存储性能数据，包括 IOPS、吞吐量、延迟、拥塞、待处理 IO 等信息（如图 9-2-91 所示）。

第 4 步，单击"Virtual SAN-后端"，可以看到 Virtual SAN 后台存储性能信息，包括 IOPS、吞吐量、延迟、拥塞、待处理 IO 等信息（如图 9-2-92 所示）。

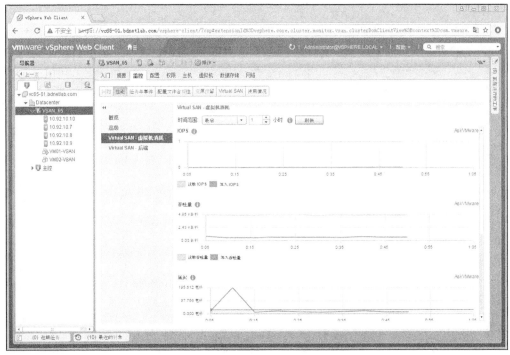

图 9-2-91　配置 Virtual SAN 性能监控之三

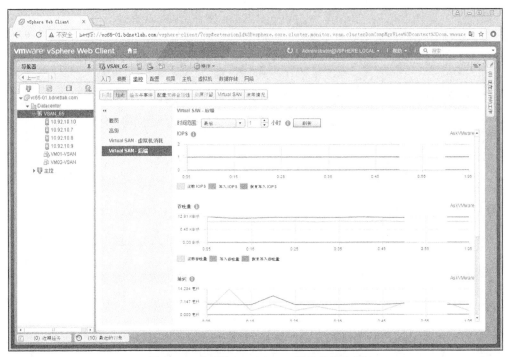

图 9-2-92　配置 Virtual SAN 性能监控之四

第 5 步，选择 ESXi 主机，与 Virtual SAN 集群相比较，ESXi 主机性能监控增加了磁盘组与磁盘选项,单击"Virtual SAN-磁盘组",可以看到该磁盘组各种 IOPS、读取命中率、吞吐量、写入缓冲区可用百分比等信息(如图 9-2-93 所示)。

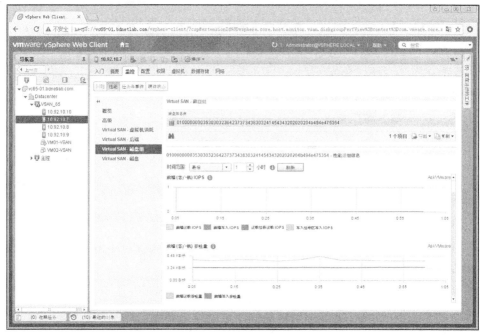

图 9-2-93　配置 Virtual SAN 性能监控之五

第 6 步,选择 ESXi 主机,单击"Virtual SAN-磁盘",可以看到该磁盘硬件层 IOPS、吞吐量、VSAN 怪延迟等信息(如图 9-2-94 所示)。

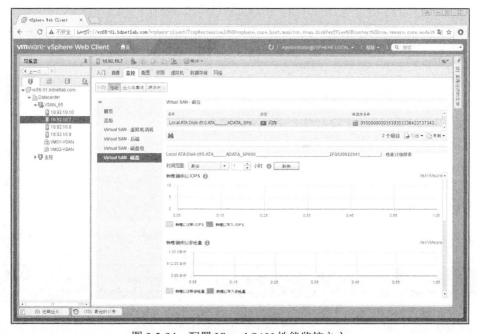

图 9-2-94　配置 Virtual SAN 性能监控之六

9.3 生产环境使用 Virtual SAN 讨论

作为 VMware vSphere 软件定义存储以及超融合 HCI 解决方案，Virtual SAN 从 2014 年开始到现在，频繁发布了多个版本。下面简单讨论生产环境应用如何使用 Virtual SAN。

9.3.1 Virtual SAN 是否能代替传统存储

Virtual SAN 是否能代替传统存储，这是一直争论的话题。VMware 官方的回复是，Virtual SAN 没有把传统存储当作对手或敌人。Virtual SAN 的思路是充分利用物理服务器存储资源，通过重新整合这些存储资源提供高性能、高可用性的存储服务。

从行业使用情况来看，Virtual SAN 与传统存储处于并行的状态，不少企业核心业务使用传统存储，虚拟桌面以及非核心业务使用 Virtual SAN。目前来说没有谁能取代谁，毕竟两种存储各有各的优势以及缺点。从发展形势看，软件定义存储以及超融合应该是未来的发展趋势。

9.3.2 生产环境使用 Virtual SAN 主机数量

生产环境中使用 Virtual SAN，表 9-1-1 中不同的 FTT 值对于 ESXi 主机数量的要求是最低值，生产环境强烈建议大于最低值。以 FTT 值为 1 为例，需要的 ESXi 主机数量为 3 台，如果在生产环境仅使用 3 台 ESXi 主机启用 Virtual SAN，Virtual SAN 可以使用，但可能发生由于组件数不够等原因导致 Virtual SAN 崩溃的情况（由于资源不足情况导致 Virtual SAN 崩溃的情况已经发生多次），所以对于 FTT 值为 1 的生产环境，至少应配置 4 台 ESXi 主机。

9.3.3 生产环境使用 Virtual SAN 网络要求

生产环境中使用 Virtual SAN，对于网络的要求分为两种情况。一种是混合架构，混合架构可以使用 1GE 网络承载 Virtual SAN 流量，但 1GE 网络不能发挥出最大效率，推荐使用 10GE 网络；另一种是全闪存架构，全闪存架构本身要求使用 10GE 网络，强制使用 1GE 网络也可以承载，但这样的使用完全不能发挥全闪存读写的优势。

9.3.4 生产环境使用 Virtual SAN 硬件兼容性要求

实验环境使用的 HP 服务器 SCSI 控制器不在 VMware 硬件兼容列表中，会出现失败以及警告信息，如图 9-2-44 所示。在生产环境中是否能使用这些不兼容性的硬件设备呢？答案是否定的。为保证 Virtual SAN 在生产环境的稳定，不推荐使用这些不兼容设备。随着 Virtual SAN 快速发展，VMware 官方增加了硬件厂商进行合作，定期更新 Virtual SAN HCL 数据库，读者可以访问 VMware 官方网站（http://partnerweb.vmware.com/service/vsan/all.json）获取 Virtual SAN HCL 数据库文件进行更新。如图 9-3-1 所示，Virtual SAN HCL 数据库已更新到 2017 年 7 月 27 日。

图 9-3-1　更新 Virtual SAN HCL 数据库

9.4　本章小结

本章对 Virtual SAN 6.5 版本进行了详细介绍，包括基础知识部分以及具体的部署。作为 VMware 软件定义存储以及超融合解决方案，Virtual SAN 具有很多的优势，国内外一些企业已经将其部署于生产环境，对于生产环境使用 Virtual SAN，推荐使用经 VMware 官方认证的服务器以及硬件。

欢迎来到异步社区!

异步社区的来历

异步社区(www.epubit.com.cn)是人民邮电出版社旗下 IT 专业图书旗舰社区,于 2015 年 8 月上线运营。

异步社区依托于人民邮电出版社 20 余年的 IT 专业优质出版资源和编辑策划团队,打造传统出版与电子出版和自出版结合、纸质书与电子书结合、传统印刷与 POD 按需印刷结合的出版平台,提供最新技术资讯,为作者和读者打造交流互动的平台。

社区里都有什么?

购买图书

我们出版的图书涵盖主流 IT 技术,在编程语言、Web 技术、数据科学等领域有众多经典畅销图书。社区现已上线图书 1000 余种,电子书 400 多种,部分新书实现纸书、电子书同步出版。我们还会定期发布新书书讯。

下载资源

社区内提供随书附赠的资源,如书中的案例或程序源代码。
另外,社区还提供了大量的免费电子书,只要注册成为社区用户就可以免费下载。

与作译者互动

很多图书的作译者已经入驻社区,您可以关注他们,咨询技术问题;可以阅读不断更新的技术文章,听作译者和编辑畅聊好书背后有趣的故事;还可以参与社区的作者访谈栏目,向您关注的作者提出采访题目。

灵活优惠的购书

您可以方便地下单购买纸质图书或电子图书,纸质图书直接从人民邮电出版社书库发货,电子书提供多种阅读格式。

对于重磅新书,社区提供预售和新书首发服务,用户可以第一时间买到心仪的新书。

用户账户中的积分可以用于购书优惠。100 积分 =1 元,购买图书时,在 里填入可使用的积分数值,即可扣减相应金额。

特 别 优 惠

购买本书的读者专享异步社区购书优惠券。

使用方法：注册成为社区用户，在下单购书时输入 S4XC5 使用优惠码 ，然后点击"使用优惠码"，即可在原折扣基础上享受全单 9 折优惠。（订单满 39 元即可使用，本优惠券只可使用一次）

纸电图书组合购买

社区独家提供纸质图书和电子书组合购买方式，价格优惠，一次购买，多种阅读选择。

社区里还可以做什么？

提交勘误

您可以在图书页面下方提交勘误，每条勘误被确认后可以获得 100 积分。热心勘误的读者还有机会参与书稿的审校和翻译工作。

写作

社区提供基于 Markdown 的写作环境，喜欢写作的您可以在此一试身手，在社区里分享您的技术心得和读书体会，更可以体验自出版的乐趣，轻松实现出版的梦想。

如果成为社区认证作译者，还可以享受异步社区提供的作者专享特色服务。

会议活动早知道

您可以掌握 IT 圈的技术会议资讯，更有机会免费获赠大会门票。

加入异步

扫描任意二维码都能找到我们：

异步社区	微信服务号	微信订阅号	官方微博	QQ 群：436746675

社区网址：www.epubit.com.cn

投稿 & 咨询：contact@epubit.com.cn